HENRY NORRIS RUSSELL

THE HR DIAGRAM

INTERNATIONAL ASTRONOMICAL UNION

UNION ASTRONOMIQUE INTERNATIONALE

SYMPOSIUM No. 80

HELD AT THE NATIONAL ACADEMY OF SCIENCES, WASHINGTON D.C.
2–5 NOVEMBER, 1977

THE HR DIAGRAM

The 100th Anniversary of Henry Norris Russell

EDITED BY

A. G. DAVIS PHILIP

Dudley Observatory and Union College

AND

D. S. HAYES

Arizona State University

D. REIDEL PUBLISHING COMPANY

DORDRECHT : HOLLAND / BOSTON : U.S.A. / LONDON : ENGLAND

1978

Library of Congress Cataloging in Publication Data

Main entry under title:

The HR Diagram.

(Symposium - International Astronomical Union ; no. 80)
Includes bibliographies and indexes.
1. HR diagrams – Congresses. 2. Stars – Congresses. 3. Russell, Henry Norris, 1877–1957. I. Russell, Henry Norris, 1877–1957. II. Philip, A. G. Davis. III. Hayes, Donald S. IV. Series: International Astronomical Union. Symposium ; no. 80.
QB815.H18 523.8 78–19125
ISBN 90-277-0905-X
ISBN 90-277-0906-8 pbk.

Published on behalf of
the International Astronomical Union
by
D. Reidel Publishing Company, P. O. Box 17, Dordrecht, Holland

Sold and distributed in the U.S.A., Canada, and Mexico
by D. Reidel Publishing Company, Inc.
Lincoln Building, 160 Old Derby Street, Hingham,
Mass. 02043, U.S.A.

Printed in The Netherlands

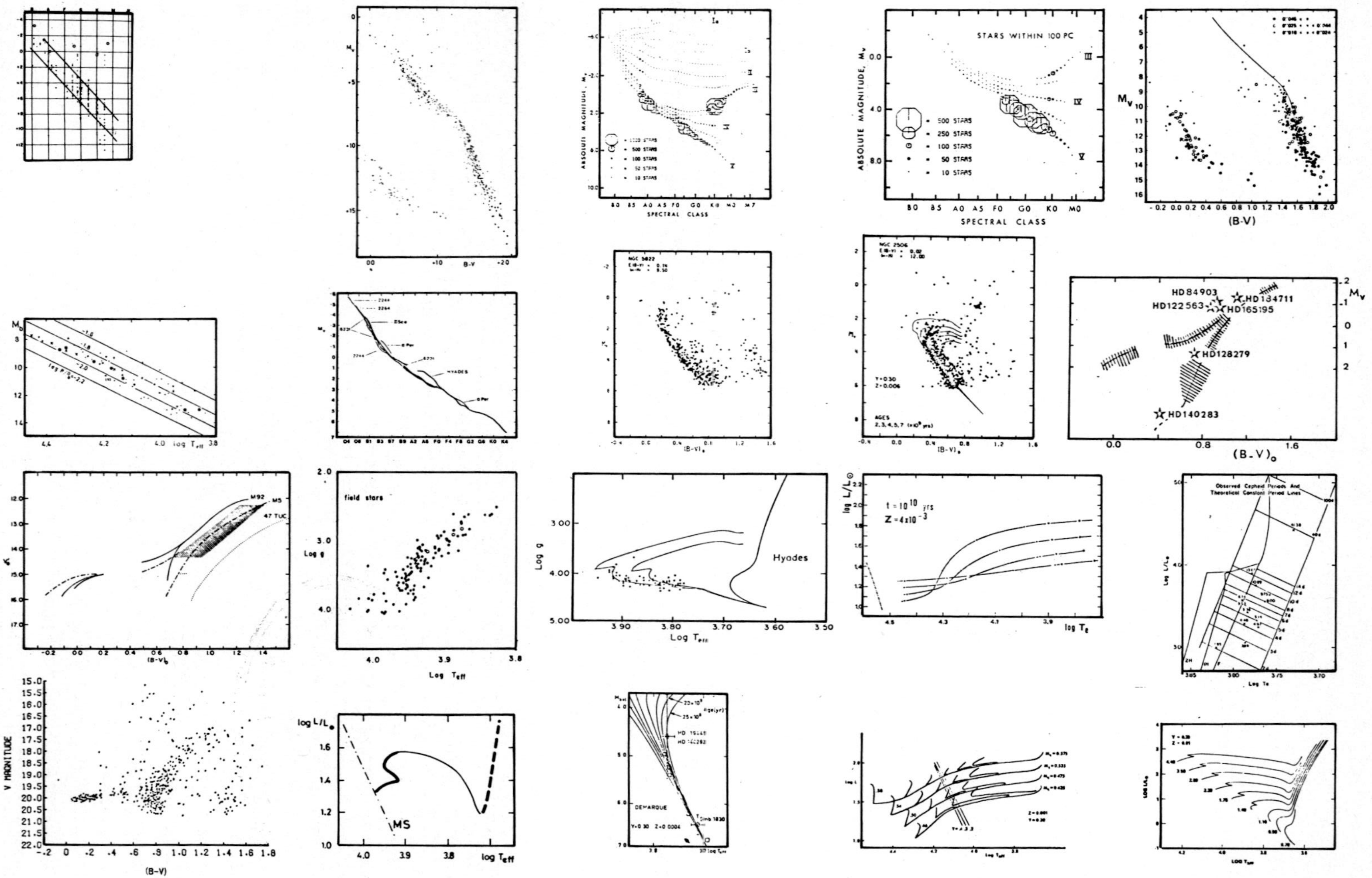

STARS WITHIN 100 PC
SPECTRAL CLASS
ABSOLUTE MAGNITUDE, M_V
500 STARS
250 STARS
100 STARS
50 STARS
10 STARS
(B-V)
$(B-V)_0$
HD84903
HD134711
HD122563
HD165195
HD128279
HD140283
Hyades
Log T_{eff}
Log g
field stars
$t = 10^{10}$ yrs
$Z = 4 \times 10^{-3}$
log $L/L_\odot$
log T_e
MS
M92
47 TUC
V MAGNITUDE
Observed Cepheid Periods And Theoretical Constant Period Lines
DEMARQUE

TABLE OF CONTENTS

V. HR DIAGRAMS, HORIZONTAL BRANCH

VI. GALAXIES AND MISCELLANEOUS SUBJECTS

POSTER SESSION

PREFACE

IAU Symposium No. 80, The HR Diagram - The 100th Anniversary of Henry Norris Russell was held on November 2-5, 1977 at the National Academy of Sciences in Washington D. C., in order to commemmorate the birth of Henry Norris Russell on October 25, 1877 and to review current problems in the use of the Hertzsprung-Russell diagram. The IAU has sponsored two previous conferences concerned mainly with the HR diagram; The Position of Variable Stars in the Hertzsprung-Russell Diagram, a colloquium held at Bamberg in 1965 and The Hertzsprung-Russell Diagram (IAU Symposium No. 10, J.L. Greenstein, ed.) held in Moscow in 1959. In 1974 a conference, Multicolor Photometry and the Theoretical HR Diagram (Dudley Obs. Report No. 9, A.G.D. Philip and D.S. Hayes, eds.) was held in Albany, N.Y.; and in 1964 a conference, Basic Data Pertaining to the Hertzsprung-Russell Diagram, was held at the Flagstaff Station of the U.S. Naval Observatory in honor of Ejnar Hertzsprung and to dedicate the 61-inch astrometric reflector. (Vistas in Astronomy Vol. 8, A. Beer and K. Aa. Strand, eds., Pergamon Press, Oxford). Volume 12 of Vistas in Astronomy, The Henry Norris Russell Memorial Volume (1970), contains a review paper on Changing Interpretations of the Hertzsprung-Russell Diagram 1910-1940, A Historical Note by B.W. Sitterly.

It was felt that there was a need for a conference on the HR Diagram at this time because of general advances in theoretical and observational investigations of various aspects of the diagram. For example, the new four meter telescopes at Kitt Peak National Observatory and Cerro Tololo Inter-American Observatory and the Anglo-Australian four meter telescope are extending studies in the HR diagram to fainter stars and to external systems, evolutionary models of the horizontal and giant branches have advanced our knowledge of the later stages of stellar evolution, and much new information concerning the chemical abundance of individual stars in stellar systems is now available. The editors are impressed with the information becoming available concerning variations in the morphology of the HR diagram, both in terms of sequences in the diagram and in terms of stellar densities along the sequences, due to parameters such as stellar age, chemical composition, mass loss, initial mass function and rotation.

A proposal sponsored by IAU Commissions 29, 35, 36, 37 and 45 was submitted to the IAU Executive Committee at the Grenoble meeting. The Executive Committee appointed the scientific organizing committee consisting of A.G. Davis Philip (Chairman), Immo Appenzeller, A.A. Boyarchuk, Pierre Demarque, Robert Dickens, Keiichi Kodaira, B. Paczynski, Stephen Strom, Sydney van den Bergh and Nolan Walborn. The local organizing committee was chaired by E.R. Dyer with Roger Bell, R.E. Berendzen, Benjamin Peery, A.G. Davis Philip, Nancy Roman, Vera Rubin and K.Aa. Strand as members.

Two hundred and seven persons from twenty different countries attended the sixteen sessions held in the superb lecture hall of the National Academy of Sciences. The chairmen of the sessions were:

I	A. G. Davis Philip	IX	Robert J. Dickens
II	Lyman Spitzer	X	Robert P. Kraft
III	Keiichi Kodaira	XI	M. Golay
IV	Immo Appenzeller	XII	George Wallerstein
V	Nolan Walborn	XIII	Vytas Straižys
VI	Philip C. Keenan	XIV	Paris Pismis
VII	Jesse L. Greenstein	XV	Pierre Demarque
VIII	W. W. Morgan	XVI	John Norris

In the four days of the symposium 79 papers were presented and 13 additional papers were posted in a poster session. Of the authors of the spoken papers, 64 were U.S., 44 were foreign and in the total were 18 women. It was gratifying that we were able to have such a high representation of both non-U.S. and women astronomers at the symposium.

The proceedings of the symposium have been printed in two volumes. The scientific sessions (III-XVI) are to be found in this volume; the first two sessions have been published separately as In Memory of Henry Norris Russell in Dudley Observatory Report No. 13, A.G. Davis Philip and David H. DeVorkin, eds. So that a complete record of the symposium will appear in this volume, a copy of the Table of Contents of the historical volume will be found on pages xvi-xvii.

The proceedings of the symposium have been edited from camera-ready manuscripts submitted by the authors. The discussion was prepared from handwritten sheets passed out to the people who participated in the discussion. Typed versions of the discussion were passed back to individuals for checking during the symposium. The whole discussion (in some cases discussion was transcribed from tape recordings) was then retyped and mailed back for final proof-reading. Author camera-ready manuscripts which were different from the elite type-face were retyped, as were papers whose format did not agree with the IAU format. A few papers that exceeded the number of allotted pages were revised by the editors. Individual authors have been made responsible for obtaining permission to copy material from other publications. We have made some effort to make the proceedings uniform throughout; for example whenever the phrase HR diagram appears it is in this form with no periods or hyphens. However, since we had a limited budget for retyping and since the value of the proceedings is much increased by rapid publication we have not corrected every difference. The order of the papers in this volume has been changed somewhat from the order in which papers were presented in the symposium. The HR diagram on the cover of this book is from the article by Houk and Fesen, Fig. 1 on page 92.

Many organizations have helped during the last year. Yale University Observatory and Louisiana State University paid for the mailing of hundreds of announcements concerning the symposium. Financial support for the meeting and preparation of the proceedings was provided by the National Science Foundation, The International Astronomical Union, Dudley Observatory and the Astronomical Society of New York.

Many people helped during the symposium and in making the volume ready for publication. There is not enough room here to thank all such persons (see Martha Liller's banquet speech in Dudley Observatory Report No. 13). We will mention a few persons who made a major contribution. E.R. Dyer, chairman of the Local Organizing Committee, did a superb job in the local arrangements and in seeing that everything ran smoothly. Mary Bongiovanni took care of the discussion sheets during the conference and did almost all the typing after the conference. Kristina Philip helped in many ways throughout the whole period.

A. G. Davis Philip

Donald S. Hayes, editors

January 28, 1978
Schenectady, New York

IN MEMORY OF HENRY NORRIS RUSSELL

TABLE OF CONTENTS

LIST OF FIGURES

All photographs are by David H. DeVorkin, courtesy of the American Institute of Physics.

LIST OF PARTICIPANTS

	Australia
Norris, John	Mt. Stromlo and Siding Spring Obs.
	Belgium
Arpigny, Claude	Universite de Liege
	Canada
Crampton, David	Dominion Astrophysical Observatory
Falk, Herbert J.	University of Western Ontario
Fitzgerald, M. Pim	University of Waterloo
Garrison, Robert F.	David Dunlap Observatory
Hogg, Helen B. Sawyer	David Dunlap Observatory
Luiken, Maike	University of Waterloo
Mitalas, Romas	University of Western Ontario
Shadick, Stanley J.	University of Western Ontario
Sreenivasan, S. Ranga	University of Calgary
Van Den Bergh, Sidney	David Dunlap Observatory
Wehlau, Amelia	University of Western Ontario
	Chile
Moreno, Adelina Gutiérrez	Universidad de Chile
Moreno, Hugo	Universidad de Chile
Walborn, Nolan R.	Cerro Tololo Inter-American Obs.
	Denmark
Hejlesen, Poul Martin	Astronomical Observatory, Copenhagen
Jørgensen, H.E.	Astronomical Observatory, Copenhagen
	England
Dickens, Robert J.	Royal Greenwich Observatory
Madore, Barry F.	University of Cambridge
Smith, R.W.	University of Cambridge
Tayler, R.J.	University of Sussex
	France
Cayrel de Strobel, G.	Observatoire de Meudon
Foy, Renaud	Observatoire de Paris-Meudon
Ochsenbein, Francois	Observatoire de Strasbourg
Perrin, M-N	Observatoire de Paris-Meudon
Spite, F.	Meudon Observatory
Spite, M.	Meudon Observatory

	India	
Parthasarathy, M.	Indian Institute of Astrophysics
	Italy
Bernacca, P.L.	Osservatorio Astrofisico Asiago
Castellani, Vittorio	University of Rome
Chiosi, Cesare	University of Padova
Renzini, Alvio	Osservatorio Astronomico, Balogna
Rosino, L.S.	University of Padova
	Japan
Kodaira, Keiichi	University of Tokyo
Sato, Koichi	International Latitude Obs. of Mizusawa
	Mexico
Mendoza, Eugenio E., V	National University of Mexico
Pismis, Paris	National University of Mexico
	Netherlands
Heck, A.	European Space Agency
Van Agt, Steven	Nijmegen University
Wesselius, Paul R.	Dept. of Space Research, Groningen
	South Africa
Balona, L.A.	South African Astronomical Observatory
Lloyd Evans, T. H. H.	South African Astronomical Observatory
	Sweden
Sinnerstad, Ulf	Stockholm Observatory
	Switzerland
Golay, M.	Observatory of Geneva
Hauck, B.	Institut d'Astronomie de l'Universite de Lausanne
Lub, J.	ESO TP Division
	U.S.A.
Abt, Helmut A.	Kitt Peak National Observatory
Aizenman, Morris L.	National Science Foundation
Ake, Thomas B.	Warner & Swasey Observatory
Albers, Henry	Vassar College

Anthony-Twarog, Barbara J.	Yale University Observatory
Assousa, George E.	Carnegie Institution of Washington
Atkinson, Robert d'E	Indiana University
Babcock, Alice	U.S. Naval Observatory
Barry, Don C.	University of Southern California
Bell, Roger A.	University of Maryland
Bidelman, William P.	Warner and Swasey Observatory
Binnendijk, L.	University of Pennsylvania
Boeshaar, Patricia C.	University of Oregon
Boggess, Albert	NASA-Goddard Space Flight Center
Boggess, Nancy	NASA Headquarters
Bond, Howard E.	Louisiana State University
Bonsack, Walter K.	University of Hawaii
Bookmyer, Beverly	Clemson University
Boyle, Robert James	Yale University
Buscombe, William	Northwestern University
Buser, Roland	California Institute of Technology
Butler, Dennis	Yale University
Canterna, Ronald W.	Louisiana State University
Carlson, John B.	University of Maryland
Chen, Peter	Case Western Reserve University
Christian, Carol A.	Boston University
Christy, James W.	U.S. Naval Observatory
Clarke, Randolph T. III	U.S. Naval Observatory
Conti, Peter S.	JILA
Cox, Arthur N.	Los Alamos Scientific Laboratory
Crawford, David L.	Kitt Peak National Observatory
Dahn, Conard C.	U.S. Naval Observatory
Demarque, P.R.	Yale University
Deming, Drake	University of Maryland
DeVorkin, David H.	American Institute of Physics
Douglass, Geoffrey G.	U.S. Naval Observatory
Drilling, John	Louisiana State University
Dunkelman, Lawrence	NASA-Goddard Space Flight Center
Dyer, Edward R., Jr.	National Academy of Sciences
Eaton, Joel A.	NASA-Goddard Space Flight Center
Edmondson, Frank K.	Indiana University
Evans, David S.	University of Texas at Austin
Fix, John D.	National Science Foundation
Flower, Phillip J.	Joint Institute for Laboratory Astrophysics
Gary, George	Howard University
Green, Louis C.	Haverford College
Greenstein, Jesse L.	California Institute of Technology
Gross, Peter G.	Warner and Swasey Observatory
Hanson, Robert B.	Yale University Observatory
Hardorp, Johannes	SUNY at Stony Brook
Harrington, R.S.	U.S. Naval Observatory
Havlen, Robert J.	University of Virginia
Hayes, Donald S.	Arizona State University

Henderson, A. Peter	University of Maryland
Henry, Richard C.	NASA Headquarters
Herbst, William	Carnegie Institution of Washington
Hoffleit, E. Dorrit	Yale University
Holdenried, Ellis R.	U.S. Naval Observatory
Houk, Nancy	University of Michigan
Humphreys, Roberta M.	University of Minnesota
Ianna, Philip A.	University of Virginia
Jackson, Peter D.	University of Maryland
Josties, F.J.	U.S. Naval Observatory
Kallarakal, V.	U.S. Naval Observatory
Keenan, Philip C.	Perkins Observatory
Kerr, Frank J.	University of Maryland
Kraft, Robert P.	Lick Observatory
Lamb, Susan A.	University of California, Los Angeles
Landolt, Arlo U.	Louisiana State University
Lesh, Janet Rountree	NASA Goodard Space Flight Center
Leung, K.C.	U.S. Department of Energy
Liebert, James W.	Steward Observatory
Liller, Martha H.	Harvard College Observatory
Lippincott, Sarah Lee	Sproul Observatory
Locanthi, Dorothy Davis	Jet Propulsion Lab
Lu, Phillip K.	Western Connecticut State College
Lukac, Carl F.	U.S. Naval Observatory
Luyten, W.J.	University of Minnesota
Manduca, Armando	University of Maryland
McClure, Robert D.	Yale University Observatory
McKee, John	University of Maryland
McNamara, Bernard	New Mexico State University
Mead, Jaylee	NASA Goodard Space Flight Center
Mengel, J.	New Haven, Connecticut
Merrill, John E.	University of Florida
Miranian, Myron	U.S. Naval Observatory
Morgan, W.W.	Yerkes Observatory
O'Brien, H.G.	University of Delaware
Pascu, Dan	U.S. Naval Observatory
Payne-Gaposchkin, Cecilia H.	Harvard College Observatory
Peery, Benjamin F., Jr.	Howard University
Penhallow, William S.	University of Rhode Island
Pesch, Peter	Case Western Reserve University
Pfeiffer, Raymond J.	Trenton State College
Philip, A.G. Davis	Dudley Observatory
Pilachowski, Catherine	University of Washington
Roman, Nancy	NASA Headquarters
Rood, Robert T.	University of Virginia
Rosendhal, Jeffrey D.	NASA Headquarters
Routly, Paul M.	U.S. Naval Observatory
Rubin, Vera C.	Carnegie Institution of Washington
Russell, Jane L.	Allegheny Observatory

Sanders, W.L.	New Mexico State University
Sanduleak, N.	Warner and Swasey Observatory
Sass, Christine A.	University of Delaware
Sass, Karl W.	University of Delaware
Schlesinger, Barry M.	Systems and Applied Sciences Corp.
Schmidt, Richard	U.S. Naval Observatory
Schommer, Robert A.	California Institute of Technology
Schwarzschild, Martin	Princeton University Observatory
Sears, Richard	University of Michigan
Shipman, Harry L.	University of Delaware
Sion, Edward M.	Villanova University
Sitterly, Charlotte Moore	Washington, D.C.
Slettebak, Arne	Ohio State University
Smith, Clayton	U.S. Naval Observatory
Snowden, Michael S.	Mt. Laguna Observatory
Southall, Rita	Howard University
Spitzer, Lyman, Jr.	Princeton University Observatory
Stencel, Robert E.	NASA/JSC
Strand, K. Aa.	Washington, D.C.
Strongylis, George	University of Maryland
Stryker, Linda Lee	Yale University Observatory
Sweigart, Allen V.	NASA Goddard Space Flight Center
Tapia, S.	Lunar and Planetary Laboratory
Thonnard, Norbert	Carnegie Institution of Washington
Thorne, Muriel	NASA (FE)
Tinsley, Beatrice M.	Yale University Observatory
Trimble, Virginia	University of Maryland
Upgren, Arthur R.	Van Vleck Observatory
Van Altena, William F.	Yale University Observatory
Vila, Samuel C.	University of Pennsylvania
Wallerstein, George	University of Washington
Warren, Wayne	NASA Goddard Space Flight Center
Weis, Edward W.	Van Vleck Observatory
Westerhout, Gart	U.S. Naval Observatory
White, Nathaniel M.	Lowell Observatory
Wilcox, Gary	Warner and Swasey Observatory
Wood, Frank Bradshaw	University of Florida
Worley, Charles E.	U.S. Naval Observatory
Zinn, Robert J.	Hale Observatories

Relatives of Henry Norris Russell

Margaret Russell Edmondson	daughter
Lucy May Russell Gardner	daughter
Henry Norris Russell, Jr.	son
Mrs. Henry Norris Russell, Jr.	daughter-in-law
William A. Russell	grandson
Paige Russell	great granddaughter
Kenneth Gordon	great grandson
Cynthia Gardner MacKay	granddaughter
Henrietta Boughton	secretary to Henry Norris Russell

	U.S.S.R.
Straižys, V.	Astronomical Observatory, Vilnius
	Vatican
McCarthy, Martin F.	Vatican Observatory
	Venezuela
Fuenmayor, Francisco	Case Western Reserve Univ. (graduate student from Venezuela)
	West Germany
Appenzeller, Immo	Landessternwarte Heidelberg-Konigstuhl
Gliese, Wilhelm	Astronomisches Rechen-Institut
Kahler, Helmuth	Hamburger Sternwarte
Pel, J.W.	Max Planck Institut fur Astronomie
Schmidt-Kaler, Theodor	Ruhr-University Bochum
Seitter, Waltraut	University of Muenster
Weidemann, Volker	University of Kiel

PART I

FUNDAMENTALS OF THE HR DIAGRAM

Frank K. Edmondson and W.W. Morgan.

SOME USES OF SPECTRAL CLASSIFICATION IN HR DIAGRAMS

W. W. Morgan

Yerkes Observatory

The MK spectrum-luminosity class diagram (S-LC diagram) between types O9 and G0 is a two-dimensional array containing approximately 100 - 150 information elements; each information element is a single cell denoted by a spectral type and a luminosity class. These individual, non-dimensional cells (i.e. A0 III) can be considered as carriers of observational information toward the HR diagram. Some categories of peculiar spectra (for example, spectra resembling that of the "silicon star" θ Aur) can be attached to certain cells, and can be considered as local perturbations upon the underlying continuity of the spectrum-luminosity class sequences. By such a procedure, the S-LC diagram cells can be considered open-ended, in that progressive additions to the cells can be made indefinitely into the future. Such a procedure can incorporate successfully an unlimited variety of "normal" and "peculiar" categories of spectra - both those already known and those still to be discovered. A morphological system that is unitary thus comes into existence - a system of considerable astrophysical interest. The translation of the detailed language of this system into that of the classical HR diagram is a task for the future.

Several new examples of what could be called local perturbations in the continuity of the luminosity-class sequences have been discovered recently by Dr. H.A. Abt and the writer, while working on a revision of the Yerkes MK Spectral Atlas for stars earlier than the Sun; these include: (1) A category of B7 III - B9 III giants which have the property of relatively narrow, stellar K lines, together with broad lines of He I. Among members of this group, some (or all) of the peculiar Mn II - Hg II stars

A. G. Davis Philip and D. S. Hayes (eds.), The HR Diagram, 3–4.

are found. (2) A strong He I star has been found recently in the Orion Association; this makes a total of three such stars known there. All three of these stars lie very close to the Zero-Age Main Sequence as defined by Morton and Adams.

The investigation described here was based on features in the spectral range 3500 - 4700 Å. It is of prime importance to carry out similar investigations in other spectral regions. For this purpose, each region should be surveyed independently of the MK results for the photographic region described here. Thus, a system independent of the MK system would be devised in each case (far UV; green - red; infrared). When each such system has been completed, its morphology can then be compared as a whole with the MK system. A similar procedure should be followed for the weak-lined stars in the photographic region used here.

The new Spectral Atlas is being prepared for publication jointly by the Yerkes Observatory of the University of Chicago and the Kitt Peak National Observatory; the authors are H.A. Abt, J.W. Tapscott and Morgan. All spectrograms used in this work were obtained by Abt with the Meinel spectrographs of the Kitt Peak National Observatory and the Cerro Tololo Inter-American Observatory. The work was made possible by grants to Morgan from the National Science Foundation.

DISCUSSION

STENCEL: I'm glad you mentioned 31 Com since it and a few other early G giants (24UMa, HR 1327) exhibit obviously larger line widths at high-dispersion compared to early spectral types (e.g. F5III, GOIV, late G giants), presumably due to its position in the Hertzsprung gap. This discontinuity is in a sense similar to your discontinuity.

McNAMARA: It has been shown by Strom in NGC 2264 and myself for the Orion Nebula cluster that age spreads in young clusters can be at least 2 x 10^6 years, even during the stage of major star production. Could this age spread, extrapolated to higher luminosities explain the age differences that you have suggested for the Northern and Southern sword sections of the Orion Nebula region?

MORGAN: There seems to be strong evidence for differences in ages for different sub-clusterings in the Orion Association. At one extreme is the Orion Nebula cluster; at the other is the expanding group in the Northern region of the Association investigated by Janet Lesh.

SOME PROGRESS AND PROBLEMS IN THE UPPER LEFT OF THE HR DIAGRAM

Nolan R. Walborn

Cerro Tololo Inter-American Observatory*

1. AN EMPIRICAL LUMINOSITY CLASSIFICATION FOR EARLY O STARS

The MK system of spectral classification does not include luminosity classes for types earlier than O9. Subsequent work with higher dispersions has allowed a luminosity classification for early O stars to be proposed. One of the principal criteria at types O9-B0 is the selective, negative luminosity effect in the HeII 4686 Å absorption line; a similar effect was found in the NIII 4634-4640-4642 Å absorption lines at O9-B0. These particular lines weaken markedly with increasing luminosity, while other HeII lines maintain a constant ratio to HeI (by definition), and other NIII lines strengthen considerably. These anomalous lines are the same ones which selectively appear in emission in Of stars, and the derived luminosity classification for earlier types is based upon the hypothesis that the negative absorption effect at O9-B0 is due to emission filling by the same mechanism producing the Of phenomenon, and that hence the latter may also be identified as a luminosity effect (Walborn 1971). One observes strong 4686 Å absorption and often weak 4640 Å emission [denoted ((f))] on the main sequence to the earliest types. As luminosity increases at a given spectral type, the HeII absorption weakens while the NIII emission increases [denoted (f)] and finally the most luminous stars have both 4686 and 4640 Å strongly in emission [denoted f]. A sequence of spectra all with types very near O7, but showing the marked changes with luminosity class in 4686 and 4640 Å, is illustrated by Walborn (1973). These effects show good agreement with other spectroscopic luminosity criteria

*Supported by the National Science Foundation under contract NSF-AST 74-04128

A. G. Davis Philip and D. S. Hayes (eds.), The HR Diagram, 5-12. All Rights Reserved.

such as SiIV absorption (Conti and Alschuler 1971) and Balmer-line strength (Crawford 1975, Morrison 1975), which are however less sensitive at the earliest types.

The absolute visual magnitudes of O stars with independently known distances support the empirical luminosity classification. The Of stars have M_V's near $-7^m.0$ to the earliest types, while the hottest main sequence stars average $-5^m.5$ (Conti and Alschuler 1971; Walborn 1972, 1973). Further calibration information is desirable, however, and provides one motivation for observations of O stars in the Magellanic Clouds. Theoretical support for the luminosity classification is provided by the work of Mihalas et al. (1972) and Mihalas and Hummer (1973), who have reproduced the observed behavior of the NIII lines as a function of temperature and gravity. These results provide a good demonstration of the value of detailed morphological investigation; previously spectra in all three of the above-mentioned categories had sometimes been classified "Of", and when absolute-magnitude differences from pure absorption-line O stars were sought, of course no clear conclusions could be reached.

Following the resolution of some initial divergences, there is now excellent general agreement between this morphological work and the extensive, parallel studies of O-type spectra at coudé dispersions by P.S. Conti and collaborators (Conti and Leep 1974). The value of morphology is that it enables one to perceive more readily the forest (and its substructures) among the trees, while the quantitative approach provides the essential data for comparison with physical theory (Conti 1973a, b; Conti and Burnichon 1975).

In summary, I think the Of stars may now be removed from the category of "peculiar stars" where they are often found. They are simply the early O-type supergiants, and their spectra represent the behavior of stellar atmospheres and certain atomic species at the relevant temperatures and gravities (although complete physical interpretations await more detailed, future theoretical models). However, while refined morphology may remove some classes of stars from the peculiar category, it invariably reveals at the same time more subtle kinds of peculiarity, to some examples of which I'll now turn.

2. EXTENDED-ENVELOPE PHENOMENA

2.1 Onfp Stars

The principal peculiarity is a combined emission/absorption structure in HeII 4686 Å, which could be described as an absorption with emission wings, or as a centrally reversed emission (Walborn 1972, 1973). As implied by the notation, these stars invariably have broadened absorption lines suggestive of

rapid rotation. This category has been denoted Oef by Conti and Leep (1974), who inferred a relationship to the Be phenomenon.

2.2 Of?p Stars

The defining characteristic is CIII 4647-4650-4651 Å emission comparable in strength to NIII 4634-4640-4642 Å (Walborn 1972, 1973, and references therein). Only 3 stars (HD 108, 148937, 191612) have been so classified to date. The emission features appear excessively sharp relative to the stellar absorption lines, and there are other suggestions of a shell phenomenon rather than an extension of the stellar atmosphere as in normal Of stars. The interrogative in the notation is intended to express doubt that these are normal O-type supergiants. There is evidence that they are massive, Population I objects, although the CIII emission property is shared by some Of-type nuclei of planetary nebulae (Heap 1977, Thackeray 1977). HD 148937 is associated with expanding nebulosity apparently ejected from the star (Catchpole and Feast 1970, Pismis 1974). Binary motion has been reported for HD 108 by Hutchings (1975), while Conti et al. (1977) did not find variable velocity for HD 148937.

2.3 OIafpe Stars

The pre-eminent peculiarity is a pronounced P Cygni spectrum. These stars are definitely highly luminous supergiants. Conti et al. (1977) concluded that the well-known example HD 152408 in Scorpius OB1 is not a spectroscopic binary. Two objects recently observed in the Large Magellanic Cloud (HDE 269858 and Sanduleak -67° 266) show probable evidence of extreme envelope extent; the SiIV 4089 and NIII 4097 Å absorption lines have interstellar-like profiles at moderate resolution (Walborn 1977).

2.4 WN Stars

The necessary characteristic which distinguishes these spectra from any of the previously discussed categories is the presence of broad emission bands. Two basic questions which must be answered are (i) whether or not these objects occur only in binary systems; and (ii) what is the relationship between the high-luminosity, narrow-line WN-A stars, usually found in massive young regions such as 30 Doradus, the Carina Nebula, and Scorpius OB1, and their lower-luminosity, moderate-mass counterparts such as V444 Cygni. The Of binaries VI Cygni No. 5 and HDE 228766 may be pre-WR systems (Bohannan and Conti 1976, Massey and Conti 1977). There are clear morphological relationships

between O3-4f and WN-A spectra (Walborn 1974). However, it is essential to determine which objects are binaries and which are probably single stars, before possible evolutionary relationships can be inferred with any confidence. The O3f star HD 93129A and the WN-A stars in the Carina Nebula are currently the subjects of a radial-velocity study by P.S. Conti, V.S. Niemela and the writer.

3. CHEMICAL ANOMALIES

3.1 OBN/OBC Stars

The observational details have recently been reviewed (Walborn 1976) and need not be repeated here. Bolton and Rogers (1977) have found that many and possibly all OBN stars are spectroscopic binaries, while none of the OBC stars are members of short-period binaries. The OBN phenomena may represent relatively extreme nitrogen enrichment due to mass transfer in binary systems, while the most significant aspect of OBC spectra may be their nitrogen deficiency. The systematic, relative nitrogen deficiency among late-O and early-B supergiants in the Orion Belt and NGC 6231 (Scorpius OB1) could be interpreted as evidence that all OB supergiant atmospheres undergo evolutionary nitrogen enhancement to some degree during their lifetimes as such. A very important result was obtained by Baschek et al. (1972), who found that the spectroscopic differences between ζOri (Orion Belt) and HD 188209 (morphologically normal CNO spectrum) could be explained by an admixture of 20 to 40% CNO cycle-processed material in the atmosphere of the latter star. Dearborn and Eggleton (1977) have suggested that mass loss could produce such an effect. If an evolutionary interpretation of these phenomena is substantiated, the fact that they are observable must be reconciled with the time scales for OB supergiants. It may be of interest in this connection that some recent models (Stothers 1976, Stothers and Chin 1977) imply more extended time scales for OB supergiants. Currently high-dispersion observations of many of these stars are being obtained by J.B. Lester, to provide data for quantitative analysis.

3.2 Helium-Rich Stars

This term refers to Population I, main-sequence objects near type B2V with enhanced HeI spectra (Osmer and Peterson 1974 and references therein). The bizarre, periodic Hα emission variations in the prototype object σOrionis E (Walborn and Hesser 1976) and its peculiar light and color curves (Hesser et al. 1977) suggest a binary interpretation, as do variable shell absorption features (Groote and Hunger 1977) and periodically variable linear polarization (Kemp and Herman 1977). However, no radial-velocity curve is observable to stringent limits, indicating a very low mass for

any secondary, and an oblique-rotator model is preferred by Shore and Bolton (1977). See also Montmerle and Michaud (1976). It is fair to say that all the proposed interpretations of this puzzling object encounter some difficulties; further investigation is certainly indicated. Many of the He-rich stars have recently been found to have periodically variable He spectra (Bond and Levato 1976; Pedersen and Thomsen 1977).

3.3 Metal Deficiency in the SMC

There is a consensus among several recent spectroscopic studies of supergiants in the Small Magellanic Cloud, to the effect that they have systematically weaker metallic lines than their galactic counterparts. These studies have included types B2-3 (Osmer 1973), A (Przybylski 1972, Feast 1976), B0 (Walborn 1977), and a range of types (Dubois et al. 1977). The same phenomenon is seen in the spectrum of SMC X-1 (Sanduleak 160, Hutchings et al. 1977). The writer found that the SiIV/HeI line ratios in 2 SMC B0I spectra are similar to those in galactic stars of luminosity class IV, and some further examples have been observed during the current season. The first extremely early O star known in the SMC has also been found; it is difficult to judge metallicity in optical spectra of such hot stars, but it may well share the apparently general metal deficiency. These massive, metal-deficient stars may prove to be of considerable interest, not only for what they can tell us about the SMC, but also as representatives of a species now extinct in our Galaxy.

4. CONCLUSION

The extended-envelope and anomalous chemical phenomena discussed above are the most important puzzles presented by the spectra of the early-type stars at the present time. The questions, both explicit and implicit, are numerous. Why does one luminous, unstable object become a P Cygni star and another a WN star? The importance of information about the binary or single-star nature of each object has been emphasized throughout; before progress can be made, it is necessary to know which of these phenomena result from interactions in massive binary systems, and which are structural or evolutionary states of massive single stars. If very low-mass companions are involved, the issue may be extremely difficult to resolve. But even then, why does one interacting system become a WN and another an OBN binary - - or are they different phases of the same kind of object? The significance of these peculiar spectra should not be underestimated. Some may turn out to be mere curiosities, but others may provide keys to an improved understanding of massive stars, once the diffusion, mixing, mass-loss,

mass-transfer, or other physical mechanisms at work can be correctly identified. Until these phenomena have been explained, the possibility cannot be excluded that some aspects of current interpretations of massive stars may change, even qualitatively. It seems clear that the upper left of the HR diagram will be an area of considerable research interest for some time to come.

REFERENCES

Baschek, B., Kodaira, K. and Scholz, M. (1972). Astrophys. Lett. 12, 227.
Bohannan, B. and Conti, P.S. (1976). Astrophys. J. 204, 797.
Bolton, C.T. and Rogers, G.L. (1977). preprint.
Bond, H.E. and Levato, H. (1976). Publ. Astron. Soc. Pacific 88, 905.
Catchpole, R.M. and Feast, M.W. (1970). Observatory 90, 136.
Conti, P.S. (1973a). Astrophys. J. 179, 161.
Conti, P.S. (1973b). Astrophys. J. 179, 181.
Conti, P.S. and Alschuler, W.R. (1971). Astrophys. J. 170, 325.
Conti, P.S. and Burnichon, M.L. (1975). Astron. Astrophys. 38, 467.
Conti, P.S., Garmany, C.D. and Hutchings, J.B. (1977). Astrophys. J. 215, 561.
Conti, P.S. and Leep, E.M. (1974). Astrophys. J. 193, 113.
Crawford, D.L. (1975). Publ. Astron. Soc. Pacific 87, 481.
Dearborn, D.S.P. and Eggleton, P.P. (1977). Astrophys. J. 213, 448.
Dubois, P., Jaschek, M. and Jaschek, C. (1977). Astron. Astrophys. 60, 205.
Feast, M.W. (1976). Mon. Not. R. Astron. Soc. 174, 9P.
Groote, D. and Hunger, K. (1977). Astron. Astrophys. 56, 129.
Heap, S.R. (1977). Astrophys. J. 215, 609.
Hesser, J.E., Moreno, H. and Ugarte P., P. (1977). Astrophys. J. Lett. 216, L31.
Hutchings, J.B. (1975). Astrophys. J. 200, 122.
Hutchings, J.B., Crampton, D., Cowley, A.P. and Osmer, P.S. (1977). Astrophys. J. 217, 186.
Kemp, J.C. and Herman, L.C. (1977). Astrophys. J. 218, 770.
Massey, P. and Conti, P.S. (1977). Astrophys. J. 218, 431.
Mihalas, D. and Hummer D.G. (1973). Astrophys. J. 179, 827.
Mihalas, D., Hummer, D.G. and Conti, P.S. (1972). Astrophys. J. Lett. 175, L99.
Montmerle, T. and Michaud, G. (1976). Astrophys. J. Supp. 31, 489.
Morrison, N.D. (1975). Astrophys. J. 200, 113.
Osmer, P.S. (1973). Astrophys. J. Lett. 184, L127.
Osmer, P.S. and Peterson, D.M. (1974). Astrophys. J. 187, 117.
Pedersen, H. and Thomsen, B. (1977). Astron. Astrophys. Supp. 30, 11.
Pismis, P. (1974). Revista Mexicana Astron. Astrof. 1, 45.
Przybylski, A. (1972). Mon. Not. R. Astron. Soc. 159, 155.

Shore, S.N. and Bolton, C.T. (1977). preprint.
Stothers, R. (1976). Astrophys. J. 209, 800.
Stothers, R. and Chin, C. (1977). Astrophys. J. 211, 189.
Thackeray, A.D. (1977). Mon. Not. R. Astron. Soc. 180, 95.
Walborn, N.R. (1971). Astrophys. J. Suppl. 23, 257.
Walborn, N.R. (1972). Astron. J. 77, 312.
Walborn, N.R. (1973). Astron. J. 78, 1067.
Walborn, N.R. (1974). Astrophys. J. 189, 269.
Walborn, N.R. (1976). Astrophys. J. 205, 419.
Walborn, N.R. (1977). Astrophys. J. 215, 53.
Walborn, N.R. and Hesser, J.E. (1976). Astrophys. J. Lett. 205, L87.

DISCUSSION

SREENIVASAN: If the Onfp stars are single objects, information about their (spin) rotational velocities is of considerable significance for their evolutionary picture. Do you have information about their v sin i? Also, the resolution of their single or binary nature is of great importance for producing proper evolutionary models including mass-loss and/or mass exchange. If v sin i is appreciable, centrifugal forces must be included in the mass-loss treatment. So, such observational guidance for modeling is crucial.

DR. WALBORN PASSED THE QUESTION TO DR. CONTI: Individual rotational velocities, v sin i, of a large number of O-type stars have been published by Conti and Ebbets (Ap. J. 1977). The Oe and Oef stars, the latter being our designation of "Onfp" stars, generally have relatively the largest v sin i among the entire set. This is similar to the situation for Be stars vis a vis normal B stars. A few Oe stars have relatively sharp lines, presumably as they are "pole-on". We believe that Oe and Oef stars are related to O stars as the Be stars are to B stars.

As far as duplicity is concerned, Dr. Katy Garmany of JILA is presently studying this question for about 10 Oe and Oef stars. A very preliminary remark is that there are no large velocity amplitudes among these stars.

WALBORN: Can you give a typical value of v sin i for the majority of them?

CONTI: Oh, I'd say they're between 150 and 200 km/sec.

NORRIS: Would you comment on the incidence of helium weakness in the B star range? In particular is it still true that no helium weak objects are found with temperatures greater than that corresponding to B2.

WALBORN: I have not worked on these stars myself, but to my knowledge the statement holds true.

APPENZELLER: Can you give us an estimate of the percentage of the O stars which show chemical anomalies?

WALBORN: The extreme OBN/OBC stars comprise no more than about 5% of all OB stars I've observed. However, one possible interpretation of the observations is that all late O-early B supergiants undergo atmospheric nitrogen enrichment. In that case the N-deficient spectra may actually show "normal" abundances (i.e., those determined from B main-sequence stars) while morphologically normal supergiant spectra may be nitrogen-enhanced.

A MORE ACCURATE HR DIAGRAM FOR THE COOLER STARS

Philip C. Keenan

Perkins Observatory
Ohio State and Ohio Wesleyan Universities

Even if we confine ourselves to the classical HR diagram, postponing the transformation of spectral types and visual absolute magnitudes into basic physical variables, there are several ways to choose the data. If, for example, we want the density of points to represent the spatial density of stars, then we must concentrate on reducing sampling errors, usually at the cost of accuracy in some parts of the diagram. The Hess diagram is very useful, but an equally useful alternative is an HR diagram which includes only those stars whose positions are determined as accurately as possible. The objective is to classify the fine structure in the pattern in the luminosity-spectrum domain, and thus to define more precisely the borders of zones which are either occupied or avoided by stars of a given population. It is this pattern which is compared with theoretical evolutionary tracks and it is such a diagram based on the revised MK classification of the recent Atlas of Keenan and McNeil (1976) that I shall present here.

The greatest uncertainty is in the absolute magnitudes, and it is only in the lower part of the diagram that individual luminosities from trigonometric parallaxes can be plotted. Through the upper two-thirds of the diagram the most homogeneous are furnished by our best spectroscopic absolute magnitudes. These are obtained from the luminosity classes by calibration through mean trigonometric parallaxes up through the giant branch and from cluster and binary membership for the supergiants. The mid-part of the diagram is tied to the giant branch, since that forms a definite ridge among all stars on the cool side of the Hertz-sprung gap. Luminosity class III is defined by the center of this

A. G. Davis Philip and D. S. Hayes (eds.), The HR Diagram, 13-20.

ridge, and for the most accurately classified stars those at the center are designated IIIab. Stars on the upper edge are called IIIa, and those on the lower edge IIIb.

When we come to compute mean trigonometric parallaxes this refined classification is one means of reducing not only the mean errors of the resulting absolute magnitudes, but also the sizes of the systematic corrections that should be applied. Statistical corrections are proportional to the variance, or the square of the dispersion. Most estimates have placed the intrinsic dispersion of the giant branch at not less than $\sigma_M \approx 0^m.6$ (Ljunggren and Oja, 1966). There are difficulties in making this estimate, but at least from G8 to about K3 the revised class III stars appear to have σ_M $0^m.4$. This reduces the Malmquist correction (ΔM_M) to constant volumes of space by a factor of about 2. In this range of types I estimate the correction to be $+0^m.22$ - but it becomes larger and more uncertain in type M. The other major correction, the difference $5\langle \log p\rangle - 5 \log \langle p\rangle$, is opposite in sense but is usually larger, since it depends on $\sigma_\tau{}^2$, where σ_τ is the total dispersion due to intrinsic dispersion plus the dispersion resulting from the large mean errors in the trigonometric parallaxes. The values of $\langle M_V\rangle = M_{\langle p\rangle} + (5\langle \log p\rangle - 5 \log \langle p\rangle) + \Delta M_M$ for IIIab giants with $m_V < 5^m.0$ are collected in Table I. The values of the first correction were computed directly from the variance of the parallaxes by the approximation:

$$5 \langle \log p\rangle - 5 \log \langle p\rangle = - 5 \log c - 2.5 \log \langle (p')^2\rangle / \log \langle p'\rangle^2 ,$$

derived by Strömberg (1936). Here $\langle p'\rangle$ is the mean of parallaxes reduced to visual magnitude 4. The number of stars in each spectral group is necessarily small, averaging 15, because the number of stars brighter than the fifth magnitude with luminosity class IIIa and having nearly solar composition, is not great. There are a few for which good classification plates have yet to be taken. The eighth row of Table I is shown in Fig. 1. The scatter of the points is due to the small samples, but it is evident that simple graphical smoothing will define the luminosity of the giant branch to about $\pm 0^m.2$, which is about equal to the uncertainty due to the possible systematic errors in the system of trigonometric parallaxes (Schilt 1954, Gliese 1972, Norgaard-Nielsen 1977).

The luminosity calibration curves for classes II and IV are established mainly by interpolation, though there are a few bright giants in clusters of known distance, and a few subgiants with good individual trigonometric parallaxes.

The accuracy with which one can assign temperature types also is improved by the better separation of luminosities, and by the segregation of stars of markedly different chemical composition. For example, the comparison of MK types with (B-I) from Johnson et al. (1966) is shown in Fig. 2 for the same IIIab

TABLE 1

1977 CALIBRATION OF CLASS III BY MEAN PARALLAXES

	G7–G9	K0–K1	K2	K3–K4	K5–M0	M1–M2
$< p^1 >$ unweighted	0.0176	0.0186	0.0129	0.0185	0.0117	0.0104
$< p^1 >$ weighted	0.0192	0.0193	0.0165	0.0147	0.0108	0.0125
Mean $< p^1 >$	0.0184	0.0189	0.0147	0.0166	0.0115	0.0114
No. of Stars	15	13	14	17	17	13
$M_{<p>}$	+0.15	+0.24	−0.02	+0.11	−0.69	−0.71
5 log c	−0.32	−0.15	−0.34	−0.26	−0.45	−0.82
ΔM_n	+0.22	+0.22	+0.22	+0.22	+0.22	+0.33
M	+0.05	+0.31	−0.10	+0.07	−0.92	−1.20
Final smoothed $<M>$	+0.20	+0.18	+0.12	−0.15	−0.70	−1.36

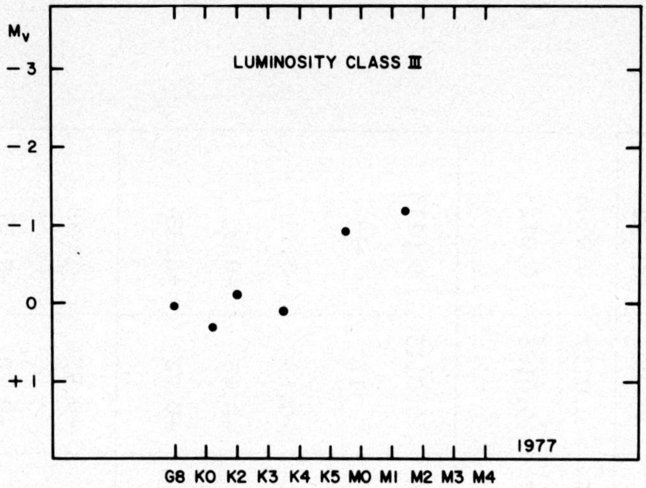

Fig. 1. Mean visual absolute magnitudes of the spectral groups, derived from trigonometric parallaxes.

giants of approximately solar composition represented in Fig. 1. The band would be appreciably broadened if IIIa and IIIb giants were included.

The resulting HR diagram is given in Fig. 3. The filled circles represent stars for which the trigonometric parallaxes are at least ten times their probable errors. The crosses show stars calibrated by their membership in binary systems, clusters or associations at reasonably well-known distances. For the stars shown as open circles the new spectroscopic absolute magnitudes were plotted. Although the diagram is intended to represent the stars of composition comparable to the Sun's, the ages necessarily increase somewhat from top to bottom. The brighter supergiants are obviously fairly young members of spiral arms, while the dwarfs must include an older disk population, but at least the obvious members of the halo population have been excluded on either kinematic or spectroscopic grounds.

The resolution in luminosity can be seen - the IIIa and IIIb giants form separate bands. The problems of classifying the most luminous supergiants are partly responsible for their apparent rarity between K0 and M0. RW Cep (HD 212466) is the coolest of the very brilliant stars not showing TiO bands and was somewhat arbitrarily placed at K0. It might as logically be called K2, which would narrow the gap and make the run of color indices more smooth. There is no doubt, however, about the real scarcity of K-type super supergiants.

A less conspicuous gap in the distribution is probably quite significant. Between G7 and K1 class II bright giants are quite rare. This deficiency is more pronounced than is shown, because

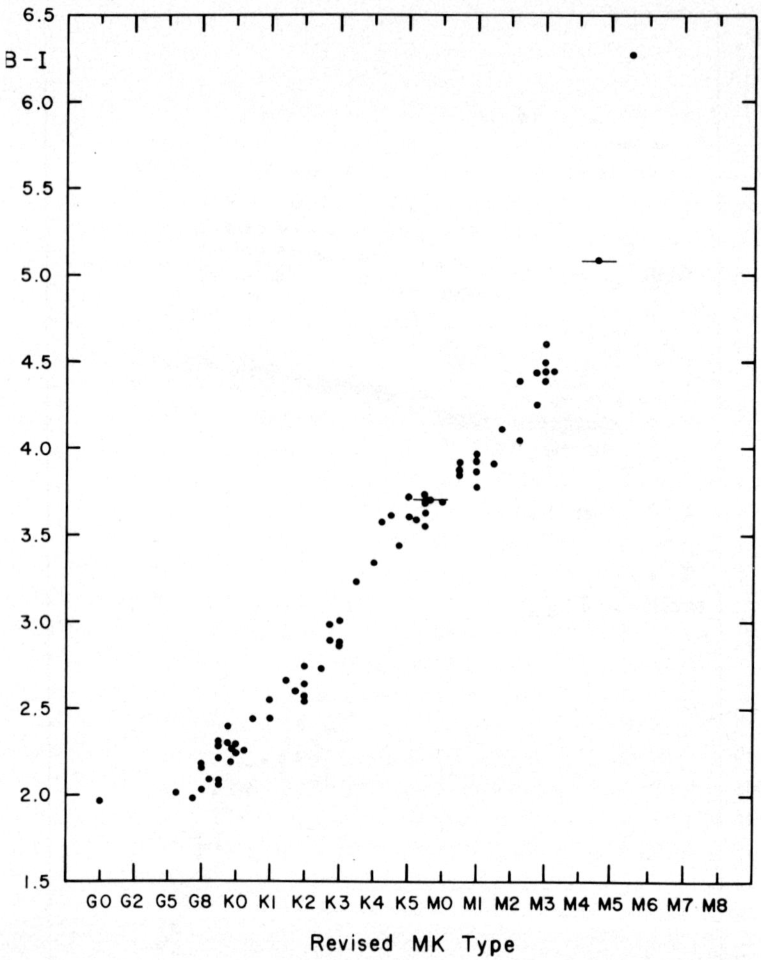

Fig. 2. (B-I) colors of IIIab giants as function of MK types. The two horizontal bars indicate stars with variable type.

---

we made an extensive search for G8 II stars in connection with the comparison of spectroscopic luminosities with Wilson's K-line absolute magnitude (Keenan and Wilson, 1977). This is the more striking since among earlier G-type stars there are a number of stars lying just above the Hertzsprung gap.

There are unrepresented kinds of stars which perhaps should be included - if we knew their individual distances. If the stars with an excess of heavy metals (barium and semibarium stars and those of type S) represent merely a particular stage in the evolution of stars having initially solar composition, then they belong in the diagram. We know statistically that they lie near the giant branch, but we do not have individual distances

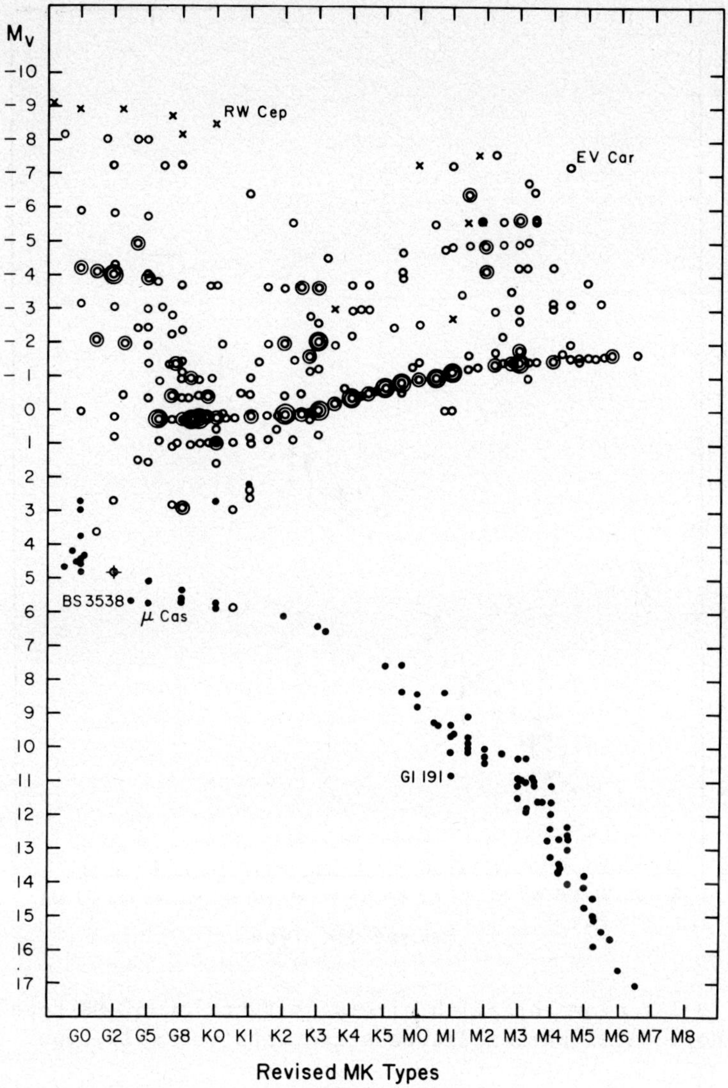

Fig. 3. HR diagram for stars of approximately solar composition. For the main sequence later than M0 most of the types are due to Boeshaar (1976).

for enough of them to make a reliable estimate of the real dispersion in their luminosities. Another very important group includes the so-called SMR stars (Spinrad and Taylor 1969), which appear actually to have only slight excesses of the iron-peak metals. Several of them, particularly BS 8924 (HD 221148) at K2.5, have notably weak lines of Ba II and Sr II. This suggests that BS 8924, at least, and possibly some others, might lie well below the giant

branch and extend the domain of the subgiants toward lower temperatures. The strength of CN in BS 8924, however, indicates a higher luminosity. The alternative interpretation is that these are giants in which practically none of the iron group elements have been converted to heavier metals by neutron fluxes. Attempts at a detailed atmospheric analysis of BS 8924 have not defined log g and the composition precisely because of difficulties in finding a satisfactory model for its atmosphere (Pilachowski, 1977).

REFERENCES

Boeshaar, P. (1976). Ph.D. Thesis, The Ohio State University.
Gliese, W. (1972). Quart. J. Roy. Astron. Soc. 13, 138.
Johnson, H.L., Mitchell, R. I., Iriarte, B. and Wisniewski, W.Z. (1966). Comm. Lunar and Planet. Lab. No. 63.
Keenan, P.C. and McNeil, R.C. (1976). An Atlas of Spectra of the Cooler Stars, Ohio State Univ. Press.
Keenan, P.C. and Wilson, O.C. (1977). Astrophys. J. 214, 399.
Ljunggren, B. and Oja, T. (1966). In IAU Symposium No. 24, Spectral Classification and Multicolor Photometry, K. Loden, L. O. Loden and U. Sinnerstad, eds. Academic Press: N.Y.
Pilachowski, C. (1977). Astron. and Astrophys. 54, 465.
Norgaard-Nielsen, H.U. (1977). Astron. and Astrophys. 59, 205.
Schilt, J. (1954). Astron. J. 59, 55.
Spinrad, H. and Taylor, B.J. (1969). Astrophys. J. 157, 1279.
Strand, K. Aa. (1958). Astron. J. 63, 152.
Strömberg, G. (1936). Astrophys. J. 84, 1936.

DISCUSSION

BELL: I would like to say how pleased I am to see Dr. Keenan's data for M_V versus spectral type. These data are needed in order to obtain stellar gravities for use in the theoretical determination of the spectral type - color - effective temperature - bolometric correction relations.

As an aside, I would also like to remark that observers who obtain log g by analyzing Fe I and Fe II line spectra could, and should, compare their values with values obtained from M_V.

MULLAN: Keenan has indicated the presence of a "gap" in the HR diagram at $M_V = -3$ or -4 and spectral types late G to late K. The left boundary of the "gap" seems to lie close to the locus of the onset of supersonic stellar winds. This locus has been determined recently in an attempt to test a speculation made by Durney (IAU Colloq. No. 19, Stellar Chromospheres, ed. S. Jordas and E. Avrett, 1973, p. 284) that the onset of supersonic stellar winds might be correlated with rapid mass loss. We used a minimum-flux corona model, and used chromospheric gas pressures (Kelch, et al. Astrophys. J. 1977, in press) to set firm upper limits on the gas pressure at the base of the corona. The locus of the onset of supersonic winds is found to agree well with the boundary of the circumstellar shell domain (Reimers, Astron. and Astrophys. **57**, 395, 1977) suggesting that Durney's speculation is probably correct. Keenan's gap also lies in roughly the same part of the HR diagram, and may therefore be correlated with rapid mass loss.

KEENAN: That's very interesting. And, of course, that's the general region where massive stars evolve very rapidly between the B-group and the M-group.

PEL: Your diagram did not cover the whole Cepheid strip, but I noticed quite a number of supergiants in the Cepheid domain. How many of these are known or suspected to be variable?

KEENAN: I've left out all Cepheid variables.

PEL: So this would mean a significant number of stable stars in the region where they should pulsate?

KEENAN: Yes, these are stars without marked variability. But some may be slightly variable.

SPECTROSCOPIC OBSERVATIONS OF LATE-TYPE HALO M DWARFS

Patricia C. Boeshaar

University of Washington

ABSTRACT

In the lower main sequence, the coolest halo population stars do not appear to be particularly subluminous in either the standard HR diagram or the (M_I, R-I) plane. Earlier type M subdwarfs (e.g., Kapteyn's Star) that lie one or more magnitudes below the main sequence exhibit enhanced hydride absorption bands relative to M dwarfs of the same temperature. From visual region (4500-7000 Å) low dispersion image tube spectrograms we find that the cool halo M dwarfs exhibit enhanced absorption in the CaOH 5530 Å band and Na I resonance lines. Easily noticeable changes in the strength of the CaH "B" band relative to that of the TiO bands was found to be more difficult to determine in halo dwarfs later than spectral type M4. Several M dwarfs with old disk type motions have spectral peculiarities similar to those of the halo dwarfs. In both cases, Balmer emission is not present in most stars, in the rest it is very weak. Therefore, the anomalous strength of the CaOH 5530 Å band and Na resonance lines appear to be a more conveniently utilized qualitative criterion in spectroscopically identifying very cool M dwarfs which are subluminous with respect to most main sequence stars of the same temperature class.

To be submitted to the Astrophysical Journal.

A. G. Davis Philip and D. S. Hayes (eds.), The HR Diagram, 21-22 and 405.

DISCUSSION

KODAIRA: You have told us that it is difficult to distinguish subdwarfs in HR diagrams. One of the reasons is obviously the weakening of TiO bands in the photometric passband V. The metal deficiency can probably shift a point only a little along the (B - V) axis, while the point is shifted at the same time upwards along the V axis. Thus the differences are strongly diminished. On the other hand, if you have an I versus (R - I) diagram, both the passbands I and R are rather insensitive to the variation by line-blocking. Therefore, I think you should plot stars in an I vs. (V-I) diagram, at least, in order to distinguish subdwarfs from the normal M dwarfs.

BOESHAAR: These are good points; however, many of the stars in my sample have no published I-magnitudes. Also, I did not use the absolute strength of the visual region TiO bands as the criterion in determining my temperature scale. I've compared my spectral types with Veeder's effective temperatures and find a good correlation. Note that I'm not calling these M dwarfs with enhanced Na "D" absorption subdwarfs, because we are not certain yet just what constitutes a true subdwarf for the coolest spectral types. I simply prefer to call these stars subluminous with respect to most main sequence stars of the same temperature class.

BIDELMAN: The M dwarfs present a truly serious classification problem. A priori, one might well expect substantial abundance variations in this group. If one uses only the TiO bands as a classification parameter, one is depending on a molecule containing two "heavy" elements, the strength of which may well depend on the M/H ratio. The D lines and CaH depend only on one element that may be deficient in high-velocity stars. These may well be more appropriate temperature classification parameters than the bands of TiO.

BOESHAAR: Mould has analyzed near-infrared coudé spectra of the early M-type subdwarf, Kapteyn's star, and obtained [Fe/H]$\sim$-0.5, [Ti/H]$\sim$-0.2. This is nothing spectacular compared with the abundances obtained for G-type subdwarfs. The metal hydride bands are known to be sensitive to changes in luminosity. In addition, the effect of pressure broadening in these late type stars is to strengthen the Na "D" lines.

BIDELMAN: If Mould's analysis is correct, then Kapteyn's star must be the star with the highest space velocity with that small a metal deficiency.

CAYREL de STROBEL: It's a pity that the sodium-enhanced stars in your sample cannot be given detailed analyses in order to find their sodium abundances. I remember that in the K-star region we have found a sodium-rich sample.

BOESHAAR: The spectra of certain M giants (such as 55 Peg shown on the previous slide) indicate sodium enrichment. However, I don't think that the strengthening of the Na "D" lines in the M dwarfs is solely an abundance effect. Regardless of whether the motion data indicate membership in the young disk, old disk, or halo populations,

CONTINUED ON PAGE 405

A COMPUTER METHOD FOR SPECTRAL CLASSIFICATION

I. Appenzeller, H. Zekl

Landessternwarte Heidelberg

Spectroscopic parallaxes are an important tool for determining distances of individual stars in the Galaxy. Unfortunately, for many astrophysical applications this method does not produce sufficiently accurate results for distant stars. Therefore, we started an attempt to improve the accuracy of spectroscopic parallaxes by evaluating spectroscopic temperature and luminosity criteria such as those of the MK classification spectrograms which were analyzed automatically by means of a suitable computer program. If necessary, the grain noise is reduced by Fourier filtering techniques. Then, the computer derives an approximate spectral type and the wave length scale from selected prominent spectral features. As a next step the computer searches the spectrum for adsorption features and produces an internal catalog of detectable absorption lines. Then, depending on the approximate spectral type, line ratios, line widths, and line depths of selected lines are calculated and compared to the corresponding values of standard stars. Since the investigation is still in progress, no final result on the achievable accuracy can be given here. However, preliminary results indicate that spectroscopic parallaxes derived by the computer method are at least as accurate as those derived from the conventional visual classification technique. A more detailed account of this investigation will be published elsewhere.

A. G. Davis Philip and D. S. Hayes (eds.), The HR Diagram, 23–24.

DISCUSSION

CAYREL de STROBEL: Could your computer method of spectral classification give us a temperature parameter as well as a gravity parameter?

APPENZELLER: In principle yes, provided that the temperature criteria used by the computer are calibrated correspondingly.

SHIPMAN: Does your computer program take line blends such as those that affect Hγ in the B stars into account?

APPENZELLER: If possible the computer separates blended lines by fitting line profiles to the individual components. If the lines are too close for this procedure, blends are treated as one absorption feature. (The Hγ line in early type stars can usually be separated from the adjacent weak absorption features.)

KODAIRA: How far down to faint objects can you apply this method? I ask this because a system of spectral classification must be applicable as homogeneously as possible.

APPENZELLER: This depends of course on the telescope, spectrograph and detector which you use. We use spectrograms of the same dispersion and widening as for conventional MK - classification. Thus, for a given instrument we have the same limiting magnitude.

KODAIRA: Do you mean that the method is independent of the dispersion of the spectra?

APPENZELLER: For the criteria used in the computer program we need a spectral resolution of about 2 Å. This requires a minimum dispersion of about 100 Å/mm if the spectra are recorded directly and about 40-50 Å/mm for image tube spectrograms.

LESH: How does the computer program distinguish between stellar and interstellar features, especially in regard to the K line? I should think that the presence of a strong interstellar K line in an early-type star would often lead the computer to conclude that the classification criteria are "contradictory".

APPENZELLER: If the computer recognizes the star as an early type star, the K-line is not used. In this case an interstellar K line has no effect. If the K line is so strong that the star is initially classified as "late type" the computer will not be able to assign a spectral type to the star. It will simply report that the spectrum does not fit into the classification scheme.

RED AND BLUE STARS IN THE DIRECTION OF THE LARGE MAGELLANIC CLOUD

M. Parthasarathy

Indian Institute of Astrophysics

SUMMARY

The Large Magellanic Cloud has been surveyed on ultra-low dispersion (10,000 Å mm^{-1}) plates taken with the 102 centimeter reflector at Kavalur. Many new red and blue stars were found. The LMC Bar was found to have a strong component of red stars and a very weak component of faint OB stars. In general the regions other than the bar contain very luminous OB stars, faint OB stars and a few very red stars. The 30 Doradus complex is rich in luminous red and blue stars.

1. INTRODUCTION

Study of the stellar content of the Large Magellanic Cloud is of fundamental importance to the understanding of stellar and galactic evolution. The determination of the relative number of blue, yellow, and red supergiants in clusters, associations and in the field will be of use in understanding the evolution of massive stars. From the objective prism surveys of Sanduleak (1970), Fehrenbach and Duflot (1970), and Westerlund (1960) a large number of LMC bright stars have been catalogued. In objective prism surveys, if one decreases the dispersion one can reach fainter stars and the crowding due to overlapping spectra can be reduced, but one looses information concerning each star. One must select features that are recognizable at very low dispersion. The use of ultra-low dispersion was first successfully employed by Morgan et al. (1954). At a dispersion of 30,000 Å mm^{-1} they were able to detect highly reddened OB stars

A. G. Davis Philip and D. S. Hayes (eds.), The HR Diagram, 25-28. All Rights Reserved.

in the VI Cygni association. Schulte (1956) employed this technique to identify new members of the VI Cyg association. More recently Bappu and Parthasarathy (1974, 1977) examined the ultra-low dispersion (10,000 Å mm^{-1}) technique in greater detail to study stars and galaxies.

2. OBSERVATIONS

We have employed an f/2 slitless spectrograph with a three degree quartz prism and a Schmidt camera at the Cassegrain focus of the 102 cm Ritchey Chretien reflector at Kavalur. The field is 40' in diameter. Spectra covering the wavelength range 3500 Å to 6600 Å were obtained on Kodak 103a-E emulsion. The dispersion is 10,000 Å mm^{-1}. The spectra are unwidened and are about 250 μ in length along the dispersion. From our initial survey of the 30 Doradus region (Bappu and Parthasarathy 1977), we found that this technique is undoubtedly a promising way to work to much fainter magnitudes and to detect blue and red stars in LMC. Now we have completed the survey of the regions around 30 Doradus and several regions along the bar and north of the bar.

3. CLASSIFICATION

The principal criterion for classification is the shape of the stellar spectrum and also the density in different portions of the image. Ultra-low dispersion spectra of different kinds of objects are shown in Fig. 1. On our LMC plates we have identified the stars classified by Sanduleak (1969) and Westerlund (1960) to calibrate our classification. The classification criteria are checked with the ultra-low dispersion spectra of our stars in NGC 2362 and h and χ Persei. When we compare our ultra-low dispersion technique with multicolor photographic photometry we find that the ultra-low dispersion spectra contain a number of sigificant information bits and in addition we have the advantage of exposing all parts of the spectra simultaneously for a large number of stars, with consequent saving of time.

4. RESULTS

From our survey we have detected a large number of new OB stars and red stars. The density of luminous OB stars is much higher in the LMC clusters and associations than in galactic ones. Luminous red and blue stars are very numerous near the 30 Doradus complex, but the red stars avoid the regions rich in gas and H II regions. In general the regions other than the Bar contain very luminous OB stars and also faint OB stars, and a few red stars. Stars of intermediate spectral type are not common.

In the LMC bar region the number of red stars in the magnitude interval 14-16 is low. There is a strong component of fainter red stars. From a study of two color composite photographs of LMC Walker et al. (1969) concluded that the Bar contains mostly Population II red stars. We find that in the LMC Bar the number density of stars of intermediate spectral type in the magnitude interval 16-18 is very low. Few faint OB stars (~ 16-17 mag) are seen in the bar.

Most of the blue stars and very red stars are obviously LMC members. A large percentage of faint stars of intermediate spectral type could be foreground stars. The presence of a strong component of red stars in the Bar of the LMC indicates that the earliest generation of stars were formed in the Bar. Tifft and Snell (1971), from a study of color-magnitude diagrams of the Bar regions, also find a very old component and also stars of intermediate ages.

The author is grateful to Prof. M. K. V. Bappu for his valuable suggestions and encouragement during the course of this work.

REFERENCES

Bappu, M.K.V. and Parthasarathy, M. (1974). Bull. Astron. Soc. India **2**, 36.
Bappu, M.K.V. and Parthasarathy, M. (1977a). Kodaikanal Obs. Bull. Ser. A. No. 2.
Bappu, M.K.V. and Parthasarathy, M. (1977b). Kodaikanal Obs. Bull. Ser. A. No. 8.
Fehrenbach, Ch. and Duflot, M. (1970). Astron. Astrophys. Special Suppl. Ser. L.
Morgan, W.W., Meinal, A.B. and Johnson, H.M. (1954). Astrophys. J. **120**, 506.
Sanduleak, N. (1970). Contr. Cerro Tololo Obs. No. 89.
Schulte, D.M. (1956). Astrophys. J. **123**, 250.
Tifft, W.G. and Snell, Ch. M. (1971). Mon. Not. R. Astron. Soc. **151**, 365.
Walker, M.F., Blanco, V.M. and Kunkel, W.E. (1969). Astron. J. **74**, 44.
Westerlund, B.E. (1960). Uppsala Astron. Obs. Ann. **4**, No. 7.

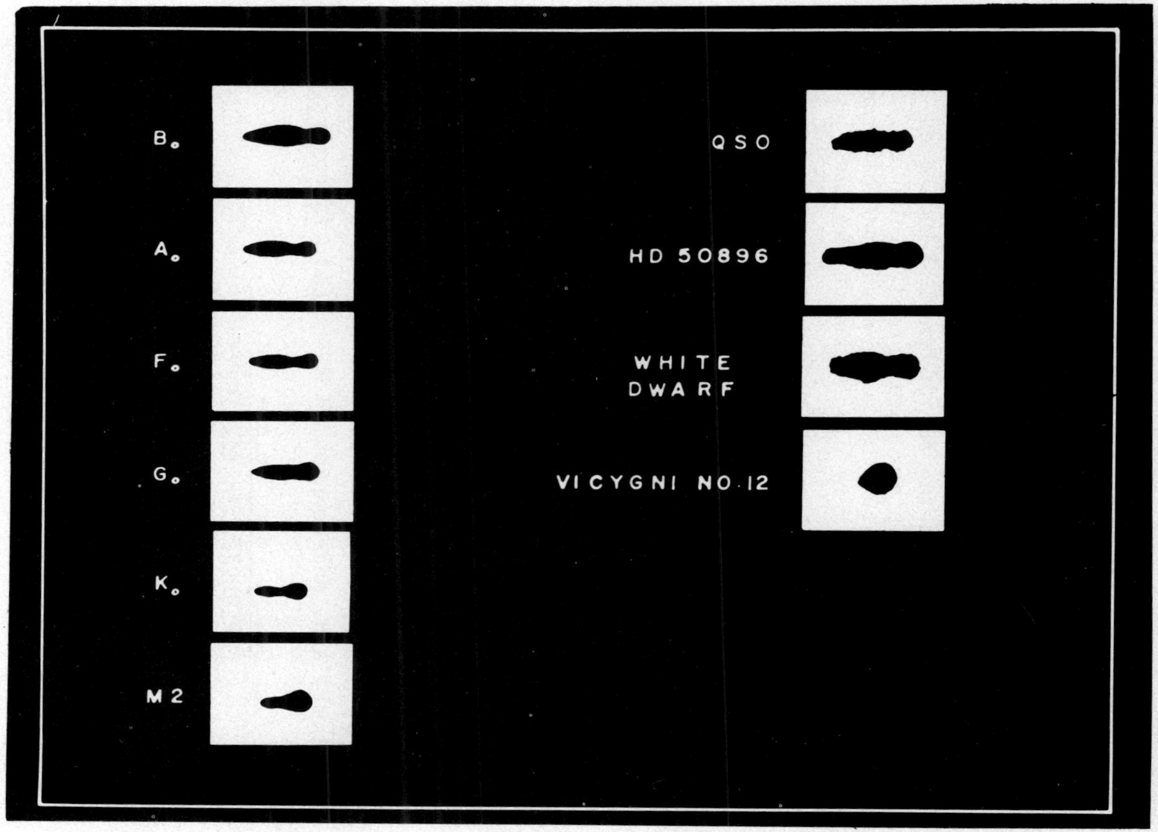

Fig. 1. Ultra-low dispersion spectra of different kinds of objects. Longer wavelengths are to the right. The discontinuity in the spectrum is due to the dip in the sensitivity of the 103 a-E emulsion in the green region.

QUANTITATIVE TWO-DIMENSIONAL CLASSIFICATIONS OF LOW DISPERSION OBJECTIVE PRISM SPECTRA

Th. Schmidt-Kaler

Astronomisches Institut Bochum

The objective prism (or grating) is the most efficient stellar spectrograph since it suffers no light losses at a slit and - compared to a slit spectrograph - only small losses in the camera optics. However, objective prism spectra have two principal disadvantages:

(1) the spectral resolution (and consequently the sharpness of spectral features) depends on the seeing, which may vary from plate to plate.

(2) quantitative measurements are difficult because of the difficulty in obtaining an exact photographic calibration; no comparison spectrum can be placed near the stellar spectra.

Therefore, low dispersion objective prism spectra have generally been used only for very rough estimates of the spectral types and for surveys for certain natural groups like the OB-stars.

In the past years Schmidt-Kaler et al. (1976) have developed methods to determine accurate, quantitative, two-dimensional classifications from objective prism spectra. We circumvented the first difficulty by restriction to equivalent widths (or similar quantities) which are - as a measure of energy - in principle independent of the variable instrumental profile. We overcame the second difficulty by constructing the characteristic curve of the plate directly from measurements of spectra on it near 4400 Å (and 3500 Å) and photoelectrically measured intensities of the same stars in the B (and U) band of the UBV system. Corrections to the broad band magnitude B (and U) have been computed to obtain the monochromatic magnitude near 4400 Å for

A. G. Davis Philip and D. S. Hayes (eds.), The HR Diagram, 29–32.

various spectral types and reddenings; these corrections are $0^m.02$ or less if the photoelectric calibration sequence is restricted to 0-F5 stars with $E_{B-V} \lesssim 1^m$.

The line-free continuum was determined by inspection of high-dispersion spectral atlasses. In the range 0-F5 we found for the dispersion 570 Å/mm of the Hamburg Schmidt plates 9 useful windows in the blue-UV region. In the case of later type stars a pseudo-continuum was defined and an equivalent area (= line depth x half width) **replaced** the equivalent widths when the wings of the lines were heavily disturbed by blends.

For a realistic test of the method we took three OB-associations on plates of the Hamburg-Cleveland survey. The internal s.e. of the equivalent widths of the Balmer lines H_γ and H_δ was about 1.5 Å (1 plate) or more exactly, for $H = \frac{1}{2}(H_\gamma + H_\delta)$,

$$\varepsilon(W_H) = 0.7 + 0.07 W_H \text{ [Å]}.$$

This is confirmed by comparisons with results from slit spectra, and with photoelectric measurements of $H\beta$.

Measures of the Balmer jump and color-indices were determined with an error of about $\pm 0^m.03$ (1 plate). On the basis of these measurements (basically the D/W_H diagram as in the Paris system and/or $(U-B)_o/W_H$) it was possible to classify the stars in the domain B0-F2 quantitatively in two dimensions with about the same accuracy as the MK-system. This statement is based on 194 stars classified in the MK-system (Fig. 1). At even lower dispersion we could use two plates taken with the thin prism at the Michigan Schmidt (1400 Å/mm) at CTIO which Dr. Duerbeck kindly put at our disposal.

The internal accuracy of the Balmer line equivalent widths was 1.6 Å and the Balmer jump turned out to be almost the same so that about the same classification accuracy is expected for B0-F2 stars. However, for a check only HD spectral types were known in sufficient number.

Considering the tremendous number of spectra on a typical Schmidt plate it is obvious that the procedure had to be automatized. For this purpose the microphotometer has been connected to an HP 2114B computer, and measurements, reductions and classification processes have been completely computerized. The total time for the classification of one spectrum is 2.5 min including visual controls, and 30 sec without these.

The collaboration of H. Unger is gratefully acknowledged.

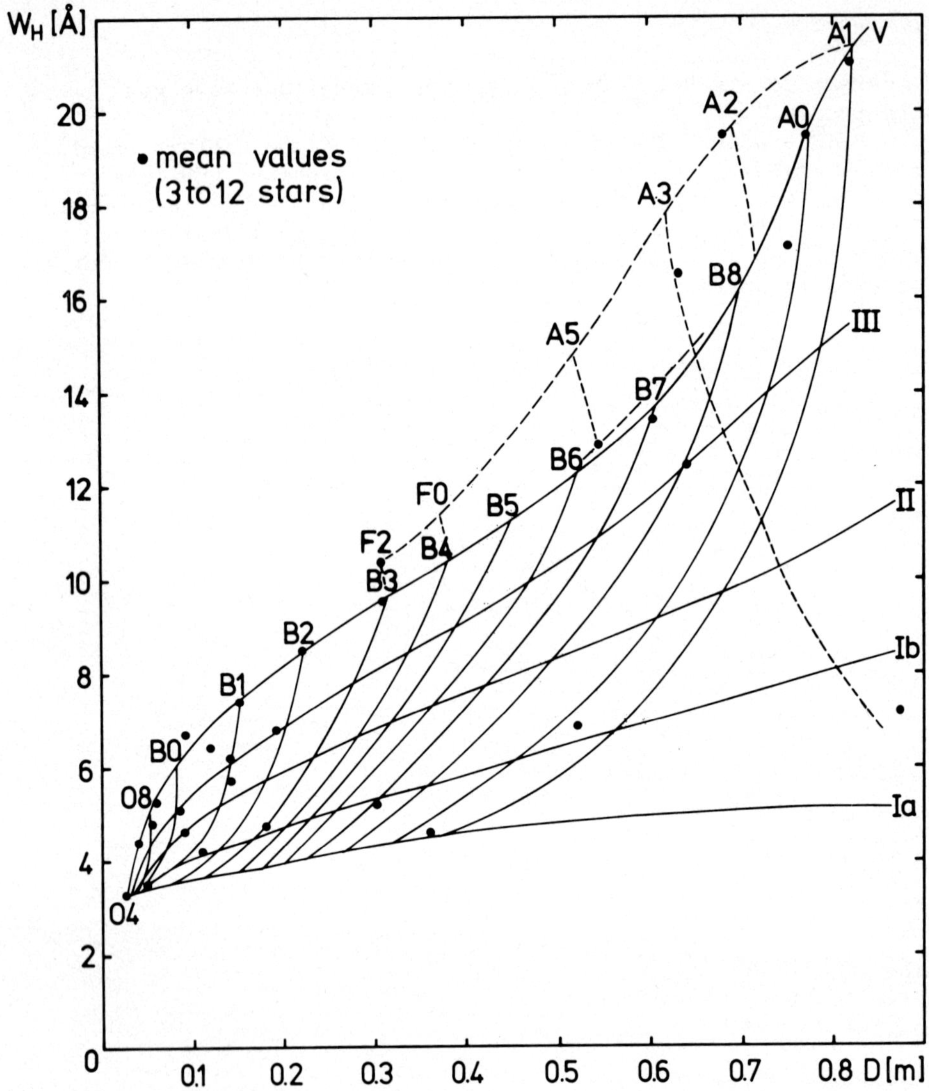

Fig. 1. Two-dimensional spectral classification of early-type stars by Balmer jump (D) and Balmer line intensity (W_H) from Hamburg Schmidt plates.

REFERENCES

Schmidt-Kaler, Th., Diaz-Santanilla, G., Rudolph, R. and Unger, H. (1976). Quantitative Klassifikation von Sternspektren mittels Objektivprismenaufnahmen. Forschungsberichte des Landes Nordrhein-Westfalen Nr. 2595, Westdeutscher Verlag, Opladen.

DISCUSSION

CAYREL de STROBEL: To what apparent magnitude have you observed?

SCHMIDT-KALER: The limiting magnitude of the Hamburg Schmidt survey plates is about 13^m. With longer exposure and less widening you can certainly reach 15^m. With the Michigan Schmidt (very low dispersion) on Cerro Tololo (very good seeing) you can reach 16^m or 17^m. The most serious problem is overlap of spectra in crowded areas.

CARBON AND LATE TYPE M STARS AS POPULATION INDICATORS

V. M. Blanco

Cerro Tololo Inter-American Observatory*

A. A. Hoag

Kitt Peak National Observatory*

M. F. McCarthy**

Vatican Observatory

The discovery of different ratios of carbon (C-type) to oxygen stars (M type) in the direction of the central regions of our own and the nearest galaxies, the Large and Small Magellanic Clouds, is the result of surveys made by Blanco, Blanco and Hoag (unpublished) who observed Baade's Window near NGC 6522 in the direction of our Galaxy's nuclear bulge and by Blanco, Blanco and McCarthy (in press) who observed selected regions in the center of the Large and Small Magellanic Clouds. We have observed each of these central regions with a combination of a grating and prism set in the converging beam of the prime focus of the CTIO 4 meter telescope (Hoag 1976). We used Eastman IV-N plates hypersensitized with silver nitrate solution; the spectral plates were exposed through a Wratten 89 filter for 60 min. We limited our spectral survey to M type stars later then M5 and to the carbon stars, using the criteria of Nassau and colleagues (1949). Thus far no S type stars have been discerned in central regions of the LMC or SMC at the dispersion used (2300 Å/mm).

The results for Baade's Window, a region 0.12 sq degree in area near NGC 6522 (α=18^{h}02^m, δ=-30°.0) are shown in Fig. 1. We note near m_c=12^m.6 a maximum in the frequency of M stars found; this value combined with an adopted distance modulus from the work of Oort and Plaut (1975) of 15^m.4 leads to a mean infrared absolute magnitude

* Operated by AURA for the National Science Foundation
**Visiting Astronomer, Cerro Tololo Inter-American Observatory

A. G. Davis Philip and D. S. Hayes (eds.), The HR Diagram, 33-38 and 405. All Rights Reserved.

M_i, of $-2^m\!.8$. With an assumed intrinsic (V-I) color of $+3^m\!.6$ this corresponds to a mean visual absolute magnitude for the M giants in the nuclear bulge region of approximately $+1^m\!.0$. There is only one C star observed in this region.

In Fig. 2 we present the counts according to m_i of late type stars for 3 central regions in the LMC. Each of these regions are 0.12 sq. deg. in area. They include the optical center ($\alpha=5^h24^m\!.5$, $\delta=-59^\circ\!.8$), the radio center ($\alpha=5^h20^m\!.2$, $\delta=-68^\circ\!.9$) and a region we have denoted Bar-west ($\alpha=5^h08^m\!.9$, $\delta=-68^\circ\!.9$) studied also by Tifft and Snell (1971). In the LMC the observed maximum frequency of late M stars occurs near $m_i=14^m\!.1$. Thus with an assumed distance modulus of $18^m\!.6$ we find a mean infrared absolute mag. $M_i=-4^m\!.6$ and a corresponding $M=-1^m\!.0$. This confirms the earlier findings by Blanco and McCarthy (1975) who used a thin-prism with the Curtis Schmidt at CTIO in a study of late type M stars. The more abundant carbon stars also show a maximum near $m_i=14^m\!.0$ in central regions of the LMC.

In Fig. 3 we display counts of late type stars grouped in magnitude intervals for the two SMC areas at $\alpha=0^h49^m\!.5$, $\delta=-73^\circ\!.4$ and at $\alpha=1^h00^m\!.0$, $\delta=-73^m\!.2$. Four late M stars were discovered in these regions, too few for a meaningful discussion of their mean absolute magnitudes. The 135 C stars show a maximum concentration near $m_i=14^m\!.4$.

One of the surprising results of this study is the apparent difference in the mean absolute magnitude derived for M giants in the nuclear bulge of our Galaxy ($M_V=+1^m\!.0$) and in the direction of the central regions of the LMC ($M_V=-1^m\!.0$). We note obvious uncertainties in the mean (V-I) colors of late type M stars. The results shown by Clube and Dawe (1978) at this Symposium point up the possibilities of revisions in the distance modulus to the Clouds and galactic center. Nonetheless the difference of 2 magnitudes in mean visual absolute magnitudes of late M stars in the two galaxies is striking.

Perhaps more puzzling in the present observations of late type stars is the difference in the three galaxies of the ratios of the two different kinds of cool stars discussed here: the carbon and late type M stars. Outstanding is the almost complete absence of carbon stars in the nuclear bulge of the Galaxy and the corresponding lack of late type M stars in the central regions of the Small Magellanic Cloud, with a 2:1 ratio of carbon to late M stars in the cool population of the Large Cloud. Parallels with population indices concerning an ordering in terms of decreasing metallicity are most attractive. Studies by Gascoigne (1969), Payne-Gaposchkin(1971), Sanduleak, MacConnell and Hoover (1972), Oort (1971), Graham (1974) and Hagen and van den Bergh (1974) have pointed out differences between the LMC, SMC and the Galaxy.

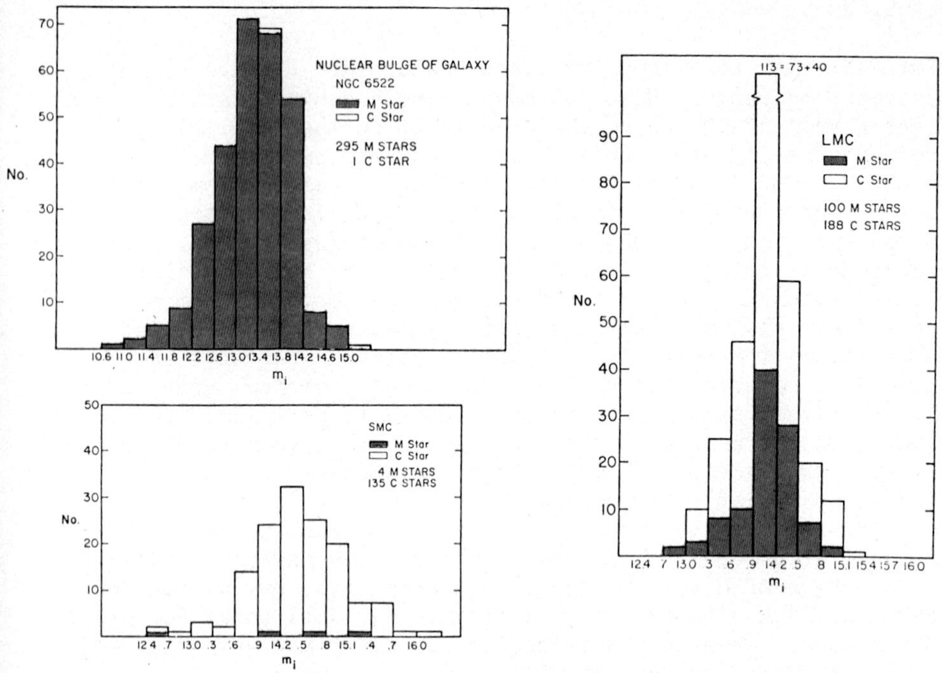

Figs. 1-3. Number of late type giants. 1, Nuclear bulge of the Galaxy; 2, Central region of LMC; and 3, Central region of SMC.

REFERENCES

Blanco, V. and McCarthy, M.F. (1975). Nature 258, 407.
Clube, S.V.M. and Dawe, J.A. (1978). In IAU Symposium No. 80, The HR Diagram, A.G.D. Philip and D.S. Hayes, eds., Reidel, Dordrecht, p.53.
Gaposchkin, C.P. (1971). The Magellanic Clouds, A.B. Muller, ed., Reidel, Dordrecht, p. 34.
Gascoigne, S.C.B. (1969). Mon. Not. Roy. Astron. Soc. 146, 1.
Graham, J. (1974). ESO/SCR/CERN: Conference on Research Programs for the New Large Telescopes, A. Reiz, ed., Geneva, p. 159.
Hagen, G. and van den Bergh, S. (1974). Astrophys. J. 189, L103.
Hoag, A.A. (1976). Publ. Astron. Soc. Pacific 88, 860.
Nassau, J.J. and Van Albada, G.B. (1949). Astrophys. J. 109, 391.
Oort, J.H. (1971). The Magellanic Clouds, A.B. Muller, ed., Reidel, Dordrecht, p. 185.
Oort, J.H. and Plaut, L. (1975). Astron. Astrophys. 41, 71.
Sanduleak, N., MacConnell, D.J. and Hoover, P.S. (1972). Nature 237, 28.
Tifft, W. and Snell, C. (1971). Mon. Not. Roy. Astron. Soc. 151, 365.

DISCUSSION

BIDELMAN: Strictly speaking, carbon stars are defined by the presence of C_2 bands. What you have found are stars showing strong infrared bands of CN, which may or may not be the same thing. There are other types of objects that show strong infrared CN.

McCARTHY: Our findings are based on low dispersion combination prism-grating plates. Certainly higher dispersion spectra are required and multicolor photometry will be of great value. Our survey reaches to $m_i = 18^m.5$ and to a visual mag near 21^m. During our observations we have secured besides the spectral plates enough direct plates to allow positional measures and magnitude determinations for charting and identification. We don't have enough plate material for precision photometry or for variable star work. When charts are published all observations will be welcomed.

PEERY: Do you know what types of carbon stars you are observing, e.g., variability characteristics such as long-period variability, or irregular variable carbon stars?

McCARTHY: At present we have a list of 280 carbon stars plus positional data (derived by Mrs. B. Blanco) and magnitude data. Our collection of direct plates is not extensive, but we can now compare our list with existing catalogues and data books for stars in the clouds. These correlations are an important next step and all suggestions of colleagues will be most welcomed.

PEL: I would like to make a radio-astronomical remark. Although radio astronomy is generally of little use in the study of the HR diagram, there is one possibility that not all stellar astronomers may be aware of: the 18-cm, OH maser-emission from long-period variables. The radio-sensitivities are good enough to observe such masers throughout the whole Galaxy, and maybe also in the Magellanic Clouds. It seems that OH-maser M stars are always oxygen-rich, and never carbon-rich. Furthermore, there are very interesting indications of correlations of the radio properties with stellar properties such as the period. I would like to urge the optical astronomers to put more effort into the identification and study of the known radio masers at optical and infrared wave-lengths, and I would like to suggest that radio astronomers, particularly at Parkes, search carefully for OH masers in globular clusters and in the Magellanic Clouds.

McCARTHY: Thank you for this suggestion. It should prove most helpful and once our charts and positional and photometric data are published could be most effective in verifying and explaining our findings.

SCHLESINGER: Do you have any information on space densities? The reason for asking is that some of your results, in particular, the increase in ratio of C/M stars going from the Galaxy to the Small Cloud, could be connected with the trend in metal abundance, decreasing from the Galaxy to the Small Cloud. When heavy element abundance goes down the Hayashi track is shifted to the blue. So some stars that would appear at M5-M9 in the Galaxy would not get as red in the clouds, but would be shifted to earlier spectral types

and not be caught in your survey. The change in ratio of C stars to M stars could then be a change in the number of M stars.

McCARTHY: Preliminary calculations by Blanco of the space density for the late M stars observed in the direction of the galactic nuclear bulge leads to about 35 stars per 10^6 pc^3 which is roughly 250 times the density observed for these stars in the solar neighborhood. The first magnitude estimates for the C and M stars in the LMC and SMC are just completed and so we can soon derive ratios for these.

Our position in giving now these first results has been to communicate our new findings as soon as they were reasonably ordered. We have tried to be aware of similarities with other studies of the Clouds and the Galaxy. We acknowledge the need of stellar models for late type giant stars and have deliberately avoided adopting a position regarding the history or evolution of these objects.

van den BERGH: These are very exciting results! One might think that the observed variations in N(C)/N(M) are due to changes in [Fe/H]. But probably life is more complicated that this because metal-poor clusters in the Clouds contain carbon stars at the tips of their red giant branches whereas not a single galactic globular is known to contain such stars. This indicates that N(C)/N(M) probably correlates with another parameter. It would be particularly important to find N(C)/N(M) in the galactic anti-center direction in which [Fe/H] is also thought to be low.

McCARTHY: I agree completely on the importance of studies in the direction of the anti-center and note that two Hoag grisms (with different dispersions and blazes) are now available for use at the KPNO 4 meter. Blanco, Hoag and I did not make any hypotheses in undertaking this study and were surprised at the C star anomaly and the marked difference in the absolute magnitude of M stars in the direction of galactic nucleus and in the center of the LMC.

LLOYD EVANS: Variable star surveys of the SMC carried out with the 74-inch telescope at Pretoria show that stars with the colors of carbon stars and in the magnitude range $16<V<18^m$ are mostly variable. The distribution of their periods is very similar to that of semi-regular carbon stars in the Galaxy.

McCARTHY: A most important desideratum will be the intercomparison of our C and M stars with known lists of variables. I shall be most anxious now that our positions have been determined to compare our lists with yours. We know that the domain of variable stars has been the most well explored in the Clouds and we have a legacy which extends from Pickering and Leavitt to the Gaposchkins and you and your colleagues from Pretoria. Confrontation of results should be fun.

WALBORN: The question is, what is this very interesting result telling us? I think it would be very difficult to understand the C to M ratio you find in the SMC in terms of the evolutionary interpretation given to these types in the Galaxy. I wonder if the result may be related to the decreasing metallicity from the galactic center through the LMC to the SMC. (1) If your SMC C stars corres-

pond to the evolutionary state of Galactic C stars, where are the SMC M stars? If the first red giant phase for metal-deficient stars occurs at higher temperatures, as suggested by Dr. Schlesinger, then your red survey may be missing them in the SMC. (2) On the other hand, one can consider the possibility that your SMC "C" stars correspond to galactic M giants in an evolutionary sense. It would be interesting to have detailed spectral syntheses for cool, very metal-deficient giants, to see what the strongest spectral features in this wavelength range should be. (Dr. van den Bergh has pointed out to me that the validity of this suggestion could be easily checked with knowledge of the colors of your stars.)

McCARTHY: The results of our observations are, I believe, telling us two things: (1) The LMC M giants with mean $m_i = 14\overset{m}{.}0$ and a distance modulus of $18\overset{m}{.}6$ have an $M_i = -4\overset{m}{.}6$ corresponding to $M_V = -1\overset{m}{.}0$, similar to the usual absolute magnitude found for the solar neighborhood. The nuclear bulge M giants with a mean $m_i = 12\overset{m}{.}6$ have a mean $M_i = -2\overset{m}{.}8$ corresponding to $M_V = +0\overset{m}{.}8$. We assume here a mean (V-I) color equal to $+3\overset{m}{.}6$ and are surprised at the apparently marked luminosity difference for M giants in the LMC and in the galactic nucleus.

(2) If we order the 3 galaxies according to decreasing metallicity as indicated by the Gaposchkins, Graham and others we would list the galactic nuclear bulge, then the LMC central region and finally (the weakest in metallicity) the SMC. This reflects the same ordering in the ratio of the number of C stars to the number of late M type stars but in a reversed sense; i.e. almost no C stars are seen in the nuclear bulge where M giants abound; C stars outnumber M stars 2:1 in the LMC central areas; in the SMC practically no M stars are found and C stars are most frequent in occurence.

As is pointed out in the printed text the argument pointing to decreasing metallicity has analogues to the findings by C.P. Gaposchkin, Thackeray, M. Burbidge, S. Gascoigne and others, who emphasize the polarity of the SMC and the Galaxy with the LMC in an intermediate state.

We need now, more than ever, models for stars in this part of the HR diagram. Carbon stars remain an astrophysical enigma.

Our magnitude work is for identification purposes. We have red and infrared plates and have established that these are red stars.

HUMPHREYS: Many red stars observed in an unpublished objective-prism survey by Blanco which covers the entire LMC have been confirmed to be M supergiants. These luminous red stars are normal when compared to the supergiants in our Galaxy. Although the stars discussed by McCarthy are giants, at least the more luminous M stars do not show any evidence for differences in their properties when compared to the same stars in our Galaxy.

McCARTHY: Blanco's earlier survey is based on 4° objective prism Curtis Schmidt plates. Our thin prism work together in 1974 was also done with the Curtis Schmidt at CTIO (Nature, Dec. 1975). Blanco's first survey has many supergiants and not the giants detected in the thin prism survey or in the current study. Walter Baade once advised

CONTINUED ON PAGE 405

MAIN SEQUENCES DEFINED BY HYADES AND FIELD STARS

Arthur R. Upgren

Van Vleck Observatory, Wesleyan University

1. INTRODUCTION

This paper reviews the main sequences defined by members of the Hyades cluster and by the field stars in the solar neighborhood. For this purpose, the discussion is limited primarily to the stars of the lower portions of the main sequence, especially those of spectral classes K and early M. There are two reasons for emphasis on the faint red dwarf stars. First, the value of a parallax depends on its size or, more accurately, on the error in parallax divided by the parallax itself. Large parallaxes of high precision occur in large numbers only for stars inhabiting the lower main sequence. Furthermore, brighter stars of earlier spectral classes are more likely to be influenced by evolutionary effects which may differ between the Hyades and field stars, and which are difficult to calibrate.

For these relatively cool red stars, it is well known that the (M_V, R-I) or (M_V, V-I) color magnitude diagram has a considerable advantage over the more conventional (M_V, B-V) diagram in determining the loci of main sequences. This is due to the linearity of the main sequence, at least over the K2-M2 spectral range, and also to the much lower sensitivity of (R-I) or (V-I) color indices to variations in metallicity than the (B-V) index, while remaining good indicators of surface temperature. This paper will concentrate on the photometric results in (R-I) since a considerable amount of observational data is now available, unlike (V-I), for both Hyades and field stars. The linearity was noted by Kron (1954, Kron et al. 1957) in his development of his R, I system of photometry, and later by Eggen and Greenstein (1965).

A. G. Davis Philip and D. S. Hayes (eds.), The HR Diagram, 39–48.

2. THE HYADES

Although photometry is a useful method of identifying members of a cluster such as the Hyades, and distinguishing them from field stars, photometry of stars which have been independently recognized by their proper motions as highly probable members, makes identification more certain since contamination by field stars becomes a problem especially at appreciable distances from the cluster center. Colors in the (B-V) and (R-I) systems are the only ones which cover a large number of the members found in the proper motion surveys by van Bueren (1952), for the stars with visual magnitude V generally brighter than 9^m, and by van Altena (1969) and Pels et al. (1975) for fainter stars. Many more stars with proper motions resembling the Hyades have been recognized, mostly through the work of Luyten (1971) but photometry is not yet available for them.

The photometric observations of Hyades members in the UBV system have been made by Johnson and his collaborators (Johnson and Knuckles 1955, Johnson et al. 1962) and extended by Upgren and Weis (1977). Observations in R and I were obtained by Eggen (1968, 1969) for a few of the Hyades identified in van Altena's proper motion study. Systematic differences between these two papers were noted and evaluated by Upgren (1974) and also by Eggen (1974) where revised values for Eggen's earlier observations are given. Upgren obtained (R-I) measures for 56 of van Altena's stars including all of those in the three papers by Eggen (Upgren 1974). Since then, Upgren and Weis (1977) have obtained BVRI photometry for the 119 newly identified members found by Pels et al. at Leiden. Some overlap occurs between the Leiden stars and the list of van Altena.

Other photometric systems have been used to measure the magnitudes of Hyades stars and to determine the distance to the cluster. The numbers of stars observed, especially those identified as members from their proper motions, is generally very small; some of these results are summarized by van Altena (1974) who shows that all but Eggen's (R-I) values lead to a Hyades distance modulus about $0^m.2$ larger than that of $3^m.05$ found by van Bueren (1952) from a convergent-point solution from proper motions. Since then, Hanson (1975) and Upgren (1974) have shown that the convergent-point method and the (R-I) sequence fit to field stars, respectively, also yield moduli of more than $3^m.2$.

3. THE FIELD STARS

The locations of main sequences defined by nearby field stars in color-magnitude diagrams are best determined from trigonometric parallaxes. However, care must be taken to avoid

systematic errors introduced by a lower limit in parallax among the stars selected. Such an error can be introduced because the observed parallaxes are, on average, spuriously too large and the high-parallax tail of the frequency distribution of more distant stars will be erroneously included. Furthermore, Lutz and Kelker (1973) have shown that a systematic error is present in any set of observed parallaxes and have calibrated the necessary corrections to absolute magnitudes derived from parallaxes, including even the largest ones. Even the rather high lower limit in parallax of 0".125, which has often been adopted in defining main sequences (e.g. Eggen 1968) introduces a systematic error of about $0^{m}.2$ according to Lutz and Kelker, since the true external mean errors in parallaxes, $\varepsilon(\pi)$, are well known to average about ±0".016.

Lutz and Kelker show that the error in absolute magnitude is dependent upon the ratio $\varepsilon(\pi)/\pi$ and cannot be calibrated whenever this ratio is larger than about 0.2. (Investigations are now in progress which aim to extend the calibration to larger ratios but they are inconclusive at the present time.) The constraint imposed by this limit reveals the great value gained in decreasing the size of parallax error. A twofold reduction in $\varepsilon(\pi)$ leads to an eightfold increase in the volume of space over which the condition of $\varepsilon(\pi)/\pi < 0.2$ holds, and hence, an eightfold increase in the number of stars available for a determination of a main sequence. In recent years, the Van Vleck Observatory has become the first to achieve such a reduction in $\varepsilon(\pi)$ and apply it to obtain parallaxes of this precision for a large number of nearby dwarf stars. For more than 100 stars from Vyssotsky's lists of dwarf stars found spectrophotometrically (Vyssotsky 1963 and references therein) the Van Vleck parallaxes show a mean external error of only ±0".008 (Upgren 1975). The mean position of the main sequence has been calculated for these stars and compares closely with the one found by Gliese (1969a, 1971) from his catalogue of nearby stars. For intervals of (R-I) the main sequence found from the Van Vleck parallaxes is given in Table I in the second column, and also in the third column after corrections are applied as determined by Lutz and Kelker. Gliese's (1971) sequence is also shown, in the last column, and is in close agreement with the previous column. The dispersion about the main sequence of dwarfs inhabiting the solar neighborhood has been shown to be about $\pm 0^{m}.4$ by Upgren (1973) after the elimination of the component of dispersion produced by errors in the parallax determinations. The intrinsic dispersion of $\pm 0^{m}.4$ is consistent with a more recent result of Veeder (1974) from a detailed analysis of bolometric magnitudes of similar stars. In his paper, Veeder notes that the intrinsic dispersion of the lower main sequence is about the same in the (M_V, B-V) color-magnitude diagram as it is in the (M_V, R-I) and (M_{bol},R-I) diagrams.

TABLE I. MEAN M_V, (R-I) RELATIONS FOR NEARBY DWARF STARS

(R-I)	Uncorrected Van Vleck	Corrected Van Vleck	Gliese
$0\overset{m}{.}35$	$7\overset{m}{.}0$	$6\overset{m}{.}6$	$6\overset{m}{.}5$
0.40	7.3	6.9	6.9
0.45	7.5	7.2	7.3
0.50	7.8	7.4	7.7
0.55	8.0	7.7	8.0
0.60	8.3	8.0	8.3
0.65	8.6	8.3	8.6
0.70	8.8	8.6	8.9
0.75	9.1	8.9	9.2
0.80	9.3	9.2	9.5
0.85	9.6	9.5	9.7
0.90	9.9	9.8	10.0
0.95	10.1	10.1	10.3
1.00	10.4	10.4	10.6
1.05	10.6	10.7	10.8

The main sequence and its dispersion are unlikely to be in error by any significant amount. However, they are derived from many red dwarfs in the solar neighborhood. Variations in the main sequence caused by differences in kinematical or chemical properties among the stars are much more difficult to measure. This is especially true for the (B-V) colors where the slope of the lower main sequence is very steep and where ultraviolet excess and line blanketing, so well known for the F and G dwarfs and subdwarfs, cannot be extended to the lower main sequence stars with confidence (see e.g. Greenstein (1965)). Among these fainter stars, the (R-I) colors are more likely to distinguish stars with differences reflecting differences in age, even though (R-I) is rather insensitive to stellar properties other than temperature, at least among the unevolved stars. The R and I indices are affected in an unknown manner by TiO blanketing since TiO bands extend beyond both of these colors. For stars later than about M0 (or (R-I) larger than about $0\overset{m}{.}7$), several investigators found high velocity dwarfs (which are likely to be old and metal deficient) to be fainter than their low-velocity counterparts of the same (R-I) color. The difference was considered by Kron (1954) to define a separate sequence. Greenstein (1965), Eggen (1968) and Gliese (1969b) have also detected the same difference, but none of these investigators included more than a few of the earlier K dwarfs which lie generally in the interval defined by $0\overset{m}{.}3 < (R-I) < 0\overset{m}{.}7$. Recently K dwarfs have been observed for (R-I) and from them Upgren (1972) found a dependence of the slope of the (M_V, R-I) main sequence upon the $|W|$ motion (the component of space motion perpendicular to the galactic plane). The slope is such that high-velocity K-dwarfs lie above the younger low-velocity dwarfs but for the

M-dwarfs the situation is reversed as Kron and others had reported. The influence of the $(U^2 + V^2)^{1/2}$ or planar motion was not as pronounced. More recently Eggen (1973) noted the same effect; his Fig. 1 shows the old-disk stars with a steeper slope than the young-disk stars with a crossover also near (R-I) = $0^m.7$ or about dM0. It is not easy to make a direct comparison between Upgren's and Eggen's conclusions, because their definitions of young and old stars differ, but the trend is the same in both papers.

The identification of main sequences belonging to different stellar populations is even less certain when the population criteria are based on spectral or chemical differences. In the same paper, Eggen (1973) notes that the K-dwarfs of the halo population (identified as such by their motion) lie about one magnitude below the (M_V,R-I) main sequence defined by young stars, specifically the Hyades. But among disk K-dwarfs of all ages, stars with greatly differing metal abundances reflected by ultraviolet excesses all define the same main sequence. Another indicator of age differences among K and early M dwarfs is the intensity of the emission component of the H and K lines of ionized calcium. Wilson and Woolley (1970) measured CaII emission intensities of 325 late main sequence dwarfs, 103 of which have recently determined Van Vleck parallaxes as well as photometric measures in the B, V, R and I colors. Fig. 1 shows a plot of these stars in the (M_V, R-I) diagram. No significant difference appears between the three groups of stars of differing CaII intensities, although the groups reveal a very definite correlation between intensity and space motion, strongly suggesting a difference in age.

Despite the fact that the late-type dwarf stars in general define nearly the same main sequence, a few are known to be clearly subluminous in both the (M_V, B-V) and (M_V, R-I) diagrams. These include some whose space motion identifies them as belonging to the halo population (halo stars make up about one percent of the total number of nearby dwarfs). But there are also a few dK-M stars whose space motions are small. Eggen's (1971) review of red subluminous stars lists several stars whose (R-I) places them in the dK range and about one magnitude below the Hyades sequence. Two recent studies confirm that not all subluminous stars are halo members. Upgren and Weis (1974) discovered an M-type subdwarf which resembles Kapteyn's star, a well known high-velocity subdwarf, but with a low space motion of only about 25 km/sec. They list it along with 20 other sdM stars; of the ten with measured radial velocity, five have a space velocity less than 40 km/sec. Hartwick _et al_. (1976) discuss ten sdM candidates and identify eight as old-disk stars from their motions and only two as halo stars. Clearly the situation of subluminosity among K and M dwarfs is still a confused one, and must await more observations for clarification.

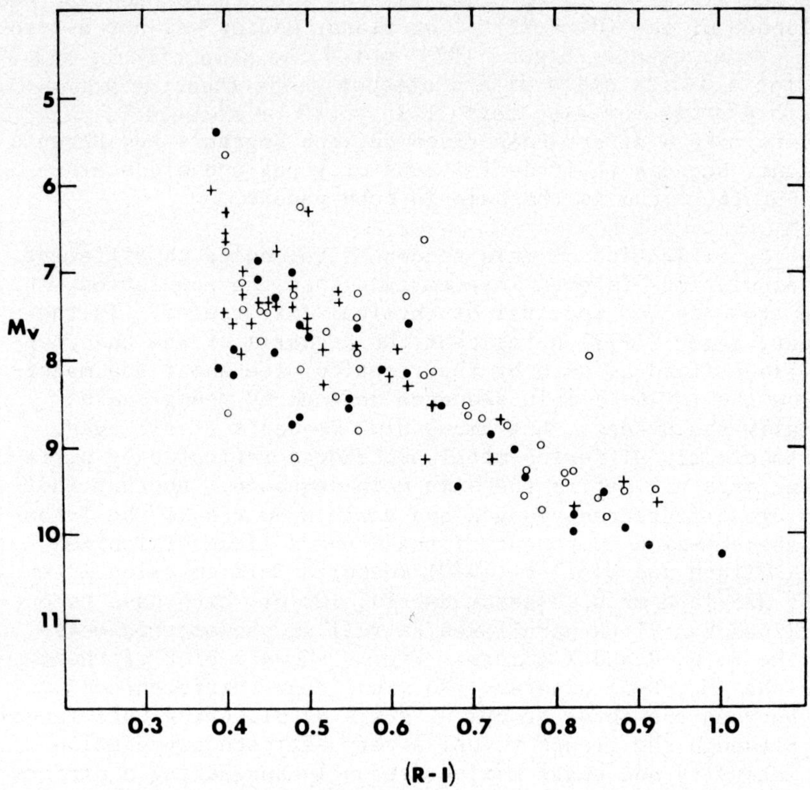

Fig. 1. The (M_V, (R-I)) diagram for stars of different CaII intensity. The 34 crosses, 39 open circles and 30 filled circles represent stars with intensities of +6 to +2, +1 to 0 and -1 to -4, respectively.

---

4. COMPARISON OF HYADES AND FIELD-STAR SEQUENCES

Figs. 2 and 3 show the (V, B-V) and (V, R-I) color-magnitude diagrams for all 119 stars identified from their motions as Hyades members by Pels et al. (1975) along with 25 additional stars claimed by van Altena (1969) to be Hyades members also from proper motion data. The photometry is taken from Upgren (1974) and Upgren and Weis (1977) who include the photometry of Johnson et al. (1962) along with their own and take an average when both are available. Every star has the same symbol in both figures, except that some obvious non-members indicated by crosses in Fig. 2 do not appear in Fig. 3 since they were not observed for (R-I). The stars lying above the main sequence by an average in both diagrams of $0^m.5$ or more are shown as open circles and are most likely to be binary

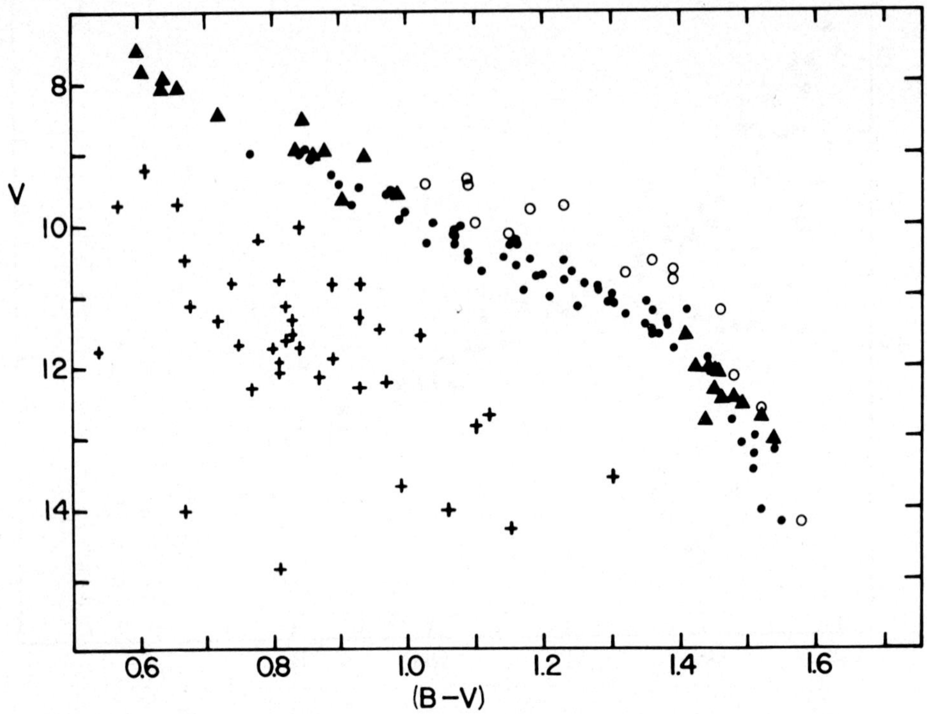

Fig. 2. The (V, B-V) diagram for stars identified as Hyades members from their proper motions. Filled circles and triangles are members according to Pels et al. and van Altena, respectively; open circles are probable binary members; crosses are non-members.

---

members although a few could be foreground stars. The 62 filled circles and 25 triangles, representing the remaining members from Pels et al. and van Altena, respectively, comprise the best representation of the Hyades main sequence in both diagrams.

A linear least-squares solution was made for the 72 stars lying on the main sequence in Fig. 3 which fall within the range $0^{m}.32 < (R-I) < 1^{m}.08$, (after excluding the probable binary members). The resulting relation is $V = 7^{m}.84 + 5.68\ (R-I)$, almost identical to the relation of $V = 7^{m}.85 + 5.67\ (R-I)$ found in an identical way (Upgren 1974) from only 28 stars which then had photometric observations because they were taken from van Altena's (1969) list alone. The addition of many more members from the list of Pels et al. (1975) does not preceptibly change the Hyades main sequence, and indicates that the distance modulus of $3^{m}.22 \pm 0^{m}.04$, derived in the 1974 paper from the 28 van Altena stars, is not in need of revision.

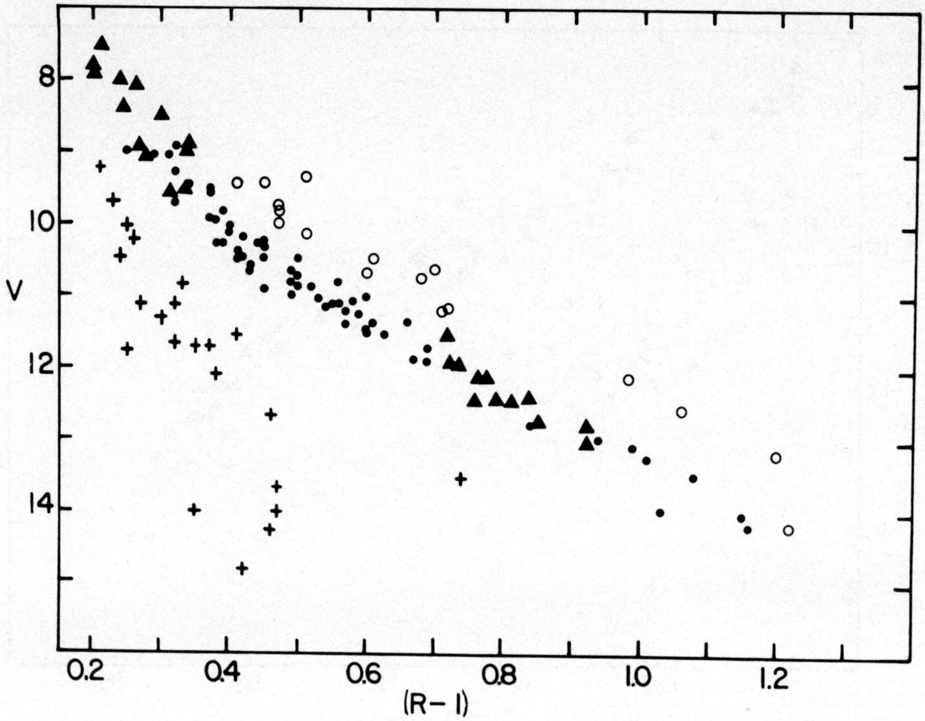

Fig. 3. The (V, (R-I)) diagram for stars identified as members from their proper motions. The symbols are defined in the same way as in Fig. 2. Each star has the same symbol as it has in Fig. 2.

This study was supported by the National Science Foundation through research grant no. AST74-00758 A03.

REFERENCES

Eggen, O.J. (1968). Astrophys. J. Suppl. 16, 49.
Eggen, O.J. (1969). Astrophys. J. 158, 1109.
Eggen, O.J. (1971). In IAU Symposium No. 42, White Dwarfs, W.J. Luyten, ed., D. Reidel, Dordrecht, p. 8.
Eggen, O.J. (1973). Astrophys. J. 182, 821.
Eggen, O.J. (1974). Publ. Astron. Soc. Pacific 86, 697.
Eggen, O.J. and Greenstein, J.L. (1965). Astrophys. J. 141, 83.
Gliese, W. (1969a). Veröeffentl. Astron. Rechen-Inst. Heidelberg, No. 22.
Gliese, W. (1969b). In Proccedings of the Symposium on Low-Luminosity Stars, S.S. Kumar, ed., Gordon and Breach, New York, p. 41.

Gliese, W. (1971). Veröeffentl. Astron. Rechen-Inst. Heidelberg, No. 24.
Greenstein, J.L. (1965). In Stars and Stellar Systems, Vol. V, Galactic Structure, A. Blaauw and M. Schmidt, eds., University of Chicago Press, Chicago, p. 361.
Hanson, R.B. (1975). Astron. J. 80, 379.
Hartwick, F.D.A., Crampton, D. and Cowley, A.P. (1976). Astrophys. J. 208, 776.
Johnson, H.L. and Knuckles, C.F. (1955). Astrophys. J. 122, 209.
Johnson, H.L., Mitchell, R.I. and Iriarte, B. (1962). Astrophys. J. 136, 75.
Kron, G.E. (1954). In Proceedings of the Third Berkeley Symposium on Mathematical Statistics and Probability, University of California, Berkeley, p. 39.
Kron, G.E., Gascoigne, S.C.B. and White, H.S. (1957). Astron. J. 62, 205.
Lutz, T.E. and Kelker, D.H. (1973). Publ. Astron. Soc. Pacific 85, 573.
Luyten, W.J. (1971). The Hyades, University of Minnesota, Minneapolis.
Pels, G., Oort, J.H. and Pels-Kluyver, H.A. (1975). Astron. and Astrophys. 43, 423.
Upgren, A.R. (1972). Astron. J. 77, 745.
Upgren, A.R. (1973). Astron. J. 78, 79.
Upgren, A.R. (1974). Astrophys. J. 193, 359.
Upgren, A.R. (1975). In Multicolor Photometry and the Theoretical HR Diagram, A.G.D. Philip and D.S. Hayes, eds., Dudley Obs. Rept. No. 9, p. 437.
Upgren, A.R. and Weis, E.W. (1975). Astrophys. J. 197, L53.
Upgren, A.R. and Weis, E.W. (1977). Astron. J. 82, 978.
van Altena, W.F. (1969). Astron. J. 74, 2.
van Altena, W.F. (1974). Publ. Astron. Soc. Pacific 86, 217.
van Bueren, H.G. (1952). Bull. Astron. Inst. Neth. XI, 385.
Veeder, G.J. (1974). Astron. J. 79, 1056.
Vyssotsky, A.N. (1963). In Stars and Stellar Systems, Vol. III, Basic Astronomical Data, K. Aa. Strand, ed., University of Chicago Press, Chicago, p. 192.
Wielen, R. (1974). In IAU Highlights of Astronomy, G. Contopoulos ed., D. Reidel, Dordrecht, p. 395.
Wilson, O. and Woolley, R. (1970). Monthly Notices Roy. Astron. Soc. 148, 463.

DISCUSSION

WEIDEMANN: Did you consider that even in (R-I) there may be differential blanketing effects shifting the sequences if there are abundance differences between the Hyades and the field stars (cf. Koester and Weidemann, Astron. Astrophys. 25, 473, 1973)?

UPGREN: For the young field stars, the difference should be very small.

LUYTEN: Concerning the alleged 115 NEW members of the Hyades Cluster announced by Pels and Oort about half of these were well known before, including some found and announced by Oort's teacher - van Rhijn.

Secondly, I might mention that Graham Hill of the Dominion Astrophysical Observatory came and processed twenty pairs of 48" Schmidt plates in the Hyades and Pleiades region with my automatic measuring machine - we now have 45000 proper motions for stars brighter than 20^m pg with mean errors of $0\rlap{.}''007$. But, in the Hyades, the more accurate your motions the more troubles you have because you begin to see the difference in the motions of the stars in the front part and those in the back part of the cluster. Then you decide that they are members of the cluster if their motions about fit and if their magnitudes and colors agree with what is known about the main sequence in the Hyades but this really involves a circular argument if then you use these stars later to make up the main sequence in the Hyades.

UPGREN: A circular argument is avoided here because the background stars are clearly separated from the stars near the Hyades main sequence in both the (B-V) and (R-I) diagrams. The only stars which could be in doubt one way or another are the few which lie just above the main sequence, but it is most likely that these are members which are unrecognized binaries.

GREENSTEIN: The degree to which the average high-velocity star is subluminous must be affected by differential line and band blanketing; the furthest infrared index I could use (1 micron) had about one-half the ΔM as did the visual luminosity. Undoubtedly a color-color system in the red-infrared would permit removal of the differential blanketing in (R-I). But bolometric measures as analyzed by Mould leave Kapteyn's star one-magnitude subluminous.

ABSOLUTE MAGNITUDES BY STATISTICAL PARALLAXES

André Heck

Liège Institute of Astrophysics (Belgium)

Our algorithm for stellar luminosity calibrations (based on the principle of maximum likelihood) allows the calibration of relations of the type:

$$M_i = \sum_{j=1}^{N} q_j C_{ij}, \quad i = 1, \ldots, n,$$

where n is the size of the sample at hand,

M_i are the individual absolute magnitudes,

C_{ij} are observational quantities (j = 1, ..., N), and

q_j are the coefficients to be determined.

If we put N = 1 and $C_{iN} = 1$, we have $q_1 = \bar{M}$, the mean absolute magnitude of the sample. As additional output, the algorithm provides us also with the dispersion in magnitude of the sample σ_M, the mean solar motion (U,V,W) and the corresponding velocity ellipsoid (σ_u, σ_v, σ_w).

The necessary input for each star consists in the position (α,δ), the apparent magnitude (m_v), the proper motions and the corresponding errors (μ_α,$\varepsilon_{\mu\alpha}$,μ_δ,$\varepsilon_{\mu\delta}$), the radial velocity (v_r) and the eventual observational quantities. The principle of the method can be defined as follows: Which law of absolute magnitudes, the radial velocities, on one hand, and the tangential velocities, on the other hand, can be considered as resulting from the same assumed kinematical model (here a Schwarzschildian distribution with correction for the galactic rotation)? Technical details are given in Heck (1974, 1976) or in a forthcoming paper in Vistas in Astronomy.

A. G. Davis Philip and D. S. Hayes (eds.), The HR Diagram, 49–52.

The algorithm has been applied to various samples, but we will restrict this paper to a general presentation of the results obtained for RR Lyrae stars and with the indices of the uvbyβ photometry.

The most recent application of the technique to a sample of about 130 RRab Lyrae stars (Heck and Lakaye, 1977) gave the following relation between luminosity and the ΔS metallicity index introduced by Preston (1959):

$$M_V = 0^m.12\ (\pm\ 0^m.04)\ \Delta S - 0^m.33\ (\pm\ 0^m.20)$$

with $\sigma_M = 0^m.51\ (\pm\ 0^m.18)$, using Parengago's (1940) absorption law. The result for the sample used by Clube and Dawe (1977) is (compare with their paper at this symposium, page 53):

$$M_V = 0^m.16\ (\pm\ 0^m.12)\ \Delta S - 0^m.17\ (\pm\ 0^m.90)$$

using Bernicki's (1967) absorption law.

From Dr. J. Lub (European Southern Observatory), we received new data concerning Heck and Lakaye's (1977) sample, especially a set of homogeneous reddening values. The derived relation is:

$$M_V = 0^m.29\ (\pm\ 0^m.03)\ \Delta S - 1^m.18\ (\pm\ 0^m.26),$$

which is slightly different from the previous relations.

The following results were obtained for main-sequence stars, when applying the method to uvby β photometric data:

A0 - A4 type stars (Heck, 1975):

$$M_V = 1.89\ (\pm\ 3.90)\ a_R - 14.52\ (\pm\ 1.93)\ r + 1^m.62\ (\pm\ 0^m.25)$$

A4 - F0 type stars (Heck, 1977a):

$$M_V = -12.99\ (\pm\ 4.20)(b-y) - 11.74\ (\pm\ 1.29)\ [c_1] + 13^m.51\ (\pm 1^m.29)$$

F-type stars (Heck, 1977b):

$$M_V = 3.13\ (\pm\ 2.84)(b-y) - 4.32\ (\pm\ 1.21)\ [c_1] + 4^m.27\ (\pm\ 0^m.25)$$

G-type stars (Heck, 1977c):

$$M_V = -27.29\ (\pm\ 3.36)(b-y) - 33.34\ (\pm\ 6.91)\ [c_1]$$
$$+19.66\ (\pm\ 4.52)(b-y)\ [c_1] + 21^m.60\ (\pm\ 2^m.07)$$

where a_r, r, (b-y) and $[c_1]$ are indices derived from the uvby β photometry (see e.g. Strömgren, 1966). The sample of G-type stars contained also stars from other luminosity classes because there were not enough main-sequence stars to get statistically significant results. No conclusive result has been obtained for B-type stars.

BIBLIOGRAPHY

Bernicki, A. (1967). Acta Astron. 17, 165.
Clube, S.W.M. and Dawe, J.A. (1977). Submitted to Monthly Notices Roy. Astron. Soc.
Heck, A. (1974). Acad. Roy. Belg. Bull. Cl. Sci. 5ème Sér. 60, 927.
Heck, A. (1975). Astron. Astrophys. 43, 111.
Heck, A. (1976). Quelques Méthodes de Détermination de la Magnitude Absolue, ed. Centre de Données Stellaires, Strasbourg.
Heck, A. (1977a). Astron. Astrophys. 56, 235.
Heck, A. (1977b). Astron. Astrophys., in press.
Heck, A. (1977c). Submitted to Astron. Astrophys.
Heck, A. and Lakaye, J.M. (1977). Submitted to Monthly Notices Roy. Astron. Soc.
Parenago, P.P. (1940). Astron. Zh. 17, 1.
Preston, G.W. (1959). Astrophys. J. 130, 507
Stromgren, B. (1966). Ann. Rev. Astron. Astrophys. Vol.4., Annual Reviews, Palo Alto, p. 433.

DISCUSSION

CAYREL de STROBEL: How much would your maximum likelihood M_V's change if you also introduced differences in He-content?

HECK: A He-content parameter can be introduced in any calibration as an additional Cij term. The way M_V will change depends upon the sample at hand, but the determination of individual absolute magnitudes will be more precise if this parameter is relevant for the particular calibration.

RR LYRAE VARIABLE STARS AND THE DISTANCE SCALE

S. V. M. Clube and J. A. Dawe

Royal Observatory Edinburgh

A statistical parallax algorithm (Clube and Dawe, 1978a,b), using the technique of maximum likelihood, has been applied to a set of 103 'ab' - RR Lyrae stars in the solar neighborhood (r 2<kpc), using observational data from the Royal Greenwich Observatory Bulletins. A second set of 130 'ab' - RR Lyrae stars has been kindly supplied to us by Dr. A. Heck (Université de Liège) to permit a comparison between our analyses. The purpose of this investigation was:

(a) to investigate the variations of kinematical parameters and absolute luminosities of these stars as functions of Preston's index ΔS and of log (Period).

(b) to identify those RR Lyrae stars in the solar neighborhood which most closely resemble those in the galactic halo, those near the galactic center, and those in the Magellanic Clouds.

To perform this analysis, the stars were divided into limited ranges of ΔS and log (Period), but with the constraint that each range should contain at least 30 members. (The minimum population required to give statistically useful solutions. Because of this constraint, adjacent bins often conain several members in common.) The results are summarized in Fig. 1.

A. G. Davis Philip and D. S. Hayes (eds.), The HR Diagram, 53–57.

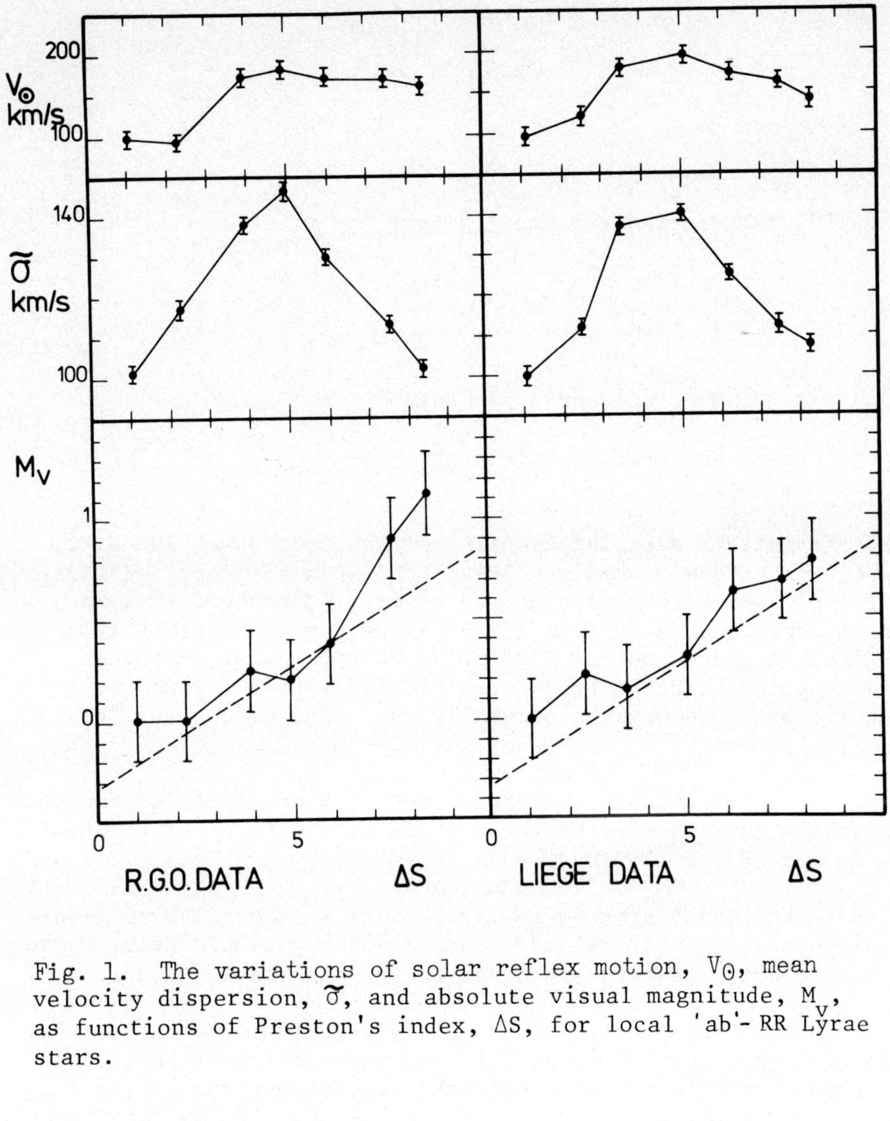

Fig. 1. The variations of solar reflex motion, $V_\odot$, mean velocity dispersion, $\tilde{\sigma}$, and absolute visual magnitude, M_V, as functions of Preston's index, ΔS, for local 'ab'- RR Lyrae stars.

$V_\odot$ is the scalar value of the solar reflex vector, $\tilde{\sigma}$ is the mean velocity dispersion (= $\{[\sigma_u{}^2 + \sigma_v{}^2 + \sigma_w{}^2]/3\}^{\frac{1}{2}}$) and M_V is the visual absolute magnitude, corrected for absorption. In the case of the R.G.O. data, Bernicki's empirical reddening law (Bernicki, 1967) has been applied to the observed color of each star, while for the Liege data a Parenago-type global formula has been used (Sharov, 1964). Both sets of data indicate the same general behavior namely, a gradual increase in the solar velocity with increasing ΔS, and a sharp maximum in the velocity dispersion at intermediate values of ΔS. There is a significant correlation between the absolute magnitude and ΔS, which is steeper in the case of the R.G.O. data. The dotted line in Fig. 1 shows the linear correlation found by Dr. Heck from his own data (Heck, private communication). The differences between our two results may arise from:

(1) differences in absorption corrections: There are indications that Bernicki's rule might overestimate the absorption by $\Delta A_v \sim 0^m.1$ (e.g. see Bernicki, 1967). However, we have chosen not to correct for this, as its effect may be counter-balanced by a correction necessary to the systematic motion of the reference stars. This amounts to $\Delta\mu = -0.007$ arcseconds p.a. in the direction of galactic rotation and appears to be confirmed by the work of Klemola and Vasilevskis (1971). When applied, such a correction would systematically make the absolute magnitude $0^m.1$ fainter for <u>both</u> sets of data.

(2) differences in the maximum-likelihood algorithms used: Both algorithms have been checked with a single set of synthetic RR Lyrae data, supplied by Dr. D. Jones of R.G.O., and their solutions in this case differ by no more than $\Delta M_v = M_v$ (Heck) $-M_v$ (CD) $= -0^m.06$ ($\Delta V = -4$kms^{-1}, $\Delta\sigma = 1$kms^{-1}).

(3) differences in stellar data: The R.G.O. data set was selected so that both proper motions and radial velocities are on uniform systems. No stars of period greater than $0^d.8$ are included, while Heck's set contains 2 stars with periods greater than $1^d.2$.

A simple interpretation of these results is that we are dealing with a mixture of at least two populations: a low-velocity component with $V \sim 50$kms^{-1}, corresponding to $\Delta S \leqslant 1$, and a high velocity component with $V \sim 180$kms^{-1}, corresponding to $\Delta S \geqslant 7$. The mean velocity variance of each population would appear to be ~ 100 kms^{-1}, the sharp peak which is manifest at $\Delta S \sim 5$ arising from the mixing of two Gaussian populations with different mean velocities.

The correlation between M_v and log (Period) is much less clear, as the periods are correlated with ΔS. Superficially, the solar reflex motion and the velocity dispersion appear to

behave with increasing period in much the same manner as they do with increasing ΔS.

Of the RR Lyrae stars investigated, we identify those with $\Delta S \geqslant 6$ as being a representative sample of the high-velocity population (though with some contamination at the 10% level). The m.l. solution for this set of stars is given in Table I.

TABLE I: M.L. SOLUTION FOR HIGH-ΔS RR LYRAE STARS

$$N = 54 \quad ; \quad \overline{\Delta S} = 7.6$$

$$M_V = 1\overset{m}{.}0 (\pm 0\overset{m}{.}2) \quad ; \quad \sigma(M_V) = 0\overset{m}{.}6 \ (\pm 0\overset{m}{.}3)$$

$$U = -2 \ , \quad V = -165 \ , \quad W = -18 \ ; \quad V_\odot = 166 \ \text{kms}^{-1}$$

$$\sigma_u = 145 \ , \quad \sigma_v = 100 \ , \quad \sigma_w = 72 \ ; \quad \tilde{\sigma} = 110 \ \text{kms}^{-1}$$

If this high-velocity sample is similar to the RR Lyrae population at the galactic center, the galactic halo and the Magellanic Clouds, and there is evidence from their respective period-frequency distributions that this is likely to be the case for at least the latter two, then the derived absolute magnitude may be used to re-calibrate the distance scale to the galactic center and to the Clouds. Using the work of Oort and Plaut (1975), and Graham (1974, 1975), we find:

$$R_o = 7.1 \ (\pm 0.7) \ \text{kpc}$$

$$\left. \begin{array}{l} \mu(\text{SMC}) = 18\overset{m}{.}6 \\ \mu(\text{LMC}) = 18\overset{m}{.}2 \end{array} \right\} \ \underline{\text{uncorrected}} \text{ for interstellar absorption.}$$

If $A_V(\text{SMC}) = 0\overset{m}{.}3$ for both the SMC and LMC, then:

$$r\ (\text{SMC}) = 46 \ \text{kpc.}$$

$$r\ (\text{LMC}) = 38 \ \text{kpc.}$$

REFERENCES

Clube, S.V.M. and Dawe, J.A. (1978a,b). submitted to Monthly Notices Roy. Astron. Soc.

Bernicki, A. (1967). Acta Astron. **17**, 165.

Graham, J.A. (1974). In IAU Symposium No. 50, Stellar Instability and Evolution, P. Ledoux, A. Noels and A.W. Rodgers, eds., D. Reidel, Dordrecht, p. 107

Klemola A.R. and Vasilevskis, S. (1971). Publ. Lick. Obs. XXII, Part III.

Oort, J.H. and Plaut, L. (1975). Astron. Astrophys. **41**, 71.

Sharov, A.S. (1964). Soviet Astron. **7**, 689.

DISCUSSION (The paper was read by Dickens)

COX: Can you tell us the numerical value of the color term in the P-L-C relation? There is theoretical evidence that the value of d log L/d log T_{eff} for constant period in the HR diagram is five rather than six from Sandage and Tammann Cepheid data.

DICKENS REFERS TO MADORE

MADORE: The resulting change in the color term in the P-L-C relation suggested by Clube's revision is quite large. From the value of 2.5, suggested by Sandage and Tammann for the (B-V) color term, Clube requires a slope closer to unity or even less.

KRAFT: Can you explain why the present result on the dependence of M_V on ΔS reverses the result found from RGO work of 10 years ago?

DICKENS: I am not familiar with the details of all this. However I suspect that a) the data sample has enlarged considerably, b) the methods of analysis are different and c) changes may have occurred in the adopted constants of the proper motion reference frame.

KRAFT: One notes that the results obtained from RR Lyraes in ω Cen show that there is no dependence of M_V on ΔS.

DICKENS: The RR Lyraes in ω Cen may well have a different chemical evolutionary history from those in the Galaxy. For example, a relatively small change in helium abundance can have a substantial effect on the luminosity at which stars of a given mass pass through the instability strip.

KRAFT: M_V increases with ΔS in Clube's work. But from stellar evolutionary theory and from Wesselink's method, one gets M_V more or less constant with ΔS, or at least a much smaller dependence.

DICKENS: It should be emphasized that only for the high ΔS group (>7) do we have a "pure" population sample. For intermediate and small values of ΔS, the statistical parallax results for M_V probably represent mixed populations and therefore any apparent correlation of M_V with ΔS must be treated with caution.

BIDELMAN: What is ΔS thought to be for the RR Lyraes in the galactic center? Is it in fact 7.5 or so?

KRAFT: I believe the few that have been done directly have a mean in the region of ΔS=3, but it must be remembered that the galactic center region contains a mixture of population types ranging from ΔS=0 to 8.

DICKENS: This means that the results for the high ΔS group (M_V = +1) are not strictly applicable to the galactic center variables.

EXTENSIONS OF THE WILSON-BAPPU EFFECT AMONG VERY LUMINOUS STARS

Robert E. Stencel

NASA/Johnson Space Center, Houston Texas

Twenty years ago, Wilson and Bappu (1957) published their observational correlation of M_V and the logarithm of the full width at half maximum of the CaII K-line central emission for G, K and M stars. The technique was stated to have $\pm$ 0^m.5 accuracy, while a tabulation of $M_V(K)$ by Wilson (1976) suggests the error may only be $\pm$ 0^m.3. This accuracy makes the approach valuable for late-type supergiants since other methods suffer from comparable errors. However, for F through M supergiants (Ia, 0), circumstellar absorption obscures the chromospheric K-line core emission and excludes such objects from the Wilson-Bappu correlation. I report here on a new class of emission lines in late-type giant and supergiant spectra that exhibit M_V correlated widths, yet are detectable among the brightest stars.

Fosbury (1971) reported a solar-like emission line at 3935 Å in Arcturus. After finding several solar-like emission lines in the Ca II H-K wings in cool giant spectra, I examined spectra across the cooler portion of the HR diagram to delineate the empirical properties of these "H-K wing emission lines." Coude´ plates for 220 bright MK standards between F0 and M8, from dwarf to supergiant, from the collections of Hale Observatories were examined. A description of the plates and the analysis can be found in published work (Stencel, 1977a, b). Several moderate strength lines, attributable to iron-peak elements, occurring between 1 - 7 Å from the H-K line centers appear in absorption in cool dwarfs, but fill-in and vanish among giants, and are present in emission lateward of K giants, G bright-giants and F supergiants (cf. Fig. 1, Stencel 1977a). Comparable lines are found in Mira and some Cepheids. The luminosity effect described can be seen

A. G. Davis Philip and D. S. Hayes (eds.), The HR Diagram, 59-62.

in Figs. 9 and 10 of the original Wilson and Bappu work (1957), and in Fig. 3 of Deutsch's (1960) work, notably just redward of the K-core emission. In summary, 24 H-K wing emission lines between 3925 - 3974 A were found which exhibit this luminosity effect (Stencel, 1977a, b).

Empirically, the H-K wing emission lines show an M_V - log FWHM correlation, although the slope of the correlation for individual H-K wing emission lines seems to depend on the square of their wavelength displacement from the CaII line centers, just as the resonance line wing opacity does. Thus, it appears that these lines arise from differing layers in the stellar atmosphere and may allow direct measure of non-thermal broadening velocity with height gradients (cf. Stencel, 1977a, b).

With this in mind, we look at the superluminous F, G, K and M stars. As a group, they show evidence for circumstellar matter and mass-loss. K-core emission is absent due to circumstellar self-absorption, but several H-K wing emission lines are unaffected due to their displacement from the line center. The wing emission at 3936 Å is a good example of this. Consider HR 2061 (α Ori, M2 Iab) and HR 8316 (μ Cep, M2 Ia). HR 2061 shows K core emission and $M_V(K) = -5^m_.5$, while HR 8316 shows no K core emission and thus has no $M_V(K)$. Based on an M_V - width relation for the 3936 Å line, HR 8316 has $M_V(3936) = -6^m_.9 \pm 0^m_.9$, in agreement with other estimates. For HR 8752 (G5V Ia0), again with no K-core emission, we find $M_V(3936) = -8^m_.2 \pm 0^m_.7$, near Sargent's (1965) value of $-9^m_.0$. For HR 9045 (ρ Cas, F8 Iap), Sandage and Tammann (1974) give $M_V = -9^m_.5$, based on uncertain membership in Cas OB5 association. I find no K-core emission, but derive $M_V(3936) = -6^m_.1 \pm 0^m_.5$. The discrepancy arises from non-membership or shell-episode spectral effects (Sargent, 1961). The point is: there seems to be a useful M_V -width effect which makes the H-K wing emission lines possible extensions of the Wilson-Bappu effect. With the use of low-light level spectral scanners at 1 Å resolution, these features may now be detectable in the brightest late-type stars of the Magellanic Clouds or of nearby galaxies.

In a slightly different spectral region, we find promise of additional M_V - width relations, and I am pleased to present the current state of this effect for the 2800 Å Mg II resonance line cores. This work is due to Y. Kondo and others with the SRL/NASA BUSS experiment. Earlier versions of Fig. 1 have been published (Kondo, et al. 1976a, b). Since the emission cores are clearly present in F stars, including the supergiants, there is some hope that the Mg II resonance lines in extremely luminous stars may reveal M_V information otherwise not available with the CaII resonance lines. Dr. Kondo and I are planning further UV observations along these lines.

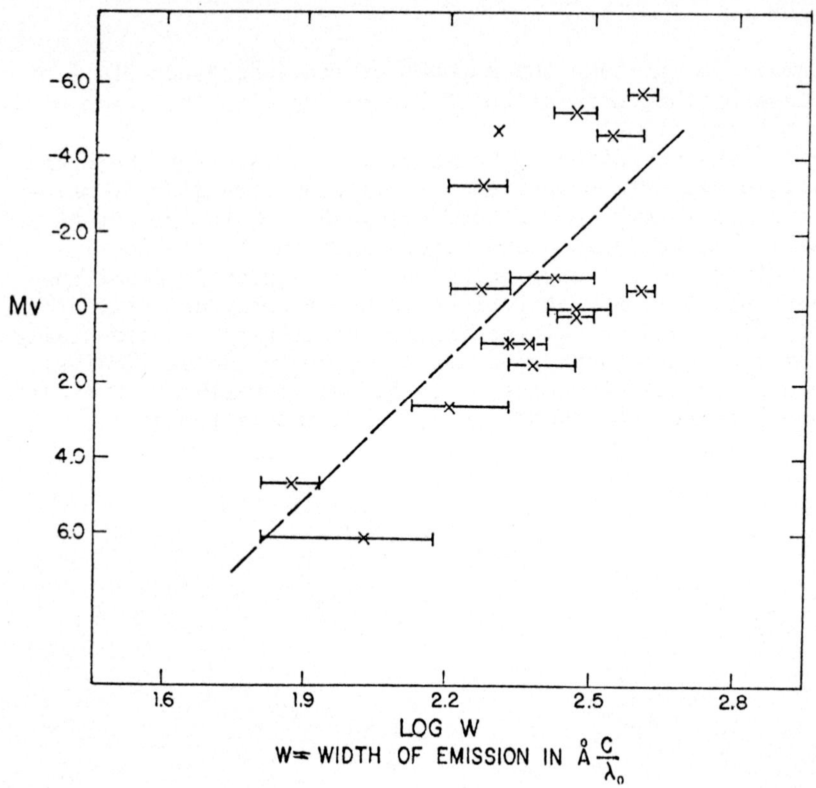

Fig. 1. Mg II core emission width versus M_V. (Kondo, Morgan and Modisette, to be published.)

REFERENCES

Deutsch, A. (1960). In Stars and Stellar Systems, Vol. VI, Stellar Atmospheres, J. Greenstein, ed., Univ. of Chicago Press, Chicago, p. 556.
Fosbury, R. (1971). Mon. Not. Roy. Astron. Soc. 155, 7P.
Kondo, Y., Morgan, T. and Modisette, J. (1976a). Astrophys. J. 207, 167.
Kondo, Y., Morgan, T. and Modisette, J. (1976b). Astrophys. J. 209, 489.
Sandage, A. and Tammann, G. (1974). Astrophys. J. 191, 603.
Sargent, W. (1961). Astrophys. J. 134, 142.
Sargent, W. (1965). Observatory 85, 33.
Stencel, R. (1977a). Astrophys. J. 215, 176.
Stencel, R. (1977b). Dissertation, Univ. Michigan. Univ. Microfilms, Ann Arbor, Order No. 77-26,369.
Wilson, O. (1976). Astrophys. J. 205, 823.
Wilson, O. and Bappu, V. (1957). Astrophys. J. 125, 661.

DISCUSSION

BUSCOMBE: Do you take any account of the suspected effect of stellar age on the correlation of luminosity with the shape of the H and K emission?

STENCEL: The age effect demonstrated by Wilson for the H and K emission arose from comparison of emission strengths of main-sequence stars in clusters of differing age. This age effect is not established for giants and supergiants due to the small sample. Since the H-K wing emission lines appear in giant and supergiant spectra, any age effect is not clearly delineated. However, this would be interesting to investigate. Incidentally, there may also be an abundance effect, e.g. in the Ba II stars, which can alter the strengths of these emission lines. R.A. Fesen and I are looking into these age and abundance effects.

STATISTICAL HR DIAGRAMS FOR ONE HUNDRED AND FIFTEEN THOUSAND PROPER-MOTION STARS

Willem J. Luyten

University of Minnesota

Since regular HR diagrams require apparent magnitudes, colors or spectra, and parallaxes, and such complete data are available for relatively few stars, there may be some advantage in making up diagrams which utilize proper motions instead of parallaxes, and are thus statistically similar to an HR diagram. The reduced proper motion, first used by Hertzsprung, is defined as $H = m + 5 + 5\log \mu$, but may also be written as $H = M + 5\log T$, where T is the tangential velocity, and is expressed in astronomical units per year. A diagram plotting H against color will thus contain the considerable dispersion in tangential velocity which is a serious disadvantage. However, this is outweighed by two practical advantages. First the one and the same person who does the proper motion survey can, and does also determine the other two quantities needed. Second, when using data obtained from such a proper motion survey one deals, statistically, with all the stars within a given distance and the results, therefore, are much more representative of the real situation in space than many HR diagrams which often contain an unrealistic preponderance of giants.

Utilizing the data obtained in the Bruce and Palomar Proper Motion Surveys I have made up such a diagram for 115,000 proper-motion stars brighter than the twenty-first photographic magnitude. It is immediately evident that white dwarfs and degenerate stars are much more numerous in space than yellow giants. Owing to the dispersion in tangential velocity and the existence of large numbers of high-velocity subdwarfs, the degenerate sequence is clearly indicated only for the bluest stars. A separate diagram has therefore been given for the degenerate stars, and this shows why relatively few yellow degenerates have been spectro-

A. G. Davis Philip and D. S. Hayes (eds.), The HR Diagram, 63–64. All Rights Reserved.

scopically confirmed thus far: the spectroscopists have limited their observations to stars which are too bright to contain more than a sprinkling of such yellow degenerates.

A further diagram is given which shows that a number of objects identified spectroscopically or photometrically as white dwarf are not degenerate at all. Using the proper-motion data to obtain corrections to my luminosity function a still further HR diagram is presented for all stars now expected to be within ten parsecs, and estimated to number about seven hundred. All these diagrams, as well as discussions of them are published in full detail in No. LI of the Proper Motion Survey with the Forty-Eight Inch Schmidt Telescope.

DISCUSSION

BUSCOMBE: H. Augensen has obtained slit spectrograms at Cerro Tololo for about 100 stars ($7<m_V<10^m$) from the LTT catalogue. Several have radial velocities more than 75 km/s relative to the Sun, and have particularly weak metallic lines.

LUYTEN: We would have expected that stars in this luminosity range would be high-velocity subdwarfs rather than fully degenerate stars.

SION: Have you constructed reduced proper motion diagrams for degenerate stars having spectroscopic identification?

LUYTEN: No, not separately, we just haven't had time to look up whether any of our 6,2000 degenerate stars had spectra determined for them.

STRAND: In your diagram of stars within 10 pc what is the evidence for the large population below 16th absolute magnitude belonging to the main sequence?

LUYTEN: These numbers were derived from my luminosity function which is still essentially the same as the one published in 1968.

STRAND: How many stars do you have at 20th magnitude with proper motions of a second of arc?

LUYTEN: I don't remember offhand but my guess is not more than eight or nine.

BIDELMAN: I would like to mention that S.G. Lee, a graduate student working with me, has determined Luyten's H vs. M relation separately for various spectral subgroups, using the stars having trigonometric parallaxes > $0''.025$ that are included in the recent Giclas northern-hemisphere proper-motion catalogue. She finds that a mean relation very similar to that determined many years ago by Luyten, using quite different input data, fits all subgroups surprisingly well. The mean tangential velocity for the various subgroups ranges between 58 and 68 km/sec.

LUYTEN: I am glad to learn that there has been no important change in one of the basic relations which is needed in the derivation of the luminosity function.

THE ABSOLUTE CALIBRATION OF THE HR DIAGRAM: FUNDAMENTAL EFFECTIVE TEMPERATURES AND BOLOMETRIC CORRECTIONS

D. S. Hayes

Arizona State University

1. INTRODUCTION

Scales of fundamental bolometric connections (B.C.) and effective temperatures (T_{eff}) as a function of spectral type or color are necessary for the comparison of observations and theory in the HR **diagram.** The basic equation defining T_{eff} may be written

$$f = (\theta^2/4)\sigma T^4_{eff} \quad (1)$$

where f is the total apparent flux, and θ is the angular diameter (θ = 2R/d). The **apparent bolometric** magnitude is

$$m_{bol} = -2.5 \log_{10} f + C = V + B.C. \quad (2)$$

The zero point constant, C, may be determined by reference to the Sun:

$$m_{bol,*} - m_{bol,\odot} = -2.5 \log_{10}(f_*/f_\odot) \quad (3)$$

We see from Eqns. (2) and (3) that the zero-point of the B.C. scale is arbitrary and that a B.C. scale may be measured by measuring f and V for a suitable sample of stars. The Earth's atmosphere blocks off a significant part of f, and interstellar hydrogen blocks the Lyman continuum, which is significant in the O-stars. The measurement of θ requires the knowledge of limb darkening, which is extremely difficult to measure, at present. Traditionally, unmeasurable portions of f have been estimated by model atmosphere calculations, as has been the limb darkening. This resort to models is "non-fundamental".

A. G. Davis Philip and D. S. Hayes (eds.), The HR Diagram, 65–76. All Rights Reserved.

Can we make truly fundamental measurements of T_{eff} and B.C.? In the strict sense, the answer is no. But the limb darkening connections to θ have a very small effect upon T_{eff}. Similarly, we have not measured θ of an O-star hot enough for the Lyman continuum flux to be a large fraction of the total flux. For cooler stars, we are now able to make measurements covering nearly all of the significant parts of the spectrum. So we can, indeed, make adequately fundamental measurements of T_{eff} and B.C.

Previous summaries of the determination of T_{eff} and B.C. are those by Kuiper (1938), Popper (1958), Harris (1963), Johnson (1966; hereafter simply "Johnson") and Code, et al. (1976; hereafter CDBB). The increase in the degree to which scales of T_{eff} and B.C. have become fundamental is clearly marked in these papers, particularly in Johnson's paper, in which extensive IR photometry is introduced, and in CDBB in which they introduce not only extensive UV photometry, but also the catalogue of θ's for hot stars measured by Hanbury Brown and his colleagues. The θ's available to Kuiper, Popper and Harris were mainly from the measurements by Michelson and Pease; there were not many of these, they were limited to K and M giants and supergiants, and some of them were of doubtful quality. In recent years, there has been a revolution in the measurement of θ's of cool stars, resulting from the introduction of speckle interferometry, the revival of amplitude interferometry, and particularly from the photometry of lunar occultations. The primary limitation still remaining relates to spectral types and luminosity classes. There is a serious lack of measurements of all cool dwarfs and of giants of middle spectral types. In this paper I use θ's of 18 K and M giants for the determination of the T_{eff}-scale for cool stars. These are only those which are most useful in the present context; it is by no means an exhaustive list of what is available.

I use Johnson's and CDBB's work wherever possible. Since Johnson's paper, much IR photometry has been published, but his mean colors and B.C.'s for cool stars have a sufficiently large data base that it does not appear profitable to revise them for this project. I regard the T_{eff} and B.C. scales for B-stars reported by CDBB to be essentially definitive. Additional UV spectrophotometry from the TD 1 satellite is available but a revision of the work by CDBB does not appear to be indicated.

2. ANGULAR DIAMETERS

I list θ's for 24 stars in Table I. Included in this list are Regulus, Vega, γ Gem, Altair and Procyon, all of which are treated by CDBB. The total fluxes for these stars, given in Table I, are not fundamental, in that they result from the

fitting of model atmospheres to the visual energy distributions. The motive for this procedure is discussed in § III. The θ's are those given by CDBB. YY Gem is included in Table I, and it is the only eclipsing system represented. A number of eclipsing systems have been analyzed for radii, but for T_{eff} we need an accurate parallax as well as photometry which would give f. Only YY Gem seemed to add useful information to this study, but it is particularly important since it is the only dwarf among the cool stars included.

The remaining 18 stars are K and M giants. I have excluded supergiants, Mira variables and carbon stars. I have not used stars whose only measurement of θ is an older one. I have used θ's corrected for darkening, and, in cases where the original source lists only an undarkened diameter, I have adopted a darkened diameter from a secondary source, most commonly Barnes and Evans (1976). My sources of θ are given in the Table.

3. INFRARED PHOTOMETRY AND THE PHOTOMETRY OF THE SUN

For the 18 giants listed in Table I, BVRIJK photometry is available, as is (V-L) for most of them, and (V-M) for a few. For YY Gem, UBVRIKL photometry is available. I have not attempted direct numerical integration of these data, preferring, instead, to interpolate B.C.'s in Johnson's tables. Each B.C. was corrected for a change in absolute calibration (see below) and for a change in zero point (see § IV). Then, using new values of V and f for the Sun (see below), f was calculated. Finally, each θ and f was used to calculate T_{eff}. The final values of these quantities are listed in Table I.

Infrared photometry in the 1-5 μ region has never been properly calibrated, although an attempt was made by Walker (1969). Johnson's (1965) calibration of UBVRIJKL is mainly based upon the solar energy distribution and assumed values for the solar IR colors. I have constructed a new absolute calibration based upon model atmospheres fitted to the visual photometry of Regulus, Vega, γ Gem, Altair and Procyon. The visual photometry is on the absolute system of Hayes and Latham (1975) with a new value for the flux at 5556Å. This latter flux is based upon the mean of the Palomar Mtn. (Oke and Schild 1970, see rediscussion by Hayes and Latham 1975), Mt. Hopkins (Hayes and Latham 1975) and Lowell Observatory (Tüg, et al. 1977) calibrations of Vega; for Vega at 5556Å I use $F_\nu = 3.52 \times 10^{-20}$ ergs/cm²/sec/Hz. The details of the calibration will be published elsewhere. The calibration requires corrections in the 10% range at some wavelengths, but the correction at K is opposite in sign to the others. Since K is located near the wavelength of maximum energy of K- and M-stars, the net effect of the change in

TABLE I

THE FUNDAMENTAL STARS

HD	Name	Type	θ_{LD}(Milliarcsec)	Source	f(ergs/cm^2/sec)	T_{eff}	B.C.
18191	RZ Ari	M6III	10.18	(7)	3.440-6	3160	-3.81
29139	α Tau	K5III	24.6	(10)	3.340-6	3590	1.21
44478	μ Gem	M3III	13.65	(7)	1.107-5	3650	2.02
47105	γ Gem	A0IV	1.39	(3)	4.569-6	9180	0.10
60179C	YY Gem	M0.5V	0.458	(1)	1.076-8	3520	1.45
61421	α CMi	F5IV-V	5.50	(3)	1.797-5	6500	0.04
86663	π Leo	M2III	5.9	(8)	1.380-6	3300	1.59
87837	31 Leo	K4III	3.55	(7)	1.076-6	4000	0.99
87901	α Leo	B7V	1.37	(3)	1.312-5	12040	0.68
91232	46 Leo	gM2	5.6	(1)	6.792-7	2950	1.74
99998	87 Leo	K4III	3.7	(1)	8.472-7	3690	1.13
112142	ψ Vir	M3III	5.86	(7)	1.770-6	3530	1.96
124897	α Boo	K2IIIp	23.1	(11)	4.875-5	4070	0.71
139663	42 Lib	K4III	2.5	(1)	4.920-7	3920	0.74
168574	HR6861	gM5	3.6	(1)	1.028-6	4150	2.82
169916	λ Sgr	K2III	4.4	(1)	2.729-6	4540	0.44
172167	α Lyr	A0V	3.24	(3)	2.836-5	9490	0.19
175775	ε^2Sgr	K1III	3.8	(7)	1.514-6	4210	0.50
187642	α Agl	A7IV-V	2.98	(3)	1.224-5	8020	0.01
196777	υ Cap	M2III	4.72	(7)	1.038-6	3440	1.68
207005	47 Cap	gM3	3.2	(1)	6.855-7	3760	2.23
216386	λ Aqr	M2III	8.2	(1)	3.733-6	3590	1.76
217906	β Peg	M2II-III	18	(1)	1.528-5	3450	1.96
219215	φ Aqr	M2III	4.9	(1)	1.923-6	3940	-1.47

Sources of θ: (1) Barnes and Evans (1976), (2) Blazit, et al. (1977), (3) Code, et al. (1976), (4) Currie, et al. (1974), (5) Gezari, et al. (1972), (6) Hanbury Brown (1968), (7) Ridgway, et al. (1977) (8) Vilas and Lasker (1977), (9) Worden (1976), (10) Ref's 4, 5, (11) Ref's 2, 4, 5, 6, 9.

absolute calibration upon the B.C. is only $0^m.01$ maximum, relative to the Sun.

I have made corrections to the visual and infrared fluxes of seven A- and late-B-stars listed by CDBB in order to correct them from the visual absolute calibration of Oke and Schild (1970) to that described above. CDBB tied their IR fluxes to the visual absolute calibration, so the correction has been applied to all $\lambda > 3400$Å. The corrections to f are small, ranging from 0.4% (α Eri, B3Vp) to 1.5% (αPsA, A3V). They have a negligible effect upon T_{eff}, but a noticeable effect upon the B.C. scale.

It is necessary to have V and (B-V) for the Sun in order to determine its B.C. and to place it into the stellar sequence of colors. Direct UBV photometry of the Sun is very difficult, and has only been tried once (Gallouët 1964). Stebbins and Kron (1957) made measurements on the six-color system, which can be transformed into V and (B-V). Otherwise, we must rely upon indirect measurements. I have re-examined all of the available evidence, and will summarize it elsewhere. For this project, I have adopted $V_\odot = -26^m.74$, $(B-V)_\odot = +0^m.65$ and $(V-R)_\odot = +0^m.54$. Measurements of the solar constant have been reviewed by Forgan (1977), whose recommended value is equivalent to f = 1.375 x 10^6 ergs/cm^2/sec which results in T_{eff} = 5786K.

4. EMPIRICAL BOLOMETRIC CORRECTIONS AND EFFECTIVE TEMPERATURES

I have summarized the final scales of B.C. and T_{eff} in Tables II and III. The B.C. scale for B-stars is that of CDBB with a shift in zero point, whereas that for stars later than the Sun is Johnson's, again with a shift in zero point. For O-stars, I have taken B.C.'s from Panagia (1973). For the A- and F-stars I have determined a new scale of B.C.'s based upon stars from CDBB corrected as described above. I have taken the zero of B.C. to be at its minimum absolute value, with the result that $B.C._\odot = -0^m.14$. For the O-stars, one must rely upon non-fundamental data, since we have θ only for ζ Oph, O9.5V. (CDBB). Furthermore, an increasing fraction of f is in the Lyman continuum, which must be calculated. I have used T_{eff}'s by Conti (1973), based upon the line spectrum interpreted by model atmospheres, with some guidance from a scale of T_{eff} published by me (Hayes 1970) which was based upon the Balmer jump. The intrinsic colors of the O-stars are not well known, and I have used my own (Hayes 1970). In Fig. 1 I show these scales of T_{eff} along with CDBB's. I also show CDBB's fundamental stars in this range, excluding supergiants. In the case of ζ Oph and δ Sco (open circles) I have used my own intrinsic colors (both stars are reddened). The dashed line with long dashes shows the line I

TABLE II

MAIN SEQUENCE STARS

TYPE	(U-V)	(B-V)	T_{eff}	B.C.
O5	$-1^{m}.48$	$-0^{m}.319$	47000K	$-4^{m}.3$
O6	1.46	.315	42000	3.9
O7	1.44	.311	38500	3.6
O8	1.41	.305	35600	3.4
O9	1.38	.298	33200	3.2
O9.5	1.35	.294	31900	3.1
B0	1.32	.286	30300	2.96
B0.5	1.28	.277	28600	2.83
B1	1.19	.26	25700	2.59
B2	1.10	.24	23100	2.36
B3	0.91	.20	18900	1.94
B5	0.72	.16	15300	1.44
B6	0.63	.14	14000	1.17
B7	0.54	.12	13000	0.94
B8	0.39	.09	11500	0.61
B9	-0.25	-0.06	10180	0.31
A0	0.00	0.00	9410	-0.15
	(B-V)	(V-R)		
B9	-0.06	0.00	10180	-0.31
A0	0.00	+0.02	9410	.15
A2	+0.06	.08	8900	.08
A5	.14	.16	8210	.02
A7	.19	.19	7920	.01
F0	.31	.30	7160	.01
F2	.36	.35	6880	.02
F5	.43	.40	6560	.03
F8	.54	.47	6190	.08
G0	.59	.50	6010	.10
G2	.63	.53	5860	.13
G5	.66	.54	5780	.14
G8	.74	.58	5580	.18
K0	.82	.64	5260	.24
K2	.92	.74	4850	.35
K5	1.15	.99	4270	.66
K7	1.30	1.15	4030	.93
M0	1.41	1.28	3880	1.21
M1	1.48	1.40	3720	1.49
M2	1.52	1.50	3600	1.75
M3	1.55	1.60	3480	1.96
M4	1.56	1.70	3370	2.28
M5	1.61	1.80	(3260)	2.59
M6	1.72	1.93	(3140)	2.93
M7	1.84	2.20	(2880)	3.46
M8	(+2.00)	(+2.50)	(2620)	-4.0

TABLE III

K AND M GIANTS

Type	(V-R)	T_{eff}	B.C.	Type	(V-R)	T_{eff}	B.C.
K0	+0^{m}77	4600K	-0^{m}42	M0	+1.23	3750K	-1^{m}28
K1	.81	4460	.48	M1	1.28	3700	1.36
K2	.84	4370	.53	M2	1.34	3640	1.52
K3	.96	4100	.60	M3	1.48	3510	1.91
K4	1.06	3950	.90	M4	+1.74	3290	-2.55
K5	+1.20	3790	-1.19				

have adopted for the transition from CDBB's scale to Conti's. **For the K- and M- giants, there are plentiful fundamental data, but I found that they do not define the shape of the T_{eff}-scale well.** As a result, I resorted to an interplay with the scale of T_{eff} for dwarfs. I have assumed that Johnson's scales of T_{eff} for dwarfs and giants correctly give the difference in T_{eff} for the two classes, at a given color. Thus, in Fig. 2 I show the fundamental giants translated to look like dwarfs. The Sun and YY Gem are also shown, and the adopted mean relation was based upon these data. An extension of the mean relation beyond the fundamental data is also shown as a lighter dashed line; it is based upon the T_{eff}'s by Greenstein, et al. (1970), Frogel, et al. (1972) and Veeder (1974). In all three cases, blackbody curves were fitted to IR photometry, with some allowance for line blocking. Also shown are the scales of T_{eff} by Johnson and by Mould and Hyland (1976) whose scale is based upon fitting model atmospheres to IR photometry.

The mean relation for dwarfs has been translated back to Teff's for giants, using Johnson's differences, and this scale is shown with the fundamental stars in Fig. 3. The mean relation defined in this way appears to be quite satisfactory, and has been adopted. Also shown in Fig. 3 is Johnson's scale for giants.

REFERENCES

Barnes, T. G. and Evans, D. S. (1976). Mon. Not. R. Astron. Soc. 174, 489.
Blazit, A., Bonneau, D., Josse, M., Koechlin, L. and Labeyrie, A. (1977). Astrophys. J. Lett., 214, L79.
Code A.D., Davis, J., Bless, R.C. and Hanbury Brown, R. (1976). Astrophys. J., 203, 417.
Conti, P. S. (1973). Astrophys. J., 179, 181.
Currie, D. G., Knapp, S. L. and Liewer, K. M. (1974). Astrophys. J., 187, 131.

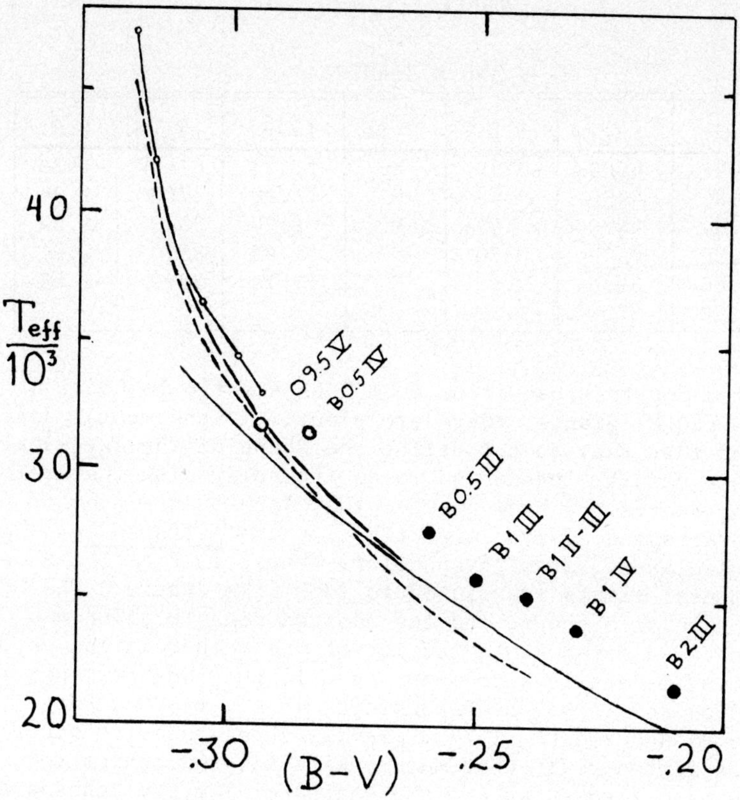

Fig. 1. The effective temperatures of hot main sequence stars. The open and closed circles are fundamental stars from Code, *et al*. 1976; the open circles denote stars plotted according to the mean colors by Hayes (1970). The full line is the mean relation from Code *et al*. (1976). The full line with small circles shows temperatures by Conti (1973) plotted with the mean colors by Hayes (1970). The line with short dashes is the mean relation by Hayes (1970); the line with long dashes is the transition line adopted for this paper.

Forgan, B. W. (1977). *Appl. Optics*, **16**, 1628.

Frogel, J. A., Kleinman, D. E., Kunkel, W., Ney, E. P. and Strecker, D. W. (1972). *Publ. Astron. Soc. Pacific*, **84**, 581.

Gallouët, L. (1964). *Ann. d'Astrophys.*, **27**, 423.

Gezari, D. Y., Labeyrie, A. and Stachnik, R. V. (1972). *Astrophys. J. Lett.*, **173**, L1.

Greenstein, J. L., Neugebauer, G., and Becklin, E. E. (1970). *Astrophys. J.*, **161**, 519.

Hanbury Brown, R. (1968). In *Ann. Rev. Astron. Astrophys.*, Vol. **6**, Annual Reviews, Palo Alto, p. 13.

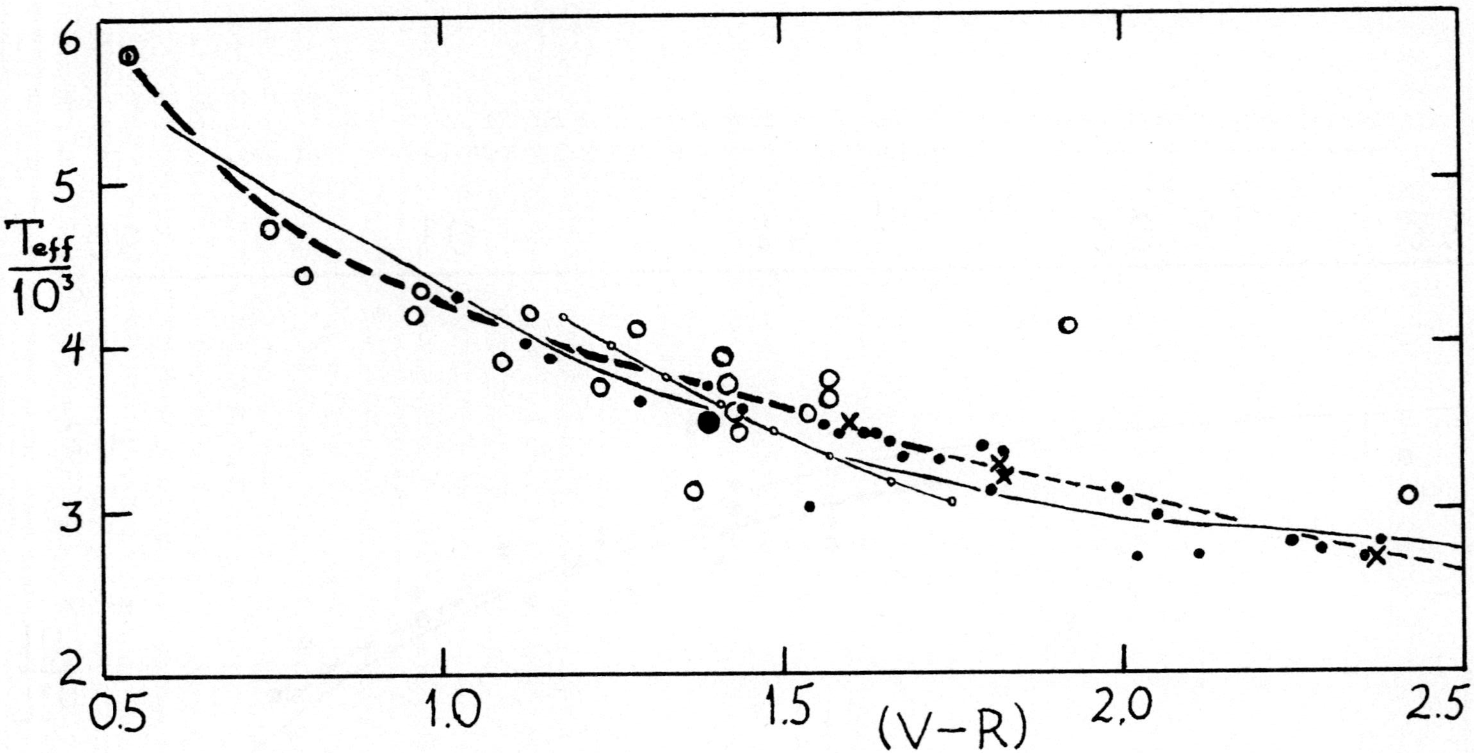

Fig. 2. The temperatures of cool dwarfs. The large open circles are fundamental giants, translated by Johnson's (1966) difference in temperatures of giants and dwarfs. The large filled circle is YY Gem, ⊙ denotes the Sun, and the dots are from Veeder (1974). Three x's are groups from Greenstein, et al. (1970); the fourth (coolest) x is Proxima Cen from Frogel, et al. (1972). The full line is from Johnson; the line with small circles is from Mould and Hyland (1976). We adopt the dashed line.

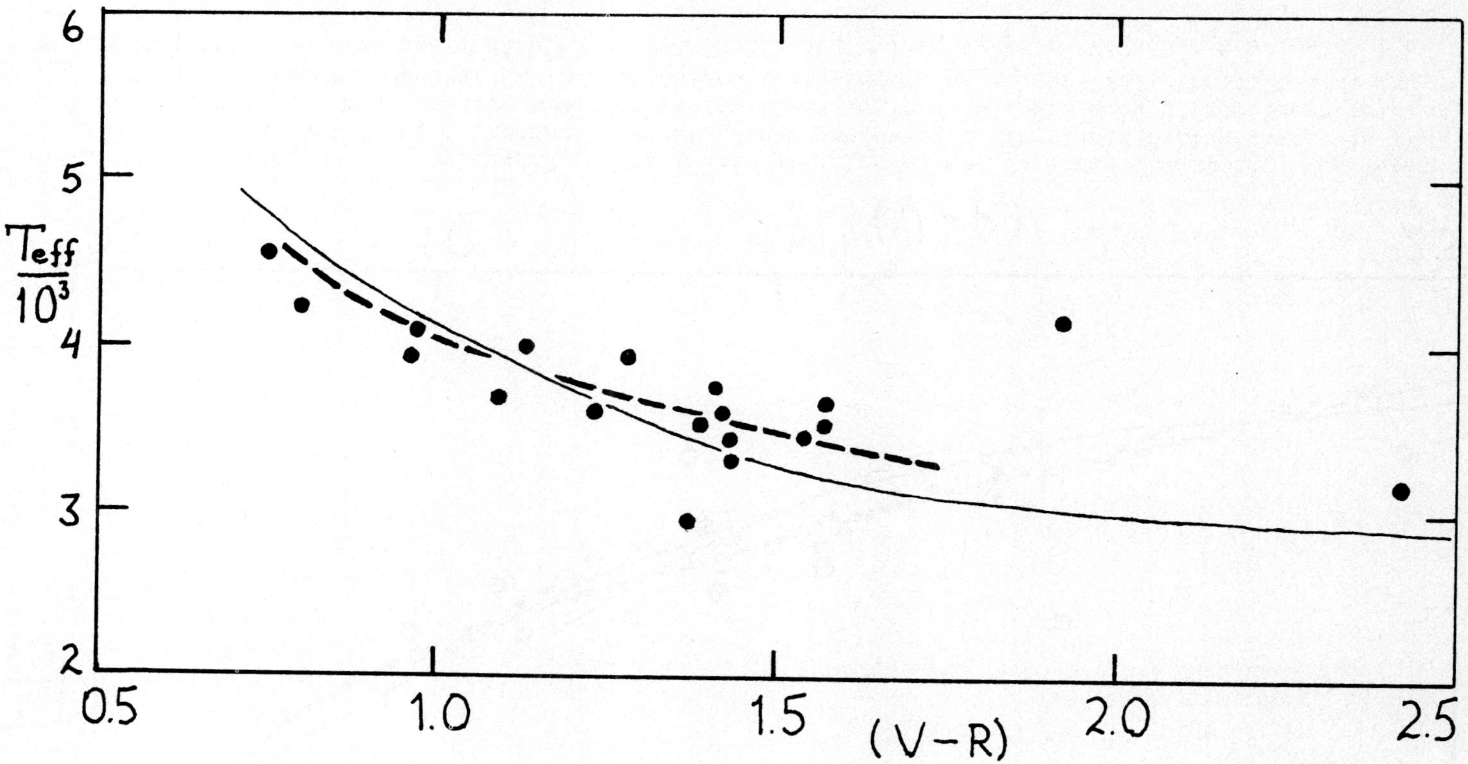

Fig. 3. The effective temperatures of K and M giants. The filled circles are fundamental stars. The full line is the mean relation for giants by Johnson (1966). The dashed line is the relation adopted for this paper.

Harris, D. L. III (1963). In Stars and Stellar Systems, Vol. III, Basic Astronomical Data, K. Aa. Strand, ed., University of Chicago Press, Chicago, p. 263.
Hayes, D. S. (1970). In IAU Symposium No. 36, Ultraviolet Stellar Spectra and Related Ground-Based Observations, L. Houziaux and H. E. Butler, eds., D. Reidel, Dordrecht, p. 83.
Hayes, D. S. and Latham, D. W. (1975). Astrophys. J., 197, 593.
Johnson, H. L. (1965). Comm. Lunar Planet. Lab., No. 53, 73.
Johnson, H. L. (1966). In Ann. Rev. Astron. Astrophys., Vol. 4, Annual Reviews, Palo Alto, p. 193.
Kuiper, G. P. (1938). Astrophys. J., 88, 429.
Mould, J. R. and Hyland, A. R. (1976). Astrophys. J., 208, 399.
Oke, J. B. and Schild, R. E. (1970). Astrophys. J., 161, 1015.
Panagia, N. (1973). Astron. J., 78, 929.
Popper, D. M. (1959). Astrophys. J., 129, 647.
Ridgway, S. T., Wells, D. C. and Joyce, R. R. (1977). Astron. J., 82, 414.
Stebbins, J. and Kron, G. (1957). Astrophys. J., 126, 266.
Tüg, H., White, N. M. and Lockwood, G. W. (1977). Astron. Astrophys. in press.
Veeder, G. J. (1974). Astron. J., 79, 1056.
Vilas, F. and Lasker, B. M. (1977). Publ. Astron. Soc. Pacific, 89, 95.
Walker, R. G. (1969). Phil. Trans. Roy. Soc. London, A, 264, 209.
Worden, S. P. (1976). Publ. Astron. Soc. Pacific, 88, 69.

DISCUSSION

GARRISON: Would you be willing to give a value for the uncertainty in (B-V) for the Sun, from your work and/or that of others?

HAYES: I estimate the uncertainty in my value to be $\pm 0\overset{m}{.}02$. Note that two other papers in this Symposium discuss this point, and they should be more definitive.

D. EVANS: I much appreciate your favorable remarks about occultation determinations of angular diameters. Things have changed since the criticism in the early days. In addition to YY Gem I would like to mention CM Dra investigated in what will become a classical study by Claud Lacy. He has also added data from a number of other eclipsing binaries.

WESSELIUS: I have compared the absolute calibrations of the ultraviolet experiments in OAO-2(WEP), TD1A(S2/68) and ANS. I find that TD1 and ANS agree quite well (within 5 to 10%) while the OAO-2(WEP) scanner data are brighter by 40% at 1550 Å and 20% at 1800 Å.

HAYES: This agrees with what I remember from the recent paper by Beeckmans in Astronomy and Astrophysics. So, it may be that the TD-1 absolute calibration is to be preferred over the OAO-2 absolute calibration. However, for the T_{eff} and B.C.- scale, Beeckmans shows that the differences which result are smaller than the uncertainties in the T_{eff} and B.C.- scale, evaluated either way.

KODAIRA: So far as I know, there is substantial scatter among the angular diameter data of individual stars. Have you used a simple average of them in deriving your temperature scale for red giants, or have you preferred some particular ones to others?

HAYES: I have taken a simple average.

FLOWER: How do your effective temperatures differ from those of Johnson's?

HAYES: My effective temperatures differ from Johnson's by up to about 200 K, as shown in Figs. 2 and 3.

FLOWER: What is the temperature difference you find between the effective temperatures of giants and dwarfs?

HAYES: The temperature difference between giants and dwarfs is between zero and about 250 K; I don't have the numbers here, but they are the same as are given by Johnson's T_{eff} - scales.

PART II

HR DIAGRAMS, THE SOLAR NEIGHBORHOOD

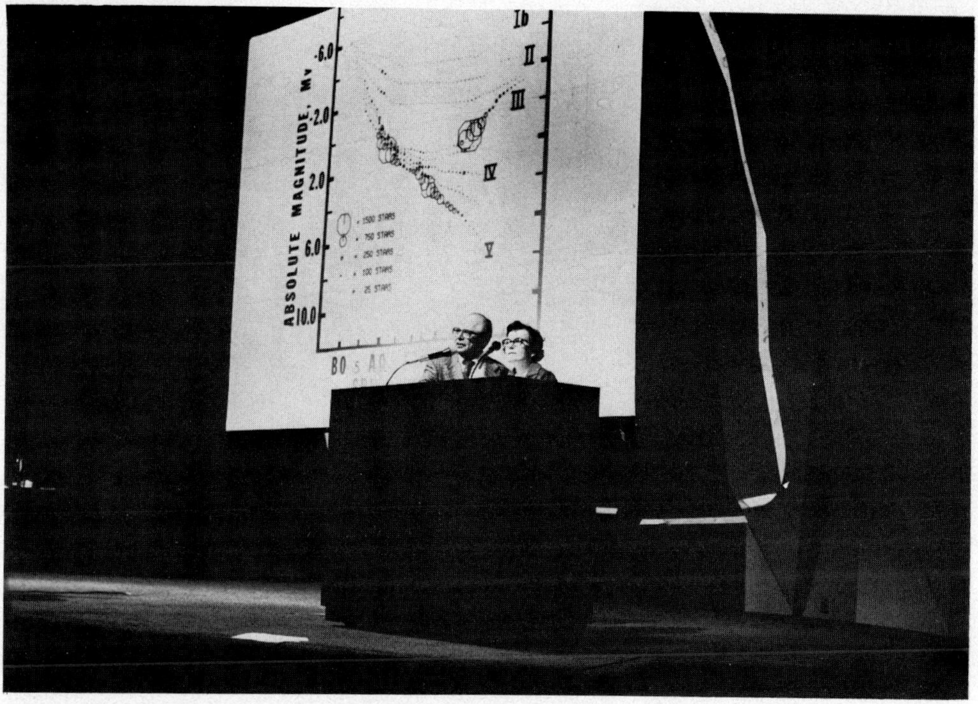

Jesse Greenstein and Nancy Houk on stage.

HERTZSPRUNG-RUSSELL DIAGRAMS AND COLOR-LUMINOSITY DIAGRAMS FOR THE STARS NEARER THAN TWENTY-TWO PARSECS

W. Gliese

Astronomisches Rechen-Institut, Heidelberg, F.R.G.

1. DATA AVAILABLE IN 1977

Observational data for stars in the solar neighborhood received before January 1969 were included in the Catalogue of Nearby Stars (Gliese, 1969) which contains more than 1800 objects, single stars and components of binaries, nearer than 22 pc. Since then more than 8 years have gone by and many new data have been published, especially new trigonometric parallaxes as well as photometric distance determinations. Today we know more than 2000 stars with parallaxes measured to be greater than 0".044. But many of these distances are still very uncertain and for a considerable number of objects, especially the faint ones, spectral classificiation is lacking or the photometry is still not satisfactory.

A discussion of relations between luminosity and other physical properties must be restricted to stars with parallaxes determined trigonometrically since this is the only primary stellar distance indicator we have. In 1977 we know nearly 1500 such objects with $\pi_t > 0''.044$ but their distances and their absolute magnitudes have been determined with very different accuracies. The numbers of these stars whose absolute magnitudes are known with standard errors between given limits is shown in Table I. The numbers of spectroscopic binaries or objects with large dispersion in their radial velocity data are given in brackets. They have been excluded from the stars in the diagrams shown in this paper. However, the number of detected spectroscopic binaries is still incomplete even in the solar neighborhood. Detailed investigations among F and G dwarfs and subgiants by Abt and Levy (1976) have implied that single stars are rare.

A. G. Davis Philip and D. S. Hayes (eds.), The HR Diagram, 79–88.

We learn from these numbers that MK types have been obtained only for somewhat more than 600 stars, mostly F, G and K dwarfs, whereas broad-band (B-V) photometry is available for about 1300 of these objects. Joy and Abt (1974) have published a classification of red dwarfs from dM0 to dM6.5 for nearly 300 of our stars. The 55 degenerates in this sample can be easily included in a (M_V,B-V) diagram.

2. EFFECTS OF THE OBSERVATIONAL ERRORS

The effects of uncertain observations on a color-luminosity diagram are demonstrated by a comparison between Fig. 1 and Fig. 2 which show the (M_V,B-V) diagrams for several hundred nearby stars. The large dots represent the stars with well-known absolute magnitudes (s.e. up to $\pm 0^m.12$). The open box at $M_V = +4^m.85$ gives the position of the Sun. The small dots in Fig. 1 show the stars with somewhat more uncertain data (s.e. of M_V from $\pm 0^m.13$ to $\pm 0^m.30$). Among them there is an insignificantly larger spread along the main sequence but no remarkable systematic difference between the regions with large and with small dots. In Fig. 2, however, the small dots represent the objects with very uncertain luminosities. Their standard errors exceed half a magnitude and, as photoelectric data are available, they are caused mainly by errors in the trigonometric parallax determinations. A striking phenomenon is the large number of small dots lying below the main sequence, especially between the limits $+0^m.5 < (B-V) < +1^m.3$ which is the region of the G and K stars. Obviously, this population does not appear in the sample of stars with accurate data.

The difference between the distribution of the stars in the two diagrams is due to the well-known systematic effects of the accidental errors in parallax measurements. The use of a lower limit to the parallaxes in this investigation produces a preference for objects with positive parallax errors and, consequently, the number of stars with apparently too low luminosities is greater

TABLE I
NUMBERS OF NEARBY STARS ($\pi_t > 0''.044$) WITH DIFFERENT STANDARD ERRORS OF THEIR ABSOLUTE MAGNITUDES

Standard errors	Number of stars (SB)	Standard errors	Sp. types MK	Sp. types Joy	Colors (B-V)	Colors (R-I)
up to $\pm 0^m.12$	126 (12)	up to $\pm 0^m.30$	216 (30)	154 (23)	525 (52)	303 (40)
±0.13 to ±0.30	428 (42)					
±0.31 to ±0.50	496 (48)	exceeding ±0.3	400 (45)	137 (11)	807 (64)	240 (7)
exceeding ±0.5	436 (45)					
Sum	1486 (147)		616 (75)	291 (34)	1332 (116)	543 (47)

than that of the objects measured to be too bright. Statistical corrections to eliminate this phenomenon vary as σ^2 where σ is the average mean error of a parallax determination in a group of objects. Therefore, this effect remains negligible for the series of stars with small errors in Fig. 1 but it is pronounced in the luminosities with standard errors exceeding half a magnitude. Conclusions drawn from the second diagram would not therefore be reliable. Further discussions in this paper are restricted to objects with standard errors in M_V not greater than $\pm 0\overset{m}{.}30$.

3. HR DIAGRAMS

The HR diagram for 186 nearby stars and for the Sun is shown in Fig. 3. The spectral types are MK classifications. Stars of luminosity classes III to VI are present: giants, subgiants, dwarfs and three subdwarfs of type G. The M-dwarf region is still sparsely populated with MK types. The scale of the abscissa has been dilated from M0 onwards so that the steepness of the main sequence remains nearly constant until M4.

The few giants, α Boo, both components of α Aur, and β Gem (HD 124897, 34029 and 62509) do not allow us to make any conclusions about luminosity class III. The diagram confirms that there is no clear dividing line between the classes IV and V, especially in the region of the early F stars.

The scatter along the main sequence shows a thickness of about one magnitude in M_V in the classes A to K. This spread is partly due to the observational errors in the trigonometric parallaxes, but a smaller part also is due to the spectral classifications (s.e. = $\pm$0.6 tenths of a spectral class; C. Jaschek and M. Jaschek, 1973; Gliese, 1973) and a small additional component is introduced by concentrating the objects only at discrete points on the abscissa. Furthermore, the material may contain some undetected spectroscopic binaries.

Even if all these effects could be eliminated a real scatter in the luminosities of main-sequence stars of the same type would be observed. The fine structure of the HR diagram of nearby F, G and K stars has been investigated by Perrin, Hejlesen, Cayrel de Strobel and Cayrel (1977) using 138 stars with fairly reliable observational data. They studied the state of evolution by means of theoretical ($\log T_{eff}$, M_{bol}) diagrams corresponding to the actual chemical composition of the stars. They concluded that the differences in metal content can account for the thickness of the observational main sequence for stars of effective temperature under 5500 K whereas the evolutionary effects become predominant for stars of effective temperature over 6000 K.

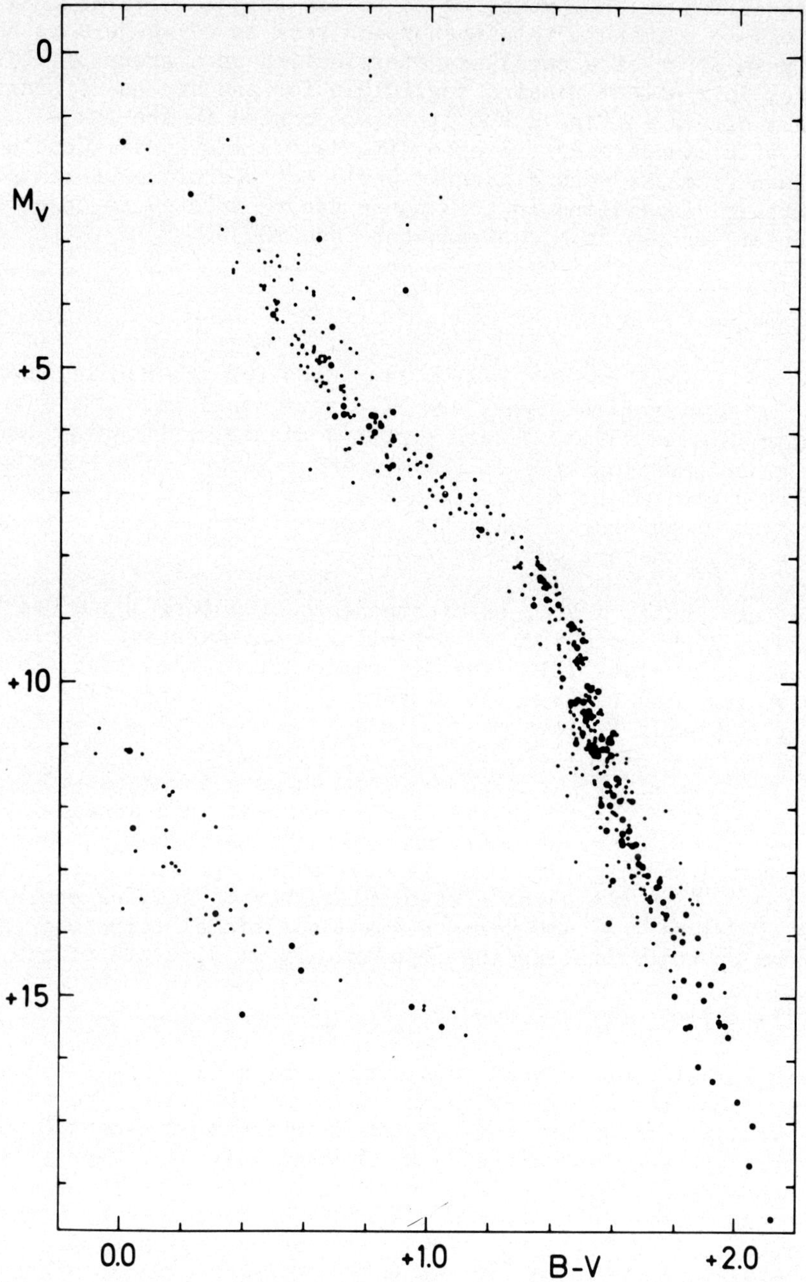

Fig. 1. Trigonometrically determined absolute visual magnitudes M_V of nearby stars <u>vs.</u> (B-V) colors. The large dots represent the stars with s.e. (M_V) up to $\pm 0\overset{m}{.}12$, the small dots represent the stars with s.e. (M_V) from $\pm 0\overset{m}{.}13$ to $\pm 0\overset{m}{.}30$.

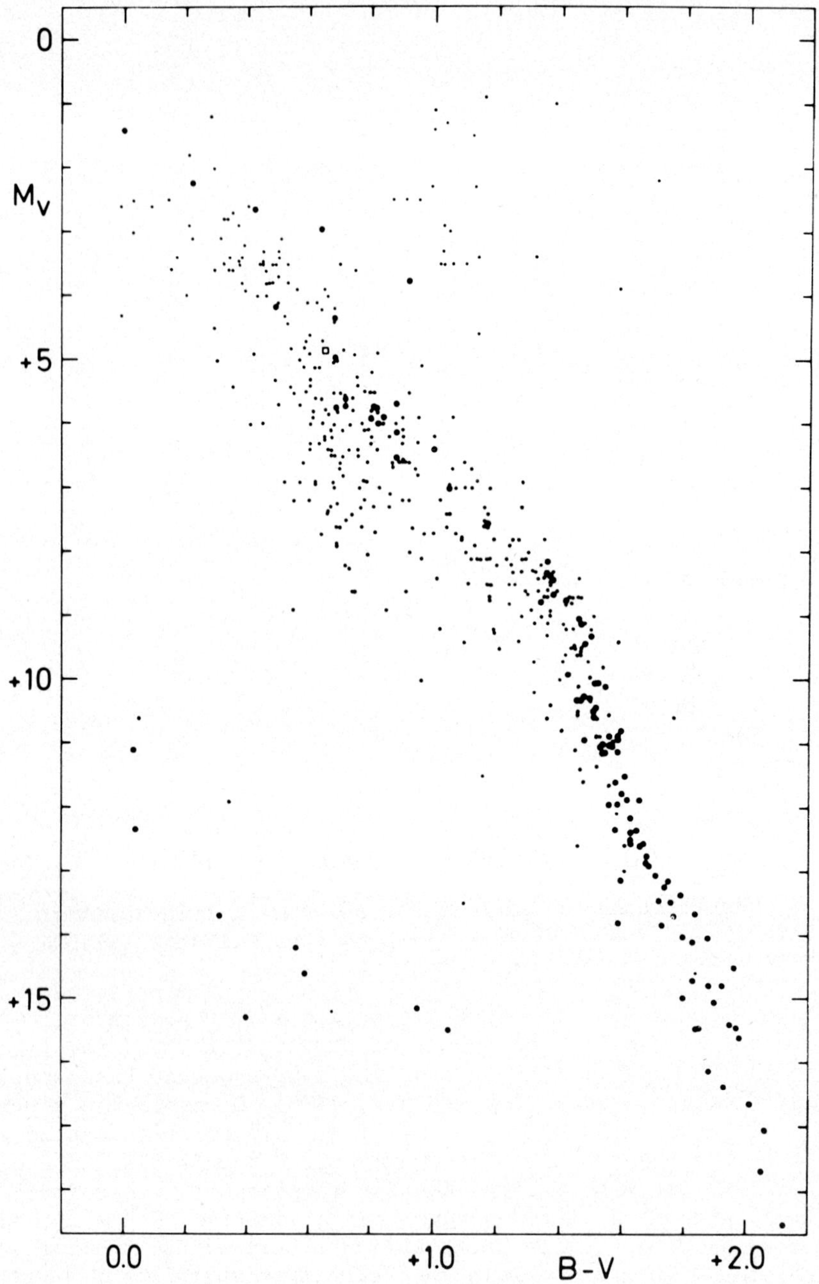

Fig. 2. Trigonometrically determined absolute visual magnitudes M_V of nearby stars vs. (B-V) colors. The large dots represent the stars with s.e. (M_V) up to $\pm 0^m.12$, the small dots represent the uncertain data with s.e. (M_V) exceeding $\pm 0^m.5$.

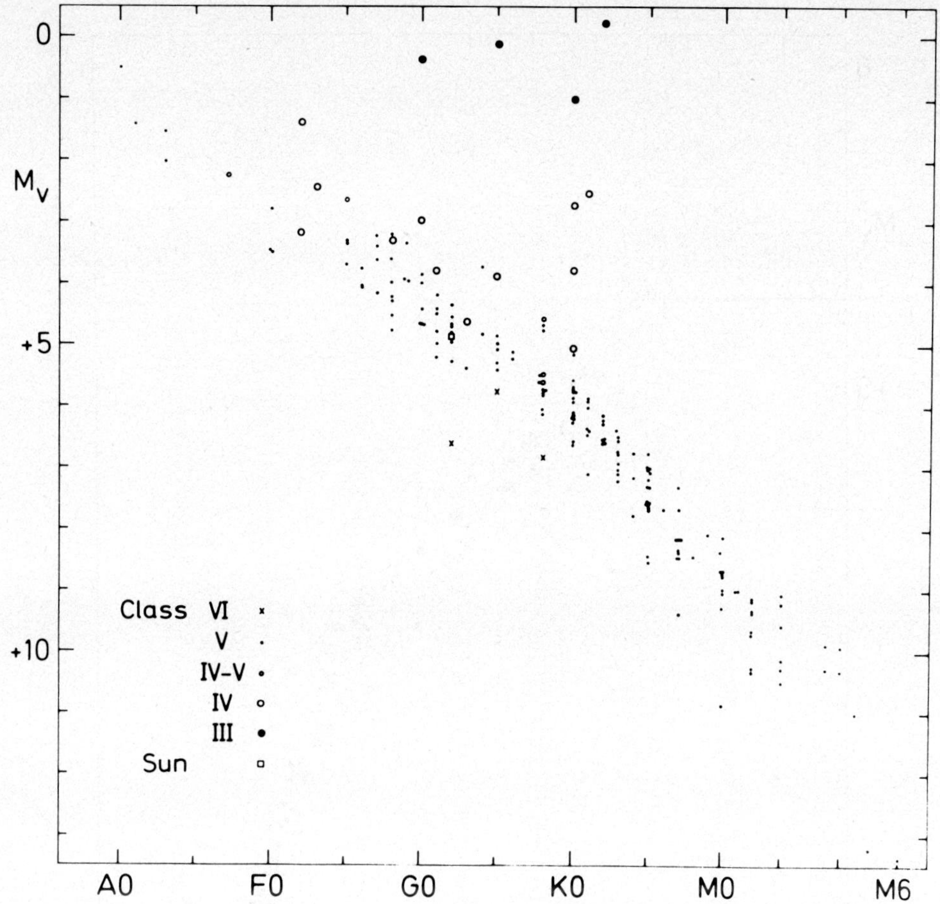

Fig. 3. Observational HR diagram for stars with trigonometric parallaxes >0".044 and with MK classification. All stars with s.e. (M_V) exceeding $\pm 0^m.30$ have been eliminated.

Observational data for such a detailed investigation are available only for a limited number of bright stars. For the red stars in the HR diagram Fig. 4 shows the main sequence based on the classification by Joy (Joy and Abt, 1974). Open circles represent the dMe stars with hydrogen emission. CaII emission can be detected in the H and K lines of nearly all M dwarfs (Wilson and Woolley, 1970) but with a large scatter in intensity. Therefore, all objects without H emission have been classified as "dM". They are plotted as dots. It is known that the CaII emission intensity is correlated with age and with the velocity dispersion of a group of red dwarfs (Wielen and Jahreiss, 1974; Wielen, 1974). Barnard's star and Ross 47 have been classed as sdM4.5; they are marked as crosses. From Fig. 4 only one conclusion can be drawn: There is an increase in the percentage of dMe stars among the latest spectral types.

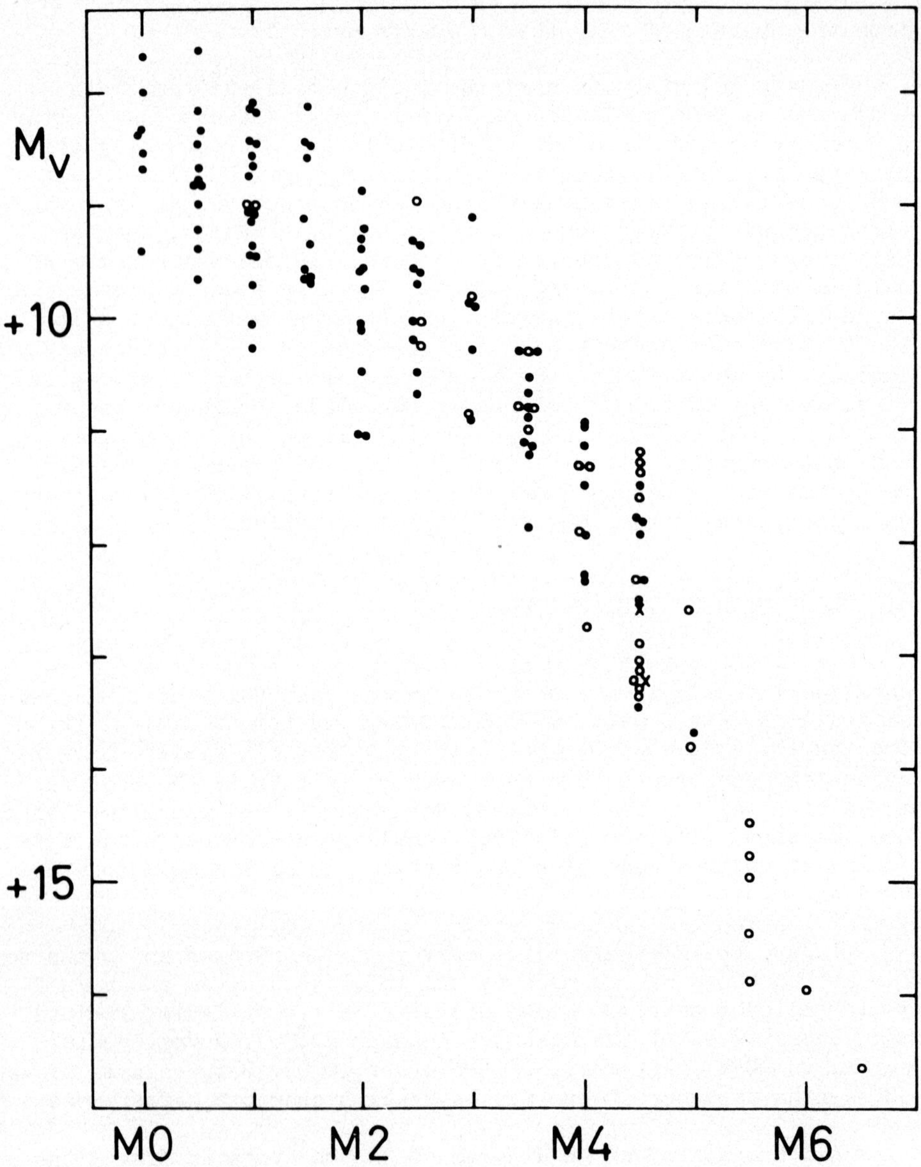

Fig. 4. Observational HR diagram for nearby red dwarfs classified by Joy (1974). The open circles represent the stars with hydrogen emission; the dots represent the other dM stars. All stars with s.e. (M_V) exceeding $\pm 0\overset{m}{.}30$ have been eliminated.

Many of these objects have proved to be flare stars; more dMe stars have been detected as flare stars as time goes on. There is a disagreement in the literature concerning their positions in a HR diagram or in various color-color diagrams. Do they deviate

from the average of the other M dwarfs or not?

A detailed discussion of the luminosities and temperatures of M dwarfs has been published by Veeder (1974) who has done infrared photometry of more than one hundred objects. Various diagrams showing "absolute bolometric magnitude versus (V-K) color" have been constructed to examine whether or not there exist systematic displacements between some groups of nearby M dwarfs, such as halo stars, disk and young disk members, hydrogen emission stars, calcium emission stars, and others. There is a considerable overlap due to the scatter of stars in each group with the result that the average displacement is not very dramatic. Halo stars are subluminous by about $0^m.2$ on the average; stars with hydrogen emission lines average about $0^m.3$ bolometric magnitudes brighter than M dwarfs without emission lines. But what is the effect of still undetected spectroscopic binaries? Do they occur more frequently among emission-line red dwarfs? Bopp and Fekel (1977) have found only four single stars among 13 BY Dra variables.

4. COLOR-LUMINOSITY DIAGRAMS

It is not possible in this short review to mention all the detailed investigations published in the last few years. MK classifications, Joy's data, UBV photometry are not sufficient for discussing all special problems. Other photometric systems have been chosen and constructed by their authors to satisfy the requirements of a well-defined problem. However, refined parameters have been measured only for relatively small groups of stars and detailed investigations have been restricted to limited regions in the HR diagram.

Today, the most extensive survey of the HR diagram in the solar neighborhood can be made by the (M_V,B-V) diagram shown in Fig. 1. It includes nearly 500 objects with fairly reliable absolute magnitudes (s.e. of the M_V not exceeding $\pm 0^m.30$). Normally, the errors in (B-V) are small, of the order $\pm 0^m.02$. Between some of the UBV series used here there may be small systematic differences.

Additional to the HR diagram of the MK types in Fig. 3 the (M_V,B-V) diagram also includes the white dwarf sequence. These objects are fairly frequent from (B-V)= $-0^m.10$ to $+0^m.70$. The search for red degenerates has not yet succeeded in detecting objects of (B-V)> $+1^m.13$ with one exception: LP 701-29 with the data ($16^m.11$ $\pm 0^m.12$, $+1^m.88$) which means that this red "white dwarf" lies on the main sequence in the (M_V,B-V) diagram (Dahn et al., 1976).

Multichannel spectrophotometry of white dwarfs with the 5 m Hale reflector by Greenstein (1976) has shown that their colors are highly correlated with their luminosities. The most useful

color is that from $1/\lambda = 2.12$ to $1.25\mu^{-1}$. It appears that most white dwarfs of a given color lie in a narrow band of luminosity.

This short review could not be complete. There are many more possibilities for drawing color-luminosity diagrams. More precise measurements are required for the nearby stars. But parallax programs have been very successful in the last few years and the steady increase of photometric data of nearby objects is noteworthy.

ACKNOWLEDGEMENTS

I am indebted to Dr. H. Jahreiss for providing me support in compiling the new observational data. Dr. Keenan graciously corrected the English text.

REFERENCES

Abt,H.A., and Levy,S.G. (1976). Astrophys. J. Suppl. 30, 273.
Bopp,B.W., and Fekel Jr.,F. (1977). Astron. J. 82, 490.
Dahn,C.C., Hintzen,P.M., Liebert,J.W., and Spinrad,H. (1976). Bull. Am. Astron. Soc. 8, 506.
Gliese,W. (1969). Veröffentl. Astron. Rechen-Inst. Heidelberg, No.22.
Gliese,W. (1973). in IAU Symposium No.50, Spectral Classification and Multicolour Photometry, C. Fehrenbach and B.E. Westerlund, eds., D.Reidel, Dordrecht, p.52.
Greenstein,J.L. (1976). Astron. J. 81, 323.
Jaschek,C., and Jaschek,M. (1973). in IAU Symposium No.50, Spectral Classification and Multicolour Photometry, C.Fehrenbach and B.E. Westerlund, eds., D.Reidel, Dordrecht, p.43.
Joy,A.H., and Abt,H.A. (1974). Astrophys. J. Suppl. 28, 1.
Perrin,M.-N., Hejlesen,P.M., Cayrel de Strobel,G., and Cayrel,R. (1977). Astron. Astrophys. 54, 779.
Veeder,G.J. (1974). Astron. J. 79, 1056.
Wielen,R. (1974). in Highlights of Astronomy, Vol.3, G. Contopoulos, ed., D.Reidel, Dordrecht, p.395.
Wielen,R., and Jahreiss,H. (1974). Mitt. Astron. Ges. 35, 214.
Wilson,O., and Woolley,R. (1970). Monthly Notices Roy. Astron. Soc. 148, 463.

DISCUSSION

D. EVANS: Because the intensity of Balmer line emission in dMe stars frequently varies with time one can envisage the possibility that some stars might at one time be classed as dMe and at another dM. I am not al all convinced that we ought to consider them separately.

GLIESE: You may be right, but the detailed discussion of the flare star phenomenon among dM or dMe stars has taken already whole sessions in variable star colloquia without solving all the problems. Certainly, it exceeds our possibilities here.

GREENSTEIN: When will the next edition of your valuable catalog appear?

GLIESE: I am just compiling a preliminary list of another 250 stars which have been observed since 1969 with parallaxes exceeding 0".044. The publication of the final Third Catalogue of Nearby Stars is planned for 1980 or 1981.

THE POSITION OF THE SUN IN THE HR DIAGRAM

Don C. Barry

University of Southern California

ABSTRACT

Since the absolute magnitude of the Sun is known accurately and the solar spectrum defines spectral type G2 V, it would appear that the position of the Sun in the HR diagram is well established. However, the color of the Sun and its metallicity relative to other stars remains controversial. It is known from theory that stellar groups of differing composition will have different main sequences.

The observed difference in the strength of the metallic lines in Coma stars relative to Hyades stars is caused by a systematic difference in [Fe/H] of less than 0.2 dex or 50%. Photographs and quantitative measures of the systematic differences between the Hyades and Coma spectra are presented. Visual comparison and quantitative analysis of the solar spectrum relative to Hyades and Coma spectra reveal that the Sun is metal rich similar to the Hyades stars rather than of normal disk metallicity as represented by Coma stars. Evidence is presented showing that the color and spectral line strengths of the Sun are more similar to stars classified G3 and G4 in the literature than G2. If the Hyades modulus is $y = 3\overset{m}{.}25$, the (B-V) color of the Sun must be $0\overset{m}{.}65$ or redder for the Sun to be on or above the Hyades main sequence.

A. G. Davis Philip and D. S. Hayes (eds.), The HR Diagram, 89-90.

DISCUSSION

HARDORP: In searching for stars whose ultraviolet absorption features match those of the Sun, I found the same as Dr. Barry: all matching stars have (B-V) = $+0^m.66$. More details can be found in a paper in the Poster-Session of this symposium, page 423.

BELL: I think it would be interesting if you were to plot similar diagrams of solar line strength versus H_β since, as you pointed out, the solar colors do depend quite sensitively on the adopted solar hydrogen line strength.

BARRY: The relation between ΣH hydrogen line strength index and H_β for Hyades and Coma cluster F stars has been published (Astrophys. J. **212**, 462, 1977). The correlation is quite tight. However, the H_β index is not useful for stars cooler than G0 because of the weakening of H_β and the strengthening of the nearby metallic lines with decreasing temperature. The solar (U-B) color reported here is not sensitive to the measured hydrogen line strength ΣH. The (B-V) color is only mildly sensitive. The colors are sensitive to the amount of line blanketing in the solar spectrum relative to other G dwarfs. If our determination that the Sun has no ultraviolet excess is assumed correct, the correlation between ultraviolet excess and the residuals in the data of those workers who found a "bluer" Sun may be used to correct their results for blanketing effects such that they are brought into agreement with the majority of workers who have found a redder Sun. Considering the results of all previous workers and our own, we find mean values of $(B-V)_{Sun} = 0^m.667$ and $(U-B)_{Sun} = 0^m.18$.

BELL: I would like to ask how you obtained the solar spectrum?

BARRY: The solar results are based on two spectra of moonlight and two spectra of a satellite of Saturn. They are widened 1 mm and have a resolution of 2 Ångstroms.

CRAWFORD: I think you should really use (R-I) as your abscissa. The β vs. (R-I) relations are very good for late F and early G type stars, and are essentially independent of blanketing. Good (R-I) would be better than ΣH, β, (B-V), etc.

BARRY: I tried that with published values of (R-I), and found so much "noise" that I couldn't separate the Hyades from Coma. Crawford has new values of (R-I), and with these I might get better results.

HR DIAGRAMS DERIVED FROM THE MICHIGAN SPECTRAL CATALOGUE

Nancy Houk and Robert Fesen

University of Michigan

In this paper we present some HR diagrams constructed using data from the Michigan Spectral Catalogues. Houk (1975) has been systematically reclassifying the Henry Draper stars on the MK system, from the south pole northward. Objective-prism plates, with a reciprocal dispersion of 108 Å/mm, have been taken with the Michigan Curtis Schmidt telescope at Cerro Tololo Inter-American Observatory in Chile. The spectra are classified visually from the plates, and the results are put onto IBM cards and magnetic tape from which the catalogues are produced.

The data used for this paper are taken from both Volume 1 (Houk and Cowley, 1975) and from part of Volume 2 (Houk, 1978, in preparation). The absolute magnitudes (M_V) were taken mainly from the Landolt-Börnstein tables (Schmidt-Kaler, 1965), with a few being from Blaauw (1963). The values were smoothed and interpolated to get the intermediate luminosity types used in classification such as Ib/II and II/III. More recent work has probably changed some of the values used, but for our purposes the general pattern of the HR diagrams is what is important.

In the plots below only the higher-quality classifications are included: those assigned qualities 1 or 2 in the catalogues. Also all peculiar stars are omitted since reliable luminosity classes can not usually be assigned. Even for spectra of quality 1 and 2, O-type stars earlier than O9 are not normally assigned luminosity types, so are virtually absent from Figs. 1 and 2. Late M stars are also under-represented because of the difficulty of getting good luminosity types when the spectra are often underexposed.

A. G. Davis Philip and D. S. Hayes (eds.), The HR Diagram, 91–98.

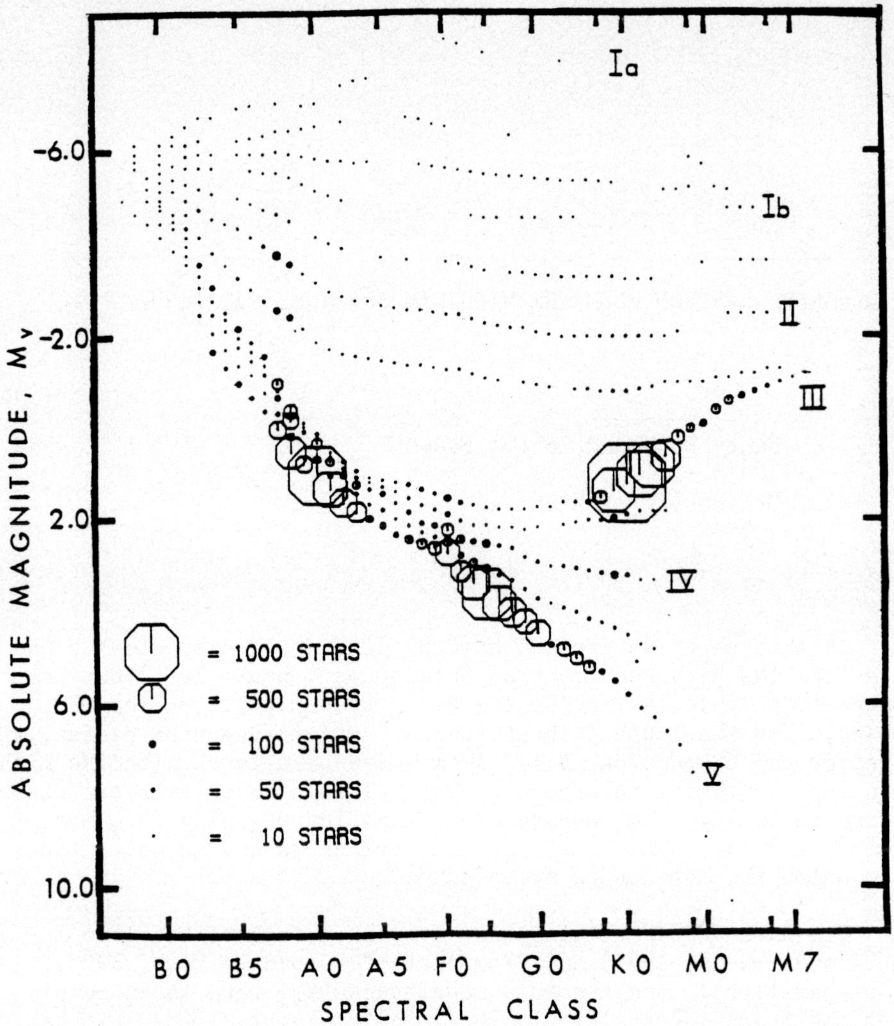

Fig. 1. HR diagram for normal stars of quality 1 and 2 in Volume 1 of the Michigan Spectral Catalogue. The unlabeled subtypes F5, G5 and K5 indicated by the shorter marks are not at the midpoint because only used spectral subtypes are included, all of them being equally spaced. Symbol size is proportional to the number of stars of the particular type. Examples are given in the key.

The plot for Volume 1 is shown in Fig. 1 (also on cover of this book). It contains the HD stars between $\delta = -90^{\circ}$ and $-53^{\circ}.0$. This diagram includes more than 24,000 stars -- about 2/3 of the total in Volume 1, the others being of too low quality or peculiar.

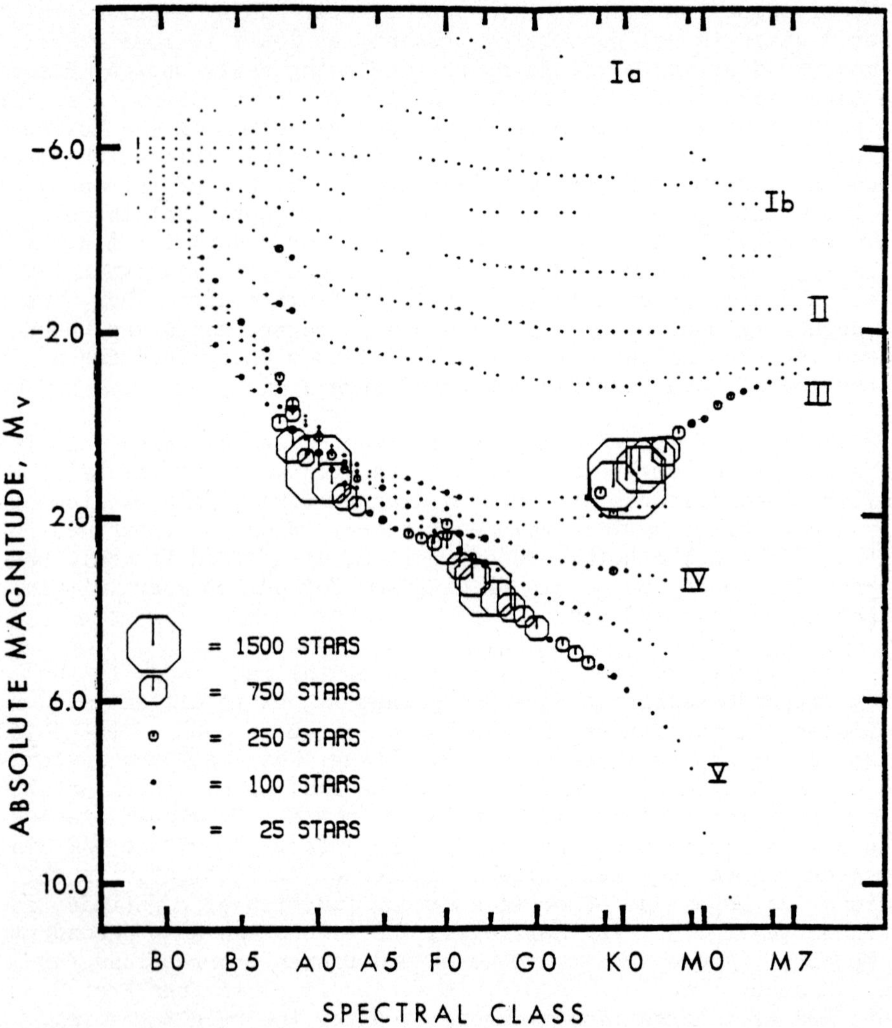

Fig. 2. HR diagram for normal stars of quality 1 and 2 in Volume 1 plus part of Volume 2 of the Michigan Spectral Catalogue.

Note that K0 III stars are most numerous, A0 V next, and then F5V. A similar pattern is seen in Fig. 2 which adds the stars that were on tape in August for Volume 2, bringing the total stars on this plot to 36,382. For the Volume 2 sample alone the A0 V types slightly exceed the number of K0 III's, but this will change when the omitted regions far from the galactic plane are added.

Perhaps the most surprising feature is the paucity of main sequence stars from A4 to A9, evident in Volumes 1 and 2. This is not an artifact of the classification scheme: a plot versus

(B-V) or temperature shows this same effect. This deficiency of later A stars is not new, being present, at least to some extent, in most HR diagrams of field stars, including early ones by Russell. The large amount of data here makes it more conspicuous. Also, in the past it has often been de-emphasized by adjusting the horizontal scale. We have no space to show an example here, but see the often-shown HR diagram prepared by Gillenberg, on p. 259 of Astronomy of the 20th Century (Struve and Zebergs, 1962). Note that there would be a gap if the scale were the same throughout the A's. Interestingly, both Mendoza (1956) and Böhm-Vitense and Canterna (1974) have found similar gaps in the (B-V)'s of field stars. They were not seen very convincingly in clusters, however, which may be one reason the finding was neglected. The latter paper includes a discussion of some possible causes of such a gap.

The sort of HR diagram given in Figs. 1 and 2, which contains stars down to a very rough limiting magnitude, is admittedly of limited usefulness compared with a diagram of a single cluster or of stars within a certain volume of space. When our paper was placed in the Solar Neighborhood session, we decided it would be appropriate to try to obtain HR diagrams for our HD stars within 25 pcs and within 100 pcs.

This first required an evaluation of the magnitudes in the Henry Draper Catalogue before using them to obtain distances. Photoelectric magnitudes are available for about 4500 of the 36,400 HD stars we used south of $\delta = -40^{\circ}$. Using these we found systematic errors in the HD magnitudes which depend, among other variables, on the position in the sky, especially on R.A. By visual examination of the errors lists we divided the area from -90° to -40° into 8 regions which required somewhat different corrections. In most regions the magnitude error is a strong function of magnitude, as is shown in Fig. 3. In that region the errors are also dependent on spectral type or color, so we made different corrections for the G-M stars than for the O-F stars. As is clear from Fig. 3 there are also large random errors even for the brighter stars, making the individual distances we obtain very uncertain. The corrections used, though of limited accuracy, are given in Table I. In the fourth column the Δm is in the sense (HD pg mag - p.e. B mag) and is the value Δm has at the brighter magnitude given in column 3. The slope, given in the last column, can then be used to calculate the Δm at any m. Outside the magnitude ranges given no corrections are needed.

The resulting HR diagrams for the 184 normal, quality-1-and-2 stars in Volumes 1 and 2 computed to be within 25 pcs and the 4700 stars within 100 pcs are shown in Fig. 4. Only the general picture is significant since the distance to any particular star is so uncertain. The majority of the HD stars within 25 pcs are early G dwarfs, with incompleteness setting in for later types. For HD stars within 100 parsecs note that the maximum is at G0 and there

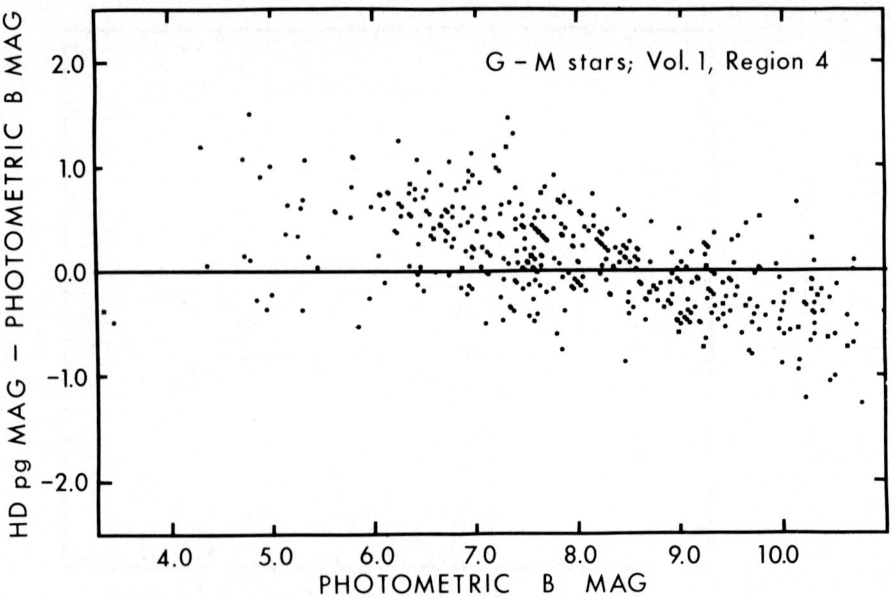

Fig. 3. Errors in HD magnitudes as a function of p.e. magnitude.

TABLE I

CORRECTIONS TO HD PG MAG AS A FUNCTION OF R.A. AND MAG

Zone	R.A. Limits	HD pg mag Range	Correction Δ mag at brtest mag	slope
Volume 1 (-90° to -53°)				
1	0^h0 - 5^h0 +21.5 - 24.0	all magnitudes	no correction	
2	5.0 - 9.0	6.0 - 10.0	+.50	-.20
3	9.0 - 10.5	7.0 - 10.2	+.40	-.19
4	10.5 - 21.5	O-F stars 6.5 - 10.7	+.25	-.24
		G-M stars 5.0 - 6.99	.00	+.35
		G-M stars 7.0 - 10.7	+.50	-.27
Volume 2 (-53° to -40°)				
1	0^h0 - 8^h0	5.8 - 8.49	+.50	-.26
	+17.5 - 24.0	8.5 - 11.5	-.20	+.20
2	8.0 - 12.0	8.5 - 11.5	-.30	.00
		O-F stars 6.2 - 8.49	+.40	-.26
		G-M stars 6.2 - 8.49	+.70	-.44
3	12.0 - 15.0	3.0 - 6.49	-.30	+.23
		6.5 - 11.5	+.70	-.30
4	15.0 - 17.5	3.0 - 6.49	-.20	+.23
		6.5 - 7.99	+.80	-.80
		8.0 - 10.5	-.20	.00

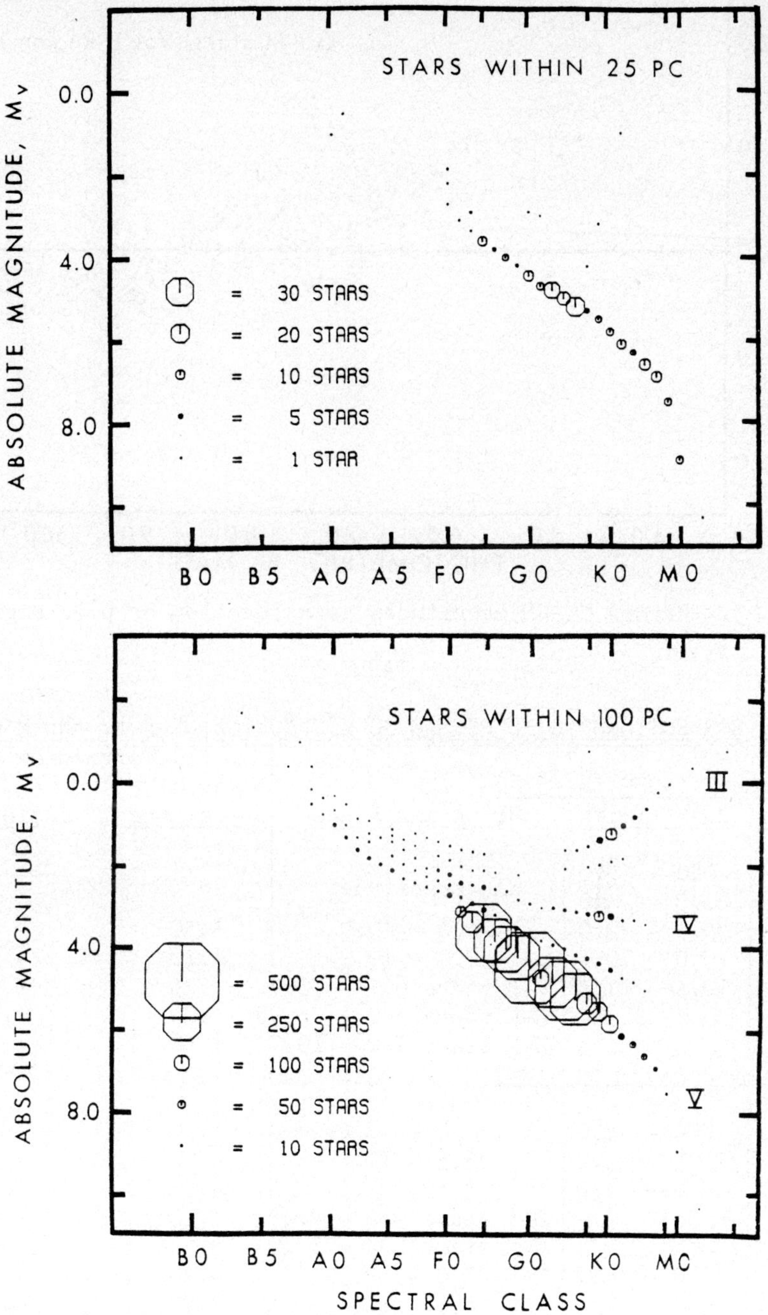

Fig. 4. HR diagrams for HD stars from Volumes 1 and 2 of the Michigan Spectral Catalogue computed to be within 25 pcs and 100 pcs.

is a quite abrupt cutoff at F3, the earlier-type stars mainly being further away than 100 pcs. In considering total numbers of stars it should be remembered that only 2/3 of the HD stars in Volume 1 and even fewer in Volume 2 were part of the reference population used to obtain Fig. 4 since poor quality, peculiar, and unprocessed (i.e., not yet on tape) stars were excluded.

HR diagrams constructed from stars divided into galactic-latitude groups might well be of interest. We did not look into this because Ochsenbein's paper, later withdrawn, was to have been along those lines. After Volume 2 of the Michigan Spectral Catalogue is complete other such correlations and statistical studies will be undertaken.

We would like to thank F. Ochsenbein and the Centre de Donnees Stellaires (Stellar Data Center) at Strasbourg for providing a tape of the available photoelectric magnitudes for HD stars south of $\delta = -40^{\circ}$. Thanks also to S.W. McCuskey for helpful discussions and suggestions. We are especially grateful to the National Science Foundation for their continuing support of this work through a series of grants.

REFERENCES

Blaauw, A. (1963). In Stars and Stellar Systems, Vol. III, Basic Astronomical Data, K. Aa. Strand, ed., Univ. of Chicago Press, Chicago, p. 383.

Böhm-Vitense, E. and Canterna, R. (1974). Astrophys. J. **194**, 629.

Houk, N. and Cowley, A.P. (1975). University of Michigan Catalogue of Two-Dimensional Spectral Types for the HD Stars, Vol. **1**, Dept. of Astronomy, Univ. of Michigan, Ann Arbor.

Houk, N. (1978). University of Michigan Catalogue of Two-Dimensional Spectral Types for the HD Stars, Vol. 2, Dept. of Astronomy, University of Michigan, Ann Arbor.

Mendoza, E.E. (1956). Astrophys. J. **123**, 54.

Schmidt-Kaler, Th. (1965). In Landolt-Börnstein New Series, Group 6, Vol. **1**, Astronomy and Astrophysics, H.H. Voigt, ed., Springer-Verlag, N.Y., p. 298.

Struve, O. and Zebergs, V. (1962). Astronomy of the 20th Century, The MacMillan Co., N.Y., p. 259.

DISCUSSION

UPGREN: I would like to direct this question to you and to Dr. Gliese. The numbers of stars in each spectral class which you find in your Volume I to be closer than 20 parsecs could be extrapolated to cover the whole sky since the nearby stars should be uniformly distributed. This would give expected numbers of stars closer than 20 parsecs in each class earlier than M. Have the numbers of stars of each class in Gliese's catalog been compared to these to detect the degree of incompleteness in each spectral class?

HOUK: I have not done this but am sure it would be very incomplete, at least from mid-G later. Extrapolating over the whole sky, I expect to find about 1100 stars within 25 parsecs, lower than Gliese's figures.

GLIESE: I use the Michigan Spectral Catalogue, Vol. 1 to search for late type nearby stars. Does it seem necessary to calibrate the mean relation between Michigan objective prism spectral types the absolute magnitudes for main sequence stars by trigonometric parallaxes?

HOUK: It might be a good idea. I don't think there are any systematic errors in that spectral region but we have not made very extensive intercomparisons.

GLIESE: It's difficult to find accurate apparent magnitudes for many of these stars which seem to be nearby objects.

HOUK: Yes, our results show rather large errors, even for stars as bright as 6th or 7th magnitude. Of the 67,000 HD stars south of declination -40°, only about 4500 have photoelectric magnitudes in the Strasbourg data file.

GLIESE: Can you give any estimate of time of the publication of further volumes of the Catalogue?

HOUK: Lets limit that to Volume 2. (Declination -53° to -40°, $\sim$ 31,000 stars.) It should be ready early in 1978. The magnetic tape version will probably be available in February; the availability of the volume depends on the printer.

GREENSTEIN: Will your catalogues and tapes include new magnitudes (determined from the spectra) and improved positions?

HOUK: We will continue to print the original HD pg magnitudes. The corrections discussed here are too uncertain and difficult to make, since they differ so much with region in the sky. Volume 2 will, however, have improved 1900 positions and precession provided by the Strasbourg Stellar Data Center. The right ascension will be given to a tenth of a second of time, the declination to one second of arc.

PART III

HR DIAGRAMS, SUBLUMINOUS STARS

Reception at the U.S. Naval Observatory, November 2, 1977.

THE HR DIAGRAM

Jesse L. Greenstein

Hale Observatories,
California Institute of Technology,
Carnegie Institution of Washington

The HR diagram is a useful shorthand locating a star in a two-coordinate space. For the astrophysicist, the y-coordinate is bolometric luminosity, M_{bol}, the x-coordinate, effective temperature, T_{eff}. Objects of given chemical composition, age (or evolutionary status) are labeled in the xy plane by mass. For an observer, y may be apparent or absolute magnitude in a certain wavelength region and x may be spectral type or color. The HR diagrams for populations differ because of age, chemical composition and stellar masses present. HR diagrams are often of mixed nature; some involve observables others derived or semi-theoretical quantities. I will display various types of HR diagrams for low-luminosity stars. For galactic or extragalactic studies the HR diagram needs a further dimension, the frequency of stars at an x,y. The mass of the Galaxy, but not its light, may be dominated by M dwarfs. HR diagrams are also interesting for their nearly empty spaces. In Fig. 1 we show as a sample, the basic results of the U.S. Naval Observatory parallax program, in which broad band (B-V) colors define the visual luminosity, M_V, on the main (MS) and degenerate (WD) sequences.

1. G-M STARS WITH THE MCSP

I report on work in progress on G-K-M dwarfs, stars of known parallax observed with the multichannel spectrophotometer (MCSP). I have observed 400 other proper-motion stars; parallax stars then calibrate photometric luminosities, $M_{1.85}$, tangential motions, T, and space motions, U,V,W. One program completed included all northern Lowell Observatory stars brighter than $16^m.5$ with motions

A. G. Davis Philip and D. S. Hayes (eds.), The HR Diagram, 101-116. All Rights Reserved.

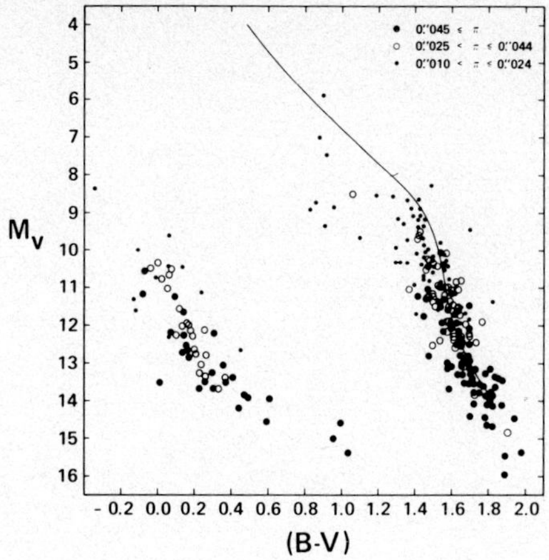

Fig. 1. The 330 USNO parallax stars, from Dahn et al. (1976) (their Fig. 2); symbols indicate size of parallax. Degenerate and main-sequence stars are well separated; (B-V) has only small amplitude for dM stars.

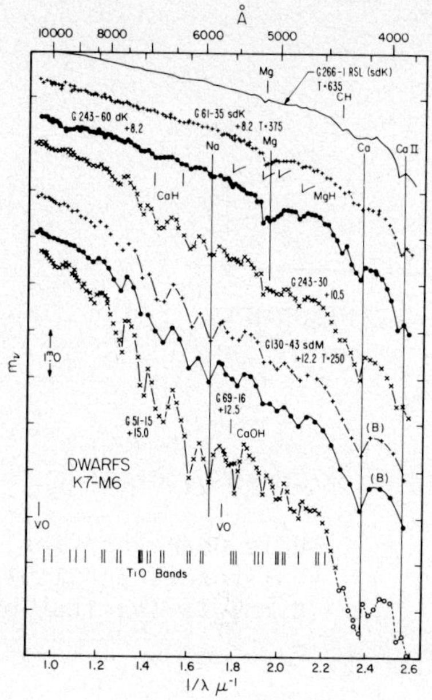

Fig. 2. MCSP output for a sample of main sequence stars; luminosities, $M_{1.85}$, and tangential motions, if high, are shown. The two curves labeled B are at lower resolution (80/160 Å). At the top is a "red subluminous star", i.e., a weak-lined K star of unknown parallax; its color gives an extreme space motion.

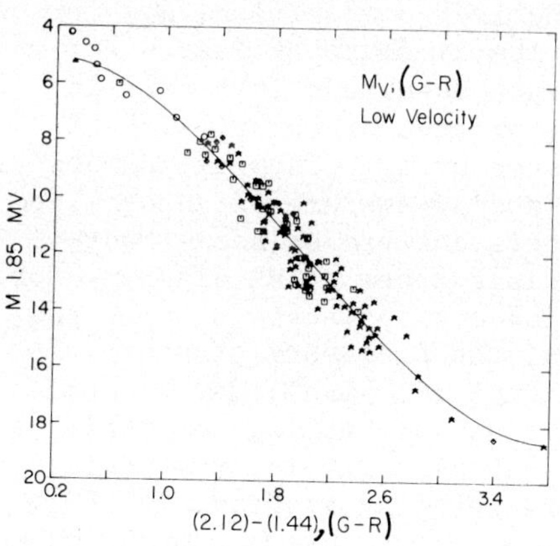

Fig. 3. Main-sequence stars of known parallax with T < 100 km s^{-1}, showing MCSP color, (G-R), [2.12] - [1.44] and $M_{1.85}$. Note the difference of shape of the lower main sequence from that in Fig. 1; symbols in Table I.

exceeding 1".0 yr^{-1} (Giclas, Burnham and Thomas 1971). An additional 50 fainter stars (to $21^m.5$) from the Luyten Palomar Schmidt Survey (1969, 1970, 1974, 1976) have been observed. About one-third are degenerate; stars were selected either for "bluish" color or large reduced proper motion. The true numbers are distorted, weighted towards high-velocity stars; this is more than compensated by the rarity of high velocity among parallax stars.

The MCSP provides magnitudes or fluxes from 11500 to 3100 Å. The accuracy, $\approx \pm 0^m.03$ is less than broadband, far less than Strömgren photometry. However, 31 photomultipliers observe a star simultaneously, with sky subtraction, so that colors are well determined. Depending on brightness, the resolution used is from 40 to 160 Å (λ < 5700 Å) and 80 to 360 Å (λ > 5700 Å). At highest resolution 155 data points are machine plotted; typical MCSP outputs are traced in Fig. 2. The information is too large to present in tabular form; for our purposes, monochromatic magnitudes, at frequencies in inverse microns, m_ν, are measured at 9 points. Magnitudes m_ν are represented by brackets or letters: I^*=[1.00]; I=[1.25]; R=[1.44]; R^*=[1.52]; G=[2.12]; B=[2.35]; U=[2.80]. The central frequencies used are related to some photoelectric systems, i.e., [1.85] is near the broadband V magnitude, at a less-absorbed position in dM stars. From O to K type stars, broad band V=[1.85]-$0^m.04$; (B-V)=[2.35]-[1.85]+$0^m.14$; (U-B)≈0.97 x ([2.80]-[2.35])-$0^m.86$; $(R-I)_{Kron}$≈1.11 x ([1.44] -[1.25] +$0^m.11$. These relations cannot be used in dM; the (R-I) relation is non-linear when $(R-I)_{Kron}$ exceeds $0^m.3$. The MCSP points were chosen by additional criteria, e.g., the strength of Ca I and TiO bands, sensitivity or freedom from TiO bands, strength of Mg I + MgH, [1.94], and Ca II + hydrogen, [2.58], not here discussed. Various color-luminosity diagrams can be derived from MCSP data.

The basic improvement in available parallaxes for faint stars was made with the new parallax reflector of the U. S. Naval Observatory at Flagstaff, constructed under the leadership of Strand (1971). Four papers with accurate parallaxes and UBV photometry have been published (Riddle et al. 1970; Routly 1972; Harrington et al. 1975; Dahn et al. 1976). These are among the first parallaxes with external errors only a small multiple of internal errors. Their small parallaxes can be used with some confidence, subject to an increase in the error of $\Delta M \approx 2.17 \Delta p/p$. At $\sigma \approx 0''.003$ and $p = 0''.019$ (a limit I have used) $\Delta M \approx 0^m.34$; even at 2σ, $\Delta M \approx 0^m.69$. Other recent parallax series from Lick (Klemola et al. 1975) and Yerkes (van Altena 1971, 1973a,b, 1975) are included, with weighted inputs. Old parallaxes have been included only if large and of high weight. The color-magnitude diagrams here presented are based on 237 G-M stars of parallax quality class 1-4. Table I shows the quality classes for the

TABLE I

QUALITY CLASSES AND SYMBOLS IN FIGURES

Q	$10^3 p$	Plot Symbol T < 150	T ≥ 150
1	67	Star	*
2	40-67	Square	X̄
3	20-39	Diamond	X
4	10-19	Triangle	crosses

parallaxes and the symbols used in the diagrams. Old parallaxes have Q increased by unity, unless > 0.″125; the distribution of Q is Q = 1, 53%; Q = 2, 34%; Q = 3, 11% and Q = 4, 2%. Fits to the color-magnitude diagrams were made separately for a sample including objects, Q = 1,2, and those Q = 3 with several determinations. As discussed below, for high-velocity stars, solutions were obtained including Q = 3,4 stars, and separately for Q = 1,2 only.

Computer handling is necessary for such a mass of data. It introduces problems for individual objects such as unresolved binaries, suspected spectroscopic binaries, flare stars and objects with lower-quality MCSP. Some binaries (> 6" separation) were resolved; unresolved pairs for which observers give $\Delta m \approx 0$ were corrected for duplicity. Unequal pairs, however, are difficult to handle and are treated as single. Some of the dispersion is caused by uncorrected duplicity (up to $0^m.75$ error); if the true error of a parallax can reach 2σ (0.″006 for USNO) at Q = 2, the dispersion introduced by the parallax is $0^m.33$. Emission line and flare stars have errors in brightness and color, e.g., visible or ultraviolet emission during mini-flares, affecting B; $H\alpha$ is located in the R^* band (6579 Å). Observations at different resolutions have been used, with little effect in G, early K and degenerate stars, but with important effects in the later dM's. These are so faint at $\nu > 1.9\ \mu^{-1}$, that we are forced to lowest resolution, (160/360 Å), and lower accuracy.

2. HR DIAGRAMS FOR PARALLAX STARS

The present data for 237 stars were computer processed to produce the diagrams separately for stars of high and low tangential velocities. Selection prevents the parallax and proper-

motion-magnitude-limited samples from containing many halo stars. If halo stars of a given M are 10^{-3} as common as old disk and Population I, the nearest example is 10 times further and 5^m fainter. Since sdM distances are small the z-direction density gradient may be neglected. I had initially planned two samples separated at T > 150 km s^{-1}, to select halo stars. The symbols are, in fact, so divided; but only 12 stars have T > 150 km s^{-1}, (Q = 1,2); Q = 3,4 makes the total 58. Only 46 stars with Q = 1,2 have T > 100 km s^{-1}.

HR diagrams are shown (Figs. 4-6) for T < 100 km s^{-1}, Q = 1,2. The fitted cubic, and dispersions, prove MCSP colors useful for calibration of stars of unknown parallax. In Fig. 7 are plotted T > 100 km s^{-1}, Q = 1,2 stars individually; the cubic obtained on adding Q = 3,4 stars is shown.

Comparison of Figs. 1 and 5, shows that B colors on the MCSP, unlike the broadband, increase rapidly as luminosities decrease. No discontinuity exists in the MS slope. Use of infrared colors give smaller range in infrared M; the large amplitude of the M_V diagram is caused by TiO and line blocking. The MS stars are clearly separated from the degenerates in Fig. 1, as they are in MCSP diagrams. In the MS broadband M_V, (B-V) diagram, +8 to +17, the slope is 13; in the $M_{2.35}$, [2.35]-[1.85] MCSP diagram the slope is 5.3. The $M_{1.85}$ is less than broadband M_V in dM stars, since [1.85] is at high flux (Fig. 2); the broadband V averages strong absorptions on a steep flux gradient. Bolometric corrections based on $M_{1.85}$ are smaller, and less composition-sensitive. The use of even broadband (R-I) will certainly improve the HR diagrams produced by the USNO program.

Details of the least-square fits will be published elsewhere. Dispersions about the cubics, σ, are about $0^m.53$ for T < 100 km s^{-1}, Q = 1,2; the $M_{1.85}$, (G-R) and $M_{2.35}$, (B-V), have $\sigma = 0^m.72$. For the T > 100 group, $\sigma \approx 0^m.57$; inclusion of Q = 3,4 gives $\sigma \approx 0^m.70$. Part of the scatter arises from uncorrected duplicity, flare stars (which cause bad scatter in the (B-V) diagram) and poor photometry in the blue-violet, where late dM stars reach 22^m, but much arises from parallax errors.

Among 179 good stars, 23 have relatively reliable Q = 3 parallaxes; 58 more with Q = 3,4 exist, of which 39 have T > 200 km s^{-1}, a few T > 500 km s^{-1}. Inclusion of Q = 3,4 gives a more genuine high-velocity sample, at the cost of larger Δp/p, and σ increases 25%. The general result is that high-velocity stars, Q = 1,2, are fainter by $\Delta M_{1.85}$, $+6^m < M_{1.85} < +10^m$. $\Delta M_{1.85}$ disappears, generally, by $+11^m.5$, and possibly forced by the cubic fit, $\Delta M_{1.85} \approx -0^m.3$ for the faintest stars. Including Q = 3,4, a purer high-velocity sample, increases $\Delta M_{1.85}$ to reach $1^m.0$,

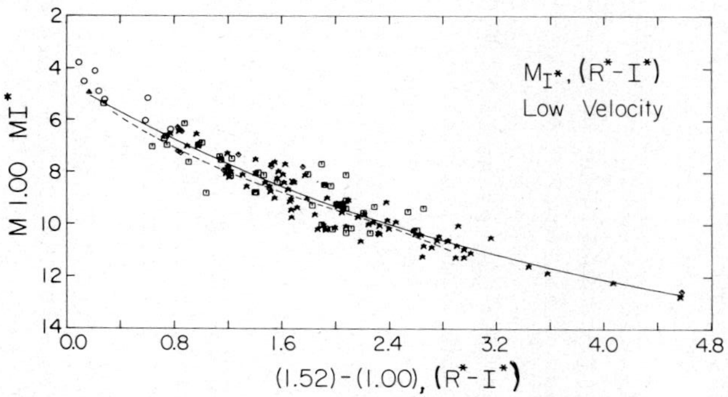

Fig. 4. HR diagram for a band-free color. $M_{1.00}$ is close to M_{bol}. High-velocity cubic dashed.

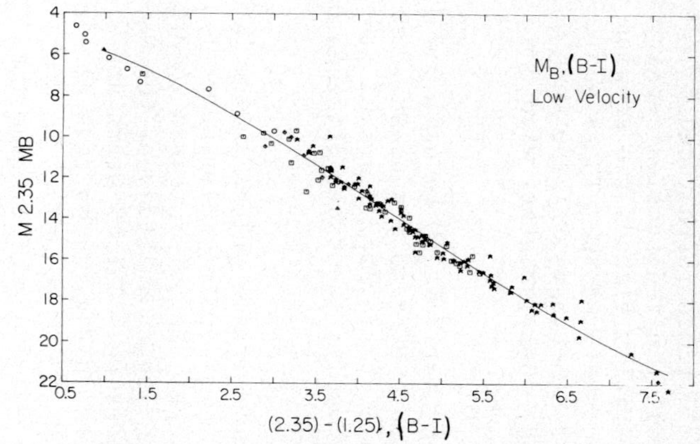

Fig. 5. Blue HR diagram, lower accuracy.

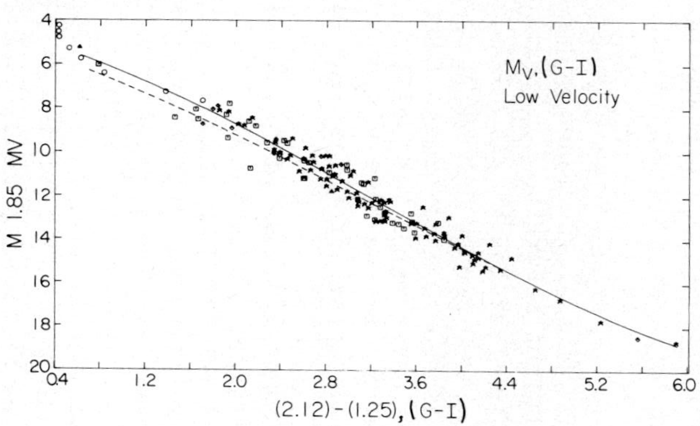

Fig. 6. (G–I) diagram, $\sigma = 0^m.51$; high velocity dashed.

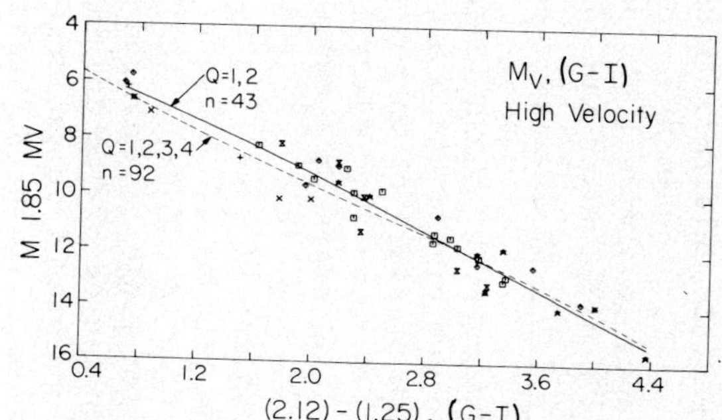

Fig. 7. Individual high-velocity stars, Q = 1,2. Lower quality solution dashed.

probably to $M_{1.85} = +12^m$. HV stars are subluminous in all MSCP colors, therefore.

Nine Hyades stars are shown in the figures as open circles; but are not used in the fitting. Eggen (1973, 74 and elsewhere) used broadband (R-I) to separate old disk from young stars (e.g., the Hyades), and from the true halo peopulation. He suggests that the Hyades contains stars above the ZAMS evolving to the MS, and brighter stars evolving upwards. The Hyades and the $M_{1.25}$ [1.44 - 1.25] results are shown in Figs. 4-6 and in Fig. 8. Four are $0\overset{m}{.}1$ bluer than the program stars, and are $0\overset{m}{.}7$ above the extrapolated, nearly straight-line, least-squares fit. Those with my (R-I) > $+0\overset{m}{.}12$ deviate by less than $0\overset{m}{.}1$.

A systematic error arises from the selection of parallax stars to observe, since stars of negative or small parallax are not included. The sample including poorer p (Q = 3,4) has $\langle p \rangle = 0\overset{''}{.}135$ for T < 100 km s^{-1}, $0\overset{''}{.}074$ for 100 < T < 200 km s^{-1} and only $0\overset{''}{.}032$ for T ≥ 200 km s^{-1}. The asymmetrical nature of the error can be estimated. For the larger parallaxes with $\sigma = \pm\, 0\overset{''}{.}003$, $\pm\, 1\sigma$ produces errors in M of $-0\overset{m}{.}23$ and $0\overset{m}{.}20$; for $\pm\, 2\sigma$ the errors are $-0\overset{m}{.}49$ and $+0\overset{m}{.}39$; for $\pm\, 3\sigma$, $-0\overset{m}{.}78$ and $0\overset{m}{.}57$. Selected by apparent, not true parallax, stars suffer a systematic dimming by $+0\overset{m}{.}02$, $+0\overset{m}{.}05$ and $+0\overset{m}{.}10$. Repeating the exercise at $\langle p \rangle = 0\overset{''}{.}20$, however, gives $\langle \Delta M \rangle = +0\overset{m}{.}03$, $+0\overset{m}{.}11$ and $+0\overset{m}{.}25$, and at $\langle p \rangle = 0\overset{''}{.}015$, $\langle \Delta M \rangle = +0\overset{m}{.}04$, $+0\overset{m}{.}19$ and $+0\overset{m}{.}49$. This effect is negligible at 1σ, nearly so at 2σ, but is comparable to the population-type luminosity difference ΔM, at 3σ. The USNO observers have not underestimated their external error by 3, nor do recent Lick and Yerkes parallaxes have errors approaching $\pm\, 0\overset{''}{.}010$. They agree well with USNO values where they overlap. Thus, we have established the degree to which high-velocity stars are subluminous, found it to disappear by mid dM, and to be independent of color baseline; i.e. it is probably not caused by line and band blocking.

Unexpectedly the high-velocity stars appear most subluminous (by at least $0\overset{m}{.}6$) at the bluest colors at which they exist in my sample (early G). If metal-poor late-type stars exist, differential line blocking makes them bluer. In banded stars, TiO would vary as the square of the metal abundances; Population II early sdM stars are, in fact, recognizable by strong MgH and CaH compared to TiO. The colors here used are as insensitive to line blocking in the red for sdG and early sdK stars, as for dG and dK stars. Yet (G-R), (R*-I*) and (R-I) agree in making our bluest "high-velocity" stars $0\overset{m}{.}6$ fainter than stars of corresponding color. The $\langle T \rangle$ = 149 km s^{-1} for the high-velocity stars, and σ = 75 km s^{-1}, i.e., does not give a truly Population II sample. The parallax sample is dominantly old-disk population, with some low-velocity Population I stars present. The F to K Hyades stars may be

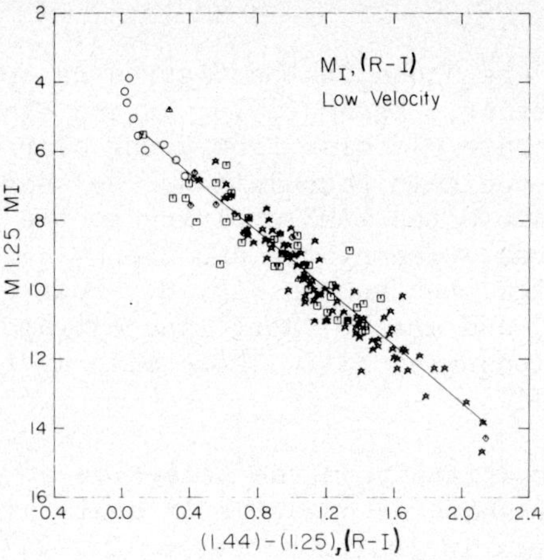

Fig. 8. The M_I, (R-I) diagram for low-velocity stars is nearly linear. The four bluest Hyades stars measured (open circles) may lie above the relation extrapolated from parallax stars, but, the few redder Hyades fit well.

---

brighter than old disk stars; the "high-velocity" stars, $T > 100$ km s^{-1}, are appreciably fainter at the blue color where evolution removes them from the Population II main sequence. It is particularly surprising that they are faintest there, rather than brighter, i.e., subgiants.

3. THE HIGH-VELOCITY POPULATION

A fundamental difference between the $T \lessgtr 100$ km s^{-1} diagrams is that the higher-velocity group lacks intrinsically faint and very red stars. In part this is a selection effect, but my continued search among fainter, redder stars with no available parallax suggests that high-velocity stars continue to be rare. The faintest low-velocity stars are $M_{1.85} = +18^m.8$ (LP619-50, Van Biesbroeck 10), near $M_{2.35} = +22^m.0$. The faintest stars with $T \approx 200$ km s^{-1}, G34-15, G260-1 (LP71-157), have $M_{1.85} = +13^m.2$, $M_{2.35} = +16^m.2$. Only seven stars have $M_{1.85} > 12^m.0$, and $T > 200$ km s^{-1}. Parallaxes for apparently fainter stars of large proper motion are clearly badly needed. I have already observed about fifty red stars from the LP and LHS catalogs using the MCSP; using the present calibration I will estimate luminosity, parallax and space motion. These stars, of course, are faint, 17^m to 19^m visual, and will be difficult objects for any parallax telescope. A few stars as red as LP619-50 and VB 10 have already been found, but none so far redder. In my catalog (July 1977) of

465 space motions a problem in the search for high-velocity late-type stars became apparent. I studied the photometric luminosity of 81 stars which lack parallax and have T ≥ 250 km s^{-1}. The median $M_{1.85} = +7^{m}.5$; only five stars have $M_{1.85} > +10^{m}.0$ and the faintest is $+11^{m}.2$. The Lowell proper-motion stars are limited to $16^{m}.5$ photographic; intrinsically faint sdM stars will be found only by systematic colorimetry beyond 20^{m}, photographic.

One example of the interesting ultra-high-velocity, so-called RSL stars has been shown in Fig. 2, G266-1 (LHS 4037) an extremely weak-lined, yellow sdK. The mean of four MCSP scans gives $M_{1.85} = +7^{m}.2$. From its proper motion and radial velocity T = 635 km s^{-1}, and the space motion is U = +223, V = -606, W = -2 km s^{-1}. The star is metal-poor and not degenerate. Photographic and SIT spectra at 3 Å resolution show sharp Ca II, Ca I, Mg I, extraordinarily strong CH, and ultraviolet Fe I. Can differential line blanketing falsify the photometric luminosity? The six different colors used predict, however, a parallax of 0"0055 ± 0"0002, so consistent that differential line blanketing cannot be important.

Such stars had been noted earlier by Eggen (1969); Greenstein had called them "Eggenites" (1969); more recently they have been called RSL (red subluminous). They exist in considerable numbers; none has a recently measured parallax. Their true location in the HR diagram is unknown. If extremely metal deficient, they may lie in a color-color diagram on the locus between normal-metal dM's and an unblanketed dM star, or one with blanketing included theoretically (Greenstein 1969; Tsuji 1969). Very metal-poor sdM's might be confused with sdK's, and late sdM's with earlier subtypes. Lack of recognition of this could increase the numbers of low-luminosity halo stars, but it would also decrease their deduced tangential velocities. Such stars may be genuinely subluminous as compared to our presently defined high-velocity group of parallax stars, but whether by $\Delta M \approx +1^{m}$ or $+3^{m}$ is not yet known. With the latter value, they could have measurable parallaxes.

4. SPECTRAL TYPES

The complex spectra and faintness of dM stars resulted in published spectral classifications being on inconsistent scales. An accurate classification is given by Boeshaar (1977) from estimated relative strengths of the bands TiO, VO, CaOH in the visual region. Color-spectral-type relations are shown in Fig. 9 for stars in common, using $(R^{*}-I^{*})$, i.e., not that of a truly Population II sample. The narrower baseline (R-I), more strongly affected by TiO, has a $1^{m}.3$ amplitude.

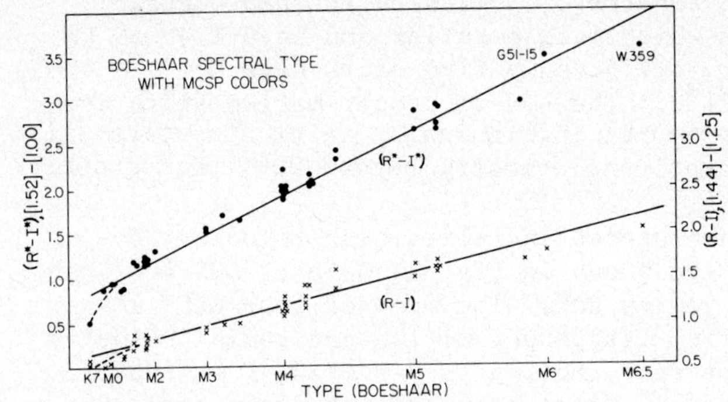

Fig. 9. Color-spectral type (P. Boeshaar's thesis). Types are on an exponential scale.

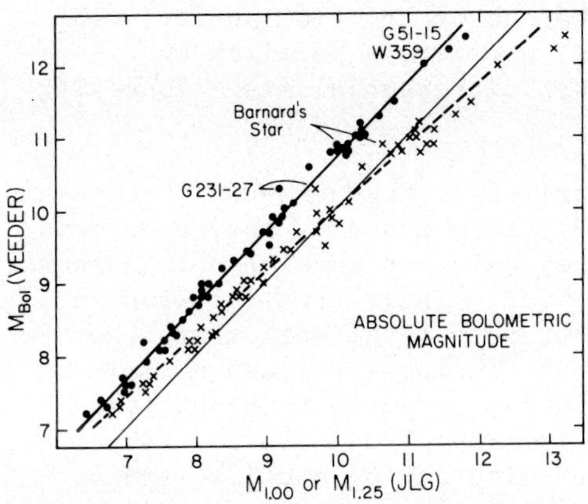

Fig. 10. Veeder's M_{bol} compared with MCSP absolute magnitudes, same parallaxes. Scatter among the faintest stars is partly observational error. A halo star deviates appreciably to the left.

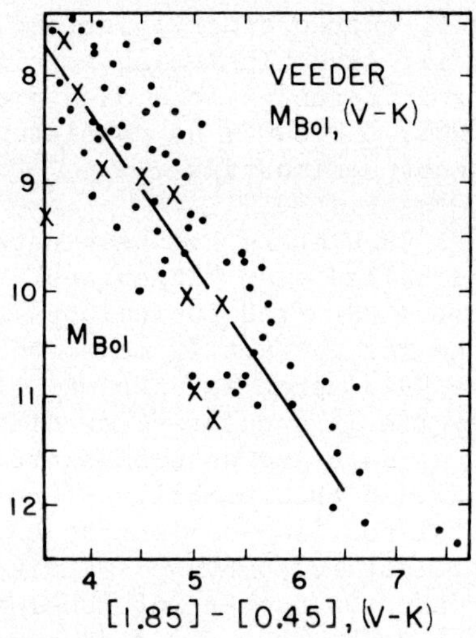

Fig. 11. Veeder's bolometric magnitudes and infrared visual color. X = halo stars.

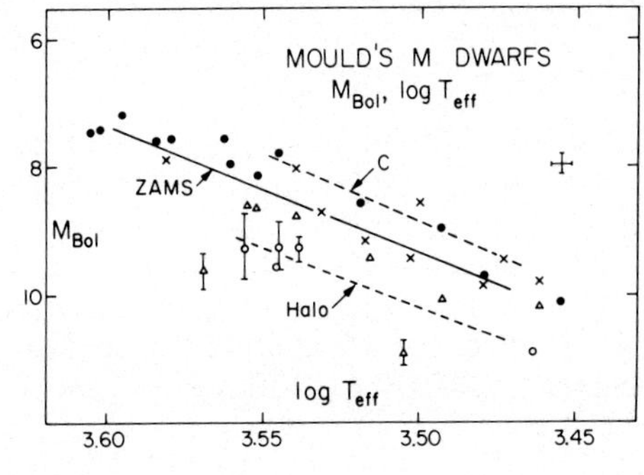

Fig. 12. M_{bol} and T_{eff} from Mould and Hyland. The upper dashed line(C) is for possible contracting stars, the lower for the halo. Filled circles, old disk; open circles & crosses halo. Kapteyn's star is the open circle at $9^m.53$, log T_{eff}=3.546.

Both relations, shown in Fig. 9, are approximately linear later than M1. Note, however, that spectral type had to be plotted on a logarithmic scale! Colors transform to spectral type, with an accuracy better than 0.5 subtype. Boeshaar has shown that a significant, precise spectral type can be determined in the red and near infrared which is well correlated with luminosity.

5. BOLOMETRIC MAGNITUDE DIAGRAMS

Consider the theoretical HR diagram giving absolute bolometric magnitude, M_{bol}, as a function of blanketed infrared colors, or, best of all, of effective temperature. Such diagrams are still remote goals but progress has been made. Infrared measurements in broadband J, K, L, M by Johnson (1965) first gave a T_{eff} scale; Greenstein, Neugebauer and Becklin (1970) redetermined lower-main-sequence radii, T_{eff}, and bolometric corrections. Veeder (1974) measured infrared colors of a number of dM stars with $p > 0''.100$. Extrapolation from broadband infrared colors to total flux is not difficult, Veeder having used a blackbody fit. His M_{bol} were available for 58 stars with MCSP data; after minor readjustment in parallaxes used we compare Veeder's M_{bol} with our near-infrared $M_{1.00}$ and $M_{1.25}$. In Fig. 10 we show how good these relations are; they seem independent of space motions or the presence of emission. The linear relations fitted are:

$$M_{bol} = 1.014\ M_{1.00} + 0^m.60,$$

$$M_{bol} = 0.863\ M_{1.25} + 1^m.41.$$

In dM stars, $M_{1.00}$ differs from M_{bol} only by a constant; $M_{1.25}$ has a 15% difference in scale.

Individual stars are shown in Fig. 10, as well as the linear relations. The M_{bol}, $M_{1.25}$ relation is significantly non-linear. The faintest, reddest stars deviate to the right suggesting that measures of I^* are depressed by TiO bands near 1 μ. Barnard's star is slightly to the left of the mean relations; although called subluminous, it is not very high velocity, $U = -141$ km s^{-1}. If subluminous it is by nearly the same amount at [1.00], [1.25] as bolometrically. One star (observed twice), G231-27, shifts much further to the left. It has strong hydride bands, a halo space motion (+372, -196, -117 km s^{-1}), and the photometric $M_{1.85}$ from different baselines are accordant. For an unknown reason it is depressed at $\lambda > 9000$ Å, as if an additional source of band absorption existed (CN and FeH are candidates). With few exceptions, dM stars are so remarkably homogeneous that we must ask: Where are the extremely metal-poor stars?

Veeder (1974) gives an HR diagram using infrared K (2.0 - 2.4 μ) at $\nu = 0.45\ \mu^{-1}$, yielding M_{bol}, (V-K) as reproduced in Fig. 11.

The straight line is: $M_{bol} = 1.12\ M_{0.45} + 1^{m}81$. As expected, the scale factor is large for $M_{0.45}$. He believes cosmic scatter to be real. There is little difference between high- and low-velocity stars, although in a true L, T_{eff} diagram, metal-poor stars lie below the normal ZAMS. There should be little effect of band blocking on M_{bol}, or on K. Band blocking could appear as extra absorption at V, i.e., as shifts to the left.

Mould's (1976) thesis at the Australian National University attempts to produce a true (M_{bol}, T_{eff}) diagram. For Kapteyn's star, with halo motion, V = -288 km s^{-1}, Mould and Hyland (1976) find a metal deficiency of a factor of 3 to 4, using blanketed model atmospheres. The star is $1^{m}0$ below the bolometric ZAMS at T_{eff} = 3500 K. They have not as yet displayed a large intrinsic luminosity deficiency for sdM stars, as shown in Fig. 12. Our ΔM for sdG or sdK stars is apparently maintained in early M, although the abundance deficiencies at the surface seem smaller than at sdG.

6. THE DEGENERATE STARS

This subject deserves nearly as much attention as the MS; my MCSP work has been mostly presented in Greenstein (1976). Only a modest increase in the data (MCSP, or Strömgren photometry) has since occurred. Many new degenerates have been discovered at the Steward and Hale Observatories. Parallaxes are badly needed for the yellow and red degenerates at the bottom of the cooling sequence. Splitting degenerate stars of known parallax into high- and low-velocity groups is not yet feasible. The main problems are such questions as (1) What is the composition sensitivity of the HR diagram? (2) Do discrete cooling sequences exist? (3) What is the spread of masses among degenerate stars? Infrared measures have not been carried out for the red degenerates to give bolometric corrections; since lines and bands are weak, model atmospheres may be sufficient to obtain M_{bol}. The extreme ultraviolet measures from space, as well as the ANS satellite program in the nearer ultraviolet provide better effective temperatures for hot stars permitting construction of an M_{bol}, T_{eff} diagram. Temperatures determined to be above 10^5 K mean that WD's exist from 10 $L_{\odot}$ to well below 10^{-4} $L_{\odot}$. Many red degenerates have been found as red and faint as any known in 1976, mostly among the Luyten Palomar survey. Most are DC metal-poor at the surface, except for LP701-29 (Dahn et al. 1977), by $0^{m}6$ the faintest known degenerate at $M_V = 16^{m}1$. But LP542/3 is a double red degenerate found by van Altena, who is measuring its parallax; Palomar spectra show both to be DC, with $M_V = 15^{m}6$ and $16^{m}0$. Thus, large composition variations are found among the faintest degenerates, unlike MS stars. From this point of view, HR diagrams for the degenerate stars are instructive. The M_V, (G-I) diagram from Greenstein (1976)

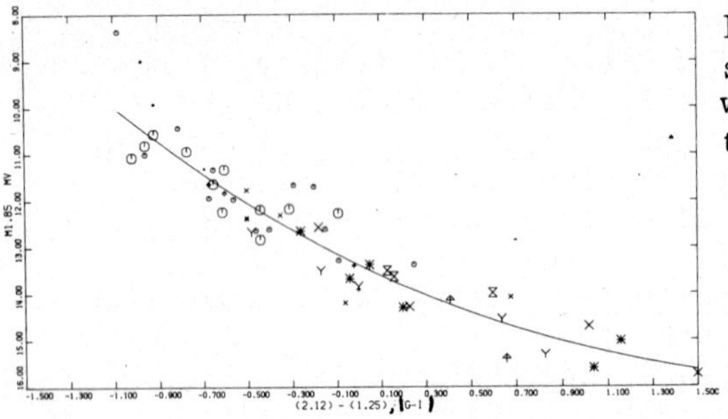

Fig. 13. M_V for degener- stars as function of a color which is relatively composition insensitive, $\sigma = 0^m.56$.

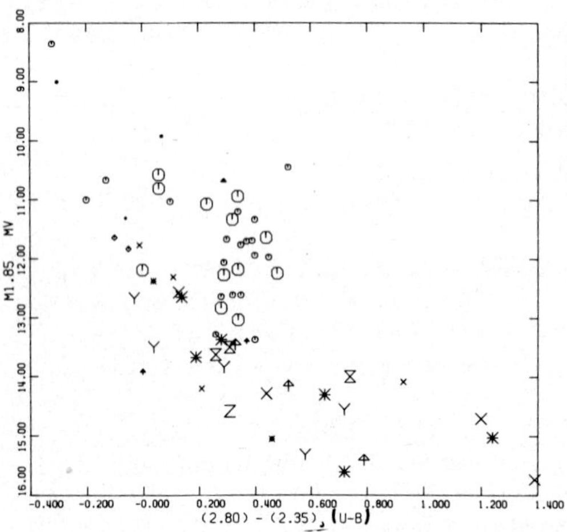

Fig. 14. M_V for WD's is poorly correlated with (U-B). The Balmer jump, absent in non-hydrogen stars, dominates and the scatter is large.

is shown in Fig. 13. The surface composition is probably irrelevant; only carbon degenerates have strong features in the MCSP [2.12] region. Only one star, EG 38 (upper right) is far off the mean curve; it is a composite, DA+dM. The symbols are: Octagons, hydrogen-rich; diamonds, helium-rich; stars, metals present; X, no lines. Compare the scatter with that in Fig. 14, where the color used is composition sensitive, i.e., it spans the Balmer jump. Colors of models for hydrogen atmospheres of increasing temperature run almost vertically at $(U-B) = 0^m.4$, to rejoin the sequence of helium dominated atmospheres near $-0^m.2$. The recognition of degenerates of differing surface composition is best done in a color-color diagram. Residual vertical scatter in a suitable HR diagram, which could be bolometric, is either a spread in mass (i.e., radius), or an effect of composition on color. This work was supported in part by the NSF, AST77-09191.

REFERENCES

Boeshaar, P. C. (1977). The Spectral Classification of M-Dwarf

Stars, preprint from unpublished Ph.D. thesis, Ohio State University.

Dahn, C. C., Harrington, R. S., Riepe, B. Y., Christy, J. W., Guetter, H. H., Behall, A. L., Walker, R. L., Hewitt, A. V., and Ables, H. D. (1976). Publ. of the U.S. Naval Obs. **20**, part 3, U. S. Government Printing Office, Washington.

Dahn, C. C., Hintzen, P. M., Liebert, J. W., Stockman, H. S., and Spinrad, H. S. (1977). in press.

Eggen, O. J. (1969). in Low Luminosity Stars, S. S. Kumar, ed., Gordon and Breach, New York, p. 3.

__________. (1973). Astrophys. J. **182**, 821.

__________. (1974). Publ. Astron. Soc. Pacific **86**, 697.

Giclas, H. L, Burnham, R. Jr., Thomas, and Thomas, N. G. (1971). The G Numbered Stars, Lowell Obs., Flagstaff. Also Lowell Obs. Bulletins.

Greenstein, J. L. (1969) in Low Luminosity Stars, S. S. Kumar, ed., Gordon and Breach, New York, p. 281.

Greenstein, J.L. (1976). Astron. J. **81**, 323.

Greenstein, J. L., Neugebauer, G., and Becklin, E. E. (1970). Astrophys. J. **161**, 519.

Harrington, R. S., Dahn, C. C., Behall, A. L., Priser, J. B., Christy, J. W., Riepe, B. Y., Ables, H. D., Guetter, H. H., Hewitt, A. V., and Walker, R. L. (1975). Publ. of U. S. Naval Obs. **24**, part 1, U. S. Government Printing Office, Washington.

Johnson, H. L. (1965). Astrophys. J. **141**, 170.

Klemola, A. R., Harlan, E. A., McNamara, B., and Wirtanen, C. A. (1975). Astron. J. **80**, 642.

Luyten, W. J. (1969). Proper Motion Survey with the 48-inch Schmidt Telescope XIX, University of Minnesota, Minneapolis. Also others of this series.

__________. (1970). White Dwarfs, University of Minnesota, Minneapolis.

__________. (1974). Proc. Nat. Acad. Sci. **71**, 4813.

__________. (1976). Catalogue of Stars with Proper Motions Exceeding 0".5 Annually, University of Minnesota, Minneapolis.

Mould, J. (1976). Ph.D. thesis, Australian National University, Canberra.

________. (1976). Astrophys. J. **210**, 402.

Mould, J., and Hyland, A. R. (1976). Astrophys. J. **208**, 399.

Riddle, R. K., Priser, J. B., Strand, K. Aa., and Riddle, R. K. (1970). Publ. of the U.S. Naval Obs. 20, part 3A-3C, U. S. Government Printing Office, Washington.

Routly, P. M. (1972). Publ. of the U.S. Naval Obs. **20**, part 6, U. S. Government Printing Office, Washington.

Strand, K. Aa. (1971). Publ. of the U.S. Naval Obs. **20**, part 1, U. S. Government Printing Office, Washington.

Tsuji, T. (1969). in Low Luminosity Stars, S. S. Kumar, ed., Gordon and Breach, New York, p. 457.

van Altena, W. F. (1971). Astron. J. 76, 932.
van Altena, W. F., and Vilkki, E. U. (1973a). Astron. J. 78, 201.
van Altena, W. F., and Stone, R. C. (1973b). Astron. J. 78, 781.
van Altena, W. F., and Vilkki, E. U. (1975). Astron. J. 80, 647.
Veeder, G. (1974). Astron. J. 79, 1056.

DISCUSSION

UPGREN: Did you correct the absolute magnitudes derived from parallaxes for the volume effect such as Lutz and Kelker have found?

GREENSTEIN: I am familiar with the correction, which Eddington first derived, and the Malmquist corrections and their effect on the mean luminosity. There exists an asymmetry, which depends on the ratio of the parallax to its true mean error. These are trivial except for the smallest parallaxes used. The evidence that high-velocity stars are subluminous by $0\overset{m}{.}6$, on the average, would disappear only if the true errors of the USNO, new Yerkes, Lick parallaxes were, in fact, three times their quoted values, which I do not believe.

THE DISTRIBUTION OF MASSES AND RADII OF WHITE-DWARF STARS

Harry L. Shipman

University of Delaware

1. INTRODUCTION AND SUMMARY

The status of determinations of white dwarf radii by model atmosphere methods is reviewed in this paper. Details will appear elsewhere (Shipman 1978). In brief, the results are that (i) the mean radius of a sample of 95 hydrogen-rich stars with parallaxes is 0.0131 $R_\odot$; (ii) the mean radius of a sample of 13 helium-rich stars is 0.011 $R_\odot$, indistinguishably different from the radius of the hydrogen-rich stars; and (iii) that the most serious limitation on our knowledge of the mean radius of white dwarfs is the influence of selection effects. An estimate of the selection effects indicates that the true mean white dwarf radius is near 0.011 $R_\odot$.

2. METHOD AND RESULTS

The calculation of white dwarf radii from model atmospheres and photometric observations is best considered in terms of the fundamental relation $f_\nu = 4\pi H_\nu (R/D)^2$ between stellar flux f_ν, monochromatic Eddington flux H_ν, stellar radius R and stellar distance D. Photometry, here from Greenstein (1976) corrected to the Hayes and Latham (1975) flux calibration where available and intermediate- or broad-band colors from McCook and Sion (1977), determines the flux f_ν and a color index. Model atmospheres (Shipman 1972, 1977) provide H_ν as a function of colors. The results are shown in Fig. 1.

A. G. Davis Philip and D. S. Hayes (eds.), The HR Diagram, 117–120.

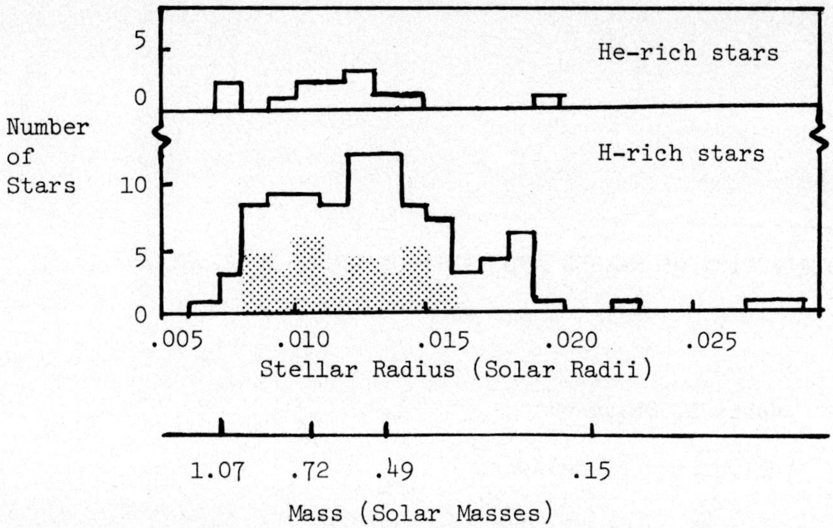

Fig. 1. Distribution of the radii of white dwarf stars. Top panel: Helium-rich stars (types DB, DC, DF, DG, λ4670, DXP). Bottom panel: Hydrogen-rich stars (type DA, with a few very cool DC and DK stars). Stars in the shaded region have high-quality distances and the radii are known to better than 10%. The Hamada-Salpeter (1961) mass-radius relation for carbon white dwarfs is shown.

Systematic errors, affecting the mean radius of the sample, come from errors in the flux calibration and from the model atmosphere determination of the H_ν vs. color relation. They are, expressed as $\Delta R/R$, about 4% for the hydrogen-rich DA stars and about 10% for the helium stars. White dwarf radii (or masses) can also be found by (i) using the five white dwarfs in binary systems, (ii) using hydrogen line profiles to estimate gravities, (iii) using the gravity dependence of the (u-v) colors for $8000 < T_{eff} < 14\,000$ K to estimate gravities, or (iv) by measuring gravitational redshifts. When method (iii) is applied with the Hayes-Latham flux calibration and convective, line-blanketed model atmospheres, updating the results of Wehrse (1975), all but method (iv) are consistent with the results of Fig. 1.

The larger white dwarf stars will, in general, be visible over greater distances and any magnitude-limited sample will tend to include more of them. Consequently the distribution of Fig. 1, averaged uncritically, may substantially overestimate the average white dwarf radius and underestimate the average white dwarf mass. Not all white dwarfs have the same radius; the

variation in radius of stars with accurate radii, in the shaded region of Fig. 1, is greater than that which can be explained by observational errors. White dwarfs in binary systems have masses that vary from 0.43 $M_\odot$ (40 Eri B) to 1.01 $M_\odot$ (Sirius B). A magnitude-limited sample will include stars from a distance $D = (4\pi H_\nu/f_\nu)^{\frac{1}{2}} R$, so that assuming that selection effects are independent of temperature, the volume sampled is proportional to the white dwarf radius R. We can then correct the observed distribution of radii N(R) by dividing N(R) by R^3 to get a corrected distribution, and then average to obtain a corrected radius of 0.0105 $R_\odot$. This procedure overestimates the selection effect because it assumes that the sample is completely magnitude-limited and that the variation in radius illustrated in Fig. 1 is all intrinsic. One can also try to eliminate the selection effect by considering only those stars hot and near enough so all stars, both large and small, would be included. This second procedure produces a mean radius of 0.0111 $R_\odot$ but the sample is small--9 stars. The agreement between the results of the two procedures argues that selection effects do occur and that the mean white dwarf radius is most likely about 0.011 $R_\odot$ rather than the 0.0131 $R_\odot$ given by the raw data.

The mean white dwarf radius, corresponding to a mass of 0.7 $M_\odot$ indicates that mass loss in the red giant stage is substantial (Weidemann 1977). One can also use this number to estimate the density of white dwarfs in the Galaxy; with Sion and Liebert's figure of 0.005 visible ($M_V < 15^m.5$) white dwarfs per cubic parsec, we have only 0.0035 $M_\odot$ pc^{-3} in the form of visible white dwarfs. Preliminary calculations (Shipman 1978) show that observational constraints on stellar birthrate functions indicate that there are fewer than 8 times as many invisible ($M_V > 15^m.5$) as visible white dwarfs, placing an upper limit of 0.03 $M_\odot$ pc^{-3} to the density of visible and invisibly cool white dwarfs.

I thank C. Sass for assistance and the National Science Foundation (AST76-20723) for financial support.

REFERENCES

Greenstein, J. L. (1976). Astron. J. 81, 323.
Hamada, T., and Salpeter, E. E. (1961). Astrophys. J. 134, 683.
Hayes, D. S., and Latham, D. W. (1975). Astrophys. J. 197, 593.
McCook, G. P., and Sion, E. M. (1977). Villanova Univ. Obs. Contr. No. 2.
Shipman, H. L. (1972). Astrophys. J.177, 723
Shipman, H. L. (1977). Astrophys. J. 213, 138.
Shipman, H. L. (1978). In preparation.
Sion, E. M. and Liebert, J. (1977). Astrophys. J. 213, 468.
Wehrse, R. (1975). Astron. Astrophys. 39, 169.
Weidemann, V. (1977). Astron. Astrophys. 59, 411.

DISCUSSION

BELL: As I understand it, you have determined the stellar radii from absolute fluxes at a particular wavelength, making use of the stellar parallaxes. You then get the masses from the mass radius relation. This then gives you the gravity. If you then determine gravities from fitting the spectral scans over a range of wavelengths, how well do they agree with the other values?

SHIPMAN: They agree quite well; we determine a mean log g = 8.0 from the separation of white dwarfs in the (u-v), (g-r) diagram.

HARRINGTON: We have measured almost 500 parallaxes at the Naval Observatory and can make a statement about the true errors. Several lines of attack indicate that we are publishing true one-sigma errors for our mean errors. Therefore, you can accept our published errors as real errors.

SHIPMAN: I'm glad to hear that.

WALBORN: What is the lowest white dwarf mass for which there is credible observational evidence?

SHIPMAN: $M/M_\odot = 0.43$.

THE HR DIAGRAM OF WHITE DWARFS

Volker Weidemann

Institut f. Theoretische Physik und
Sternwarte Universität

The lone dot in the lower left of Russell's original diagram (1914) was disregarded, since it did not fit into his suggestive scheme of giant-to-dwarf evolution. Bottlinger (1923) - in his nearly forgotten study of photoelectric color indices - seems to be the first who, with the evidence of the then three classical white dwarfs, invoked a new equation of state and clearly rejected all other possibilities of explanation. Kuiper (1941) summarized the knowledge obtained during the thirties, mainly by his spectroscopic work. A decade later Luyten (1952) presented impressive HR diagrams based on admittedly crude color indices and, for many more objects, on reduced proper motions.

The new era of photoelectric UBV photometry was opened by Harris III (1956), soon to be extended to the large surveys of Eggen and Greenstein (1965 and sequ.), for which Greenstein secured high quality spectroscopy with the 200" telescope.

Quantitative evaluation started with Greenstein (1958), and the full theory of stellar atmospheres was applied to a first sample of 22 DA stars by Weidemann (1963). The result demonstrated that white dwarfs do have the radii and surface gravities expected for completely degenerate configurations with $\mu_e = 2$. It seemed thus justified to enlarge the sample considerably by inclusion of all stars for which no distances were known but for which radii were derived from surface gravities. The resulting HR diagram is shown in Weidemann (1968).

A. G. Davis Philip and D. S. Hayes (eds.), The HR Diagram, 121-124.

New information was obtained by Strömgren photometry (Graham, 1972) which with a clearly outlined DA cooling sequence disproved the so-called two-sequence hypothesis, and moreover showed an astonishingly small range in surface gravities (Weidemann, 1971). A smaller scatter in radii has recently been confirmed by Greenstein (1976) in his evaluation of multichannel observations, on plotting visual absolute magnitudes against new color indices which are good temperature indicators.

At the same time Schulz (1977) has evaluated all available DA observations (parallaxes, UBV, uvby, Hβ equivalent widths, complete scans by Oke (1974) and Greenstein (1974)), and tried to determine best values of surface gravity, effective temperature and radius by a weighted least square analysis, on comparison with model atmospheres available in the literature. Since the results showed systematic deviations around 10000 K as well as differences between high and low temperature objects, a new set of model atmospheres has been calculated by Koester (1977). On application of the Schulz method we now obtain the HR diagram of Fig. 1 which clearly demonstrates a constant average radius all the way from the hot to the cool end of the DA sequence. Again, no indication of more than one sequence is visible. The average radius is log $R/R_{\odot}$ = -1.90$\pm$0.10, corresponding to a mass range of 0.55$\pm$0.2 $M/M_{\odot}$ on the Hamada-Salpeter mass-radius

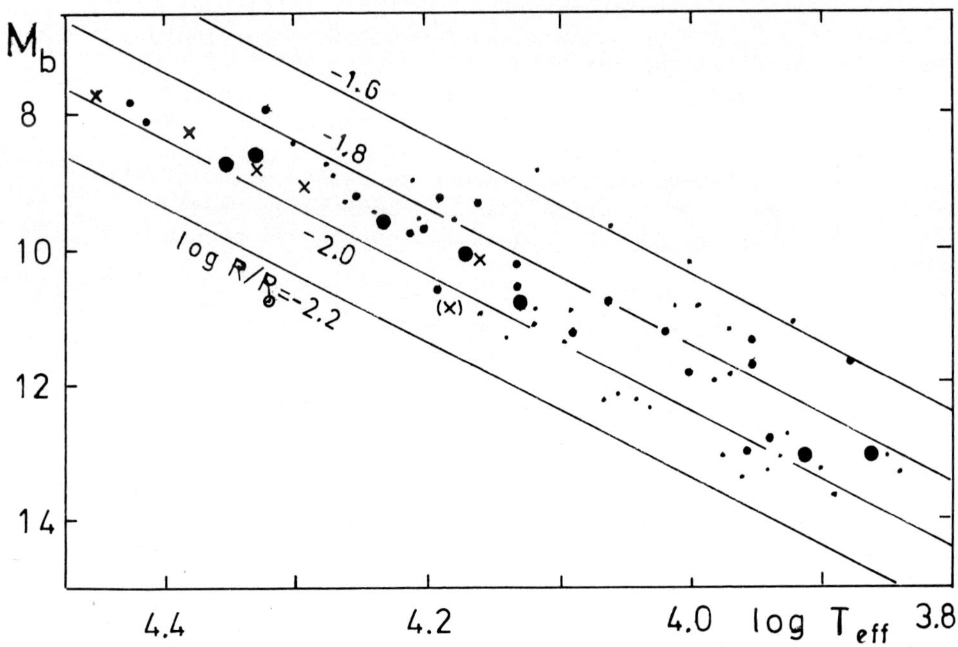

Fig. 1. HR diagram for white dwarfs of spectral type DA.
●: uvby, UBV, good parallaxes, •: ditto, distances less reliable;
·: UBV photometry only, or uncertain parallaxes; x: Hyades.

relation for C-configurations. The surface gravity scatter is even smaller, and corresponds to a mass range from 0.40 to 0.65 $M_\odot$ for two thirds of the stars for which Strömgren photometry is available. Details will be published elsewhere (Koester, Schulz and Weidemann, 1977). The narrow mass distribution of white dwarfs is the result of considerable mass loss in pre-white dwarf stages, combined with the shape of the initial mass function and effects of galactic evolution (for details see Weidemann, 1977).

The data for the hotter non-DA stars (DB, DC, 4670 Å) are still too uncertain as to establish systematic differences in gravities, masses or radii compared to the DA sample. As far as the coolest degenerates is concerned, the atmospheric composition is less clear, and models are preliminary (Shipman, 1977, Wehrse, 1977).

In this context two facts should be stressed: There seems to be no 'red deficit' if due consideration is given of selection effects (Sion and Liebert, 1977), and, second, the number of established degenerates with cooling ages larger than 2×10^9 yrs is so small as to exclude at present any direct use of degenerates as probes of the galactic past. However, there exists an indirect possibility: We can already rule out certain models of galactic evolution which are not compatible with the narrow mass distribution and the population characteristics of white dwarfs (Weidemann, 1978).

REFERENCES

Bottlinger, K.F. (1923). Veröff. Univ. Sternwarte Berline-Babelsberg 3, Heft 4.
Eggen, O. and Greenstein, J.L. (1965). Astrophys. J. **141**, 83.
Graham, J. (1972). Astron. J. **77**, 144.
Greenstein, J.L. (1958). Encycl. of Physics **50**, 161. Springer Vlg. Berlin.
Greenstein, J.L. (1974). kindly communicated.
Greenstein, J.L. (1976). Astron. J. **81**, 323.
Harris III, D.L. (1956). Astrophys. J. **124**, 665.
Koester, D. (1977). to be published.
Koester, D., Schulz, H. and Weidemann, V. (1977). in preparation.
Kuiper, G.P. (1941). Colloquium on Novae and White Dwarfs. Hermann, Paris, p. 201.
Luyten, W.J. (1952). Astrophys. J. **116**, 283.
Oke, J.B. (1974). Astrophys J. Suppl. **27**, 21.
Russell, H.N. (1914). Popular Astronomy **22**, 275.
Schulz, H. (1977). Die Massenverteilung der Weissen Zwerge vom Spektraltyp DA. Thesis, Univ. Kiel.
Shipman, H.L. (1977). Astrophys. J. **213**, 138.
Sion, E.M. and Liebert, J. (1977). Astrophys. J. **213**, 468.
Wehrse, R. (1977). Sec. Europ. Workshop on White Dwarfs, Monte Porzio, Memoria Soc. Astronomica Italiana, in press.

Weidemann, V. (1963). Zeitschr. f. Astrophysik 57, 87.
Weidemann, V. (1968). Ann. Rev. Astron. Astrophys. 6, 351.
Weidemann, V. (1971). In IAU Symposium No. 42, White Dwarfs. W.J. Luyten, ed., D. Reidel, Dordrecht, p. 81.
Weidemann, V. (1977). Astron. Astrophys. 59, 411.
Weidemann, V. (1978). In IAU Symposium No. 76, Planetary Nebulae, in press.

DISCUSSION

LIEBERT: How dependent are your interesting conclusions about the spread of gravities and radii on the use of the models by Koester rather than older calculations?

WEIDEMANN: They are very dependent on them. The old models, e.g. by Wehrse, were calculated with solar helium abundance, whereas the new models are practically pure hydrogen (H/He = 10^5, H/Metals = 10^4 solar).

LIEBERT: Then the new color-color diagram results are due more to changes in model chemical composition rather than the model programs themselves?

WEIDEMANN: Yes, partly, but there remain minor unexplained differences, e.g. in comparisons with the models of Wickramasinghe and Strittmatter.

GREENSTEIN: Did you include the objects with Strömgren photometry by Richard Green? They show a larger intrinsic scatter than did John Graham's, and might be considered to indicate a larger true variation in log g for the degenerate stars.

WEIDEMANN: Yes, Green's data were included with the exception of those at high temperatures, which show a large unexplained scatter.

SHIPMAN: The large scatter for log g values determined from any two-color diagram - Strömgren or multichannel - at high temperatures is due to the small change in observed color with variations in log g at high temperatures.

WEIDEMANN: You are right, but this has been incorporated by the weights assigned. The scatter in Green's data, however, is much larger observationally than Graham's.

ULTRAVIOLET COLORS OF SUBDWARF O STARS

Paul R. Wesselius

Kapteyn Institute

1. INTRODUCTION AND INSTRUMENT

The group of subdwarf O stars consisting of field stars and some central stars of old planetary nebulae does occupy an interesting place in the HR diagram. Greenstein and Sargent (1974) have tried to establish this place, and conclude that especially the hottest ones need ultraviolet data to improve the values of effective temperature and absolute luminosity.

I therefore observed some twenty sdO stars in the far ultraviolet using the spectrophotometer in the Netherlands' satellite ANS (van Duinen et al, 1975). It has five quite rectangular photometric bands at 1550 (50 or 150), 1800 (150), 2200 (200), 2500 (150), and 3300 (100) Å. The numbers between brackets indicate the widths of the bands in Ångstroms. These measurements could be done with the ANS-UV instrument mainly because of its small field of view (2.5 x 2.5 arc minute), a factor of more than 10 better than the aperture of comparable previous satellite instruments: The background light is much suppressed and close-by stars can be avoided.

2. DATA REDUCTION

Most objects have been observed several times. Usually the counting rates of a single observation (photomultipliers in the photon counting mode have been used) have an error (one standard deviation) of typically a few percent. No variability was detected at this level for any of the sdO stars, and therefore all observations of an object were averaged together. The reduction procedure leading to "net" count rates at a specific date, and by

A. G. Davis Philip and D. S. Hayes (eds.), The HR Diagram, 125–128.

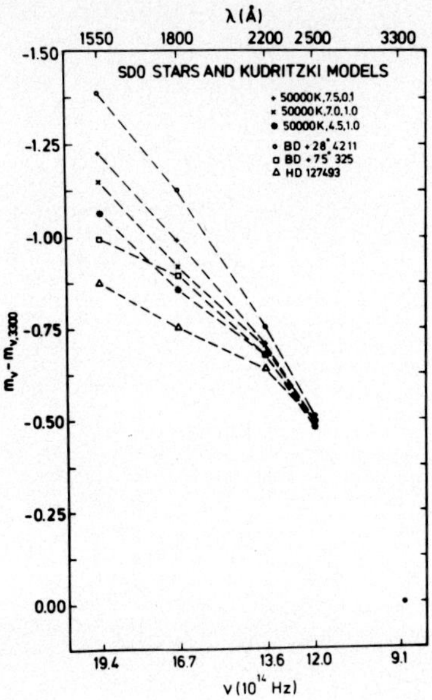

Fig. 1

For the models the numbers behind the plotting symbols indicate effective temperature, log g, and He abundance. The three sd0 stars were corrected for E(B-V) = 0.03, 0.03 and 0.07 from top to bottom.

applying an absolute calibration to absolute fluxes and magnitudes, has been described by van Duinen et al (1977).

The extinction correction poses a special problem for these stars. The extinction bump in the UV at 2200 Å (one of the ANS-UV channels has been chosen at precisely this wavelength) makes it possible for most hot objects to correct for extinction by "smoothing" the bump (see e.g. Pottasch et al, 1977). The hot sd0 stars however, can have quite high helium abundances, and consequently the absorption edge due to He II (n=3) at 2050 Å may show up. This would produce a "bump" at 2200 Å going the other way. An extensive investigation was undertaken and I believe that the problem has been solved satisfactorily. This will be discussed in a future paper.

3. COMPARISON WITH MODELS; DISCUSSION

Dr. Kudritzki kindly folded the UV fluxes of some of his sd0 models (as described in his paper of 1976) with the ANS passbands. A few models are shown in Fig. 1 along with the data on three subdwarf O stars. The He edge should be (see the models) and is (HD 127493) apparent in the ANS-UV data. The steep rise towards 1550 Å exhibited by the models and the data (BD + 28° 4211) is caused by the high gravities, and is in sharp contrast to the more gentle rise of the normal O stars. Fig. 2 shows 18 sd0 stars with some main-sequence stars from O7 to B4; it is evident that the sd0 stars are a lot more "ultraviolet" than the main-sequence stars.

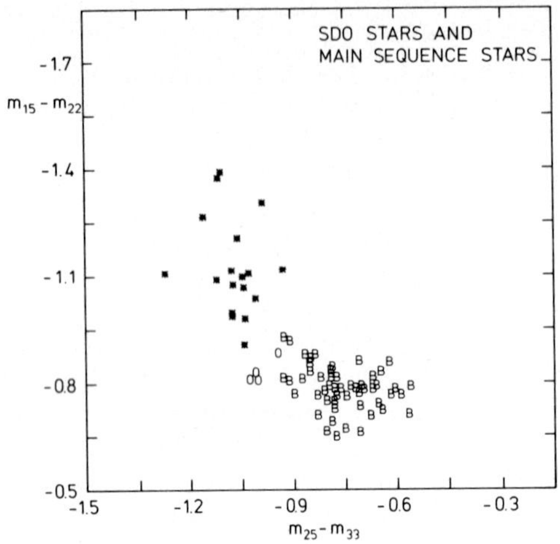

Fig. 2

Color-color plot; the stars have been corrected for interstellar extinction. 15, 22, 25 and 33 indicate the 1550, 2200, 2500 and 3300 Å band. The "star" symbols represent sdO stars, "O" and "B" refer to spectral classes O and B.

REFERENCES

Greenstein, J.L. and Sargent, A.I. (1974). Astrophys. J. Suppl. 28, 157.

Kudritzki, R.P. (1976). Astron. Astrophys. 52, 11.

Pottasch, S.R., Wesselius, P.R., Wu, C.C. and van Duinen, R.J. (1974). Astron. Astrophys. 54, 435.

van Duinen, R.J., Aalders, J.W.G., Wesselius, P.R., Wildeman, K.J., Wu, C.C., Luinge, W. and Snel, D. (1975). Astron. Astrophys. 39, 159.

van Duinen, R.J., Wesselius, P.R., Wu, C.C., Aalders, J.W.G. and Kester, D. (1977). in preparation.

DISCUSSION

SHIPMAN: What is the helium abundance of the high-helium star you mentioned (BD - 31° 4800) determined from the optical spectrum? Are there any hydrogen lines visible in its spectrum?

WESSELIUS: As far as I know this star shows very strong helium lines and no Balmer lines.

STENCEL: The Kudritzki models have a large range of gravities (log g = 4.5 to 7.5) yet do not match the observed flux range. Do you require a larger range of gravities to succeed here?

WESSELIUS: No, a larger range of effect temperatures is needed.

NORRIS: I would like to comment that your results concerning the temperatures and helium abundances of the sdO stars appear to support the suggestion originating, I think, with Strittmatter that there exists a boundary in the (log g, T_{eff}) plane hotter than which surface peculiarities, in particular helium weakness, cannot exist. According to Strittmatter this boundary is defined by the condition that mass loss removes an atmosphere on a timescale short compared to that necessary for diffusive processes to be operative. If one accepts a sequence of evolution from sdO to DA, and that sdO stars are not helium weak, then your results support the hypothesis that as the stars become cooler diffusion turns on as the stars cross the 'Strittmatter boundary' and diffusive timescales become shorter than those for mass loss. Would you comment further on the high helium abundances in the sdO stars, and on these ideas?

WESSELIUS: sdO stars are defined as showing He II 4686 Å, and all that we have anlyzed so far show large He abundances.

NORRIS: Perhaps Dr. Greenstein or Dr. Shipman could comment.

GREENSTEIN: I believe that not all subdwarf O stars need show He II. Very hot degenerate stars have no helium at the surface; hot sdO's (a) have He II, if there are many lines including ionized metals, or (b) show almost no lines, except shallow hydrogen. The hottest star in the ANS diagram, +28°4211, has no or very weak He II 4686 Å, yet is an O subdwarf. The fainter nuclei of planetary nebulae have weak He II. The description of type O by the presence of He II is near breaking down. The settling out of helium may occur as log g increases, but in addition the composition may be affected by details at the end of the planetary-nucleus stage.

WESSELIUS: Among the sdO stars I've observed there are three nuclei of planetary nebulae, and they have rather strong helium lines, and furthermore look completely similar to the other sdO stars.

WALBORN: I think the spectra of subdwarf O stars show He II lines by definition. Certainly +28°4211 has very strong 4686 Å line and +75°325 has strong 4200, 4541 Å lines as well.

GREENSTEIN: Not necessarily in high-gravity stars, which may also be helium-deficient at their surfaces.

SUBLUMINOUS STARS IN THE HR DIAGRAM

C. C. Dahn and R. S. Harrington

U. S. Naval Observatory

1. INTRODUCTION

Eggen (1969) has defined a subluminous star "... as one that is fainter by at least a factor of 10 in visual luminosity than the main sequence stars of the same (U-V)." Implicit is the suggestion that single stars might exist in the region of the HR diagram between the subdwarfs and the normal degenerate sequence. While the evolutionary calculations of Chin and Stothers (1971) indicate that stars with masses <0.2 $M_\odot$ will pass through this region following the completion of nuclear burning, these models have not yet been confirmed. We summarize Naval Observatory astrometry and photometry as it pertains to the existence of such stars.

2. PROPOSED SEQUENCE OF SUBLUMINOUS STARS

Several observational investigations have suggested the existence of a sequence of subluminous stars which apparently bridges the region between the normal degenerate sequence and the subdwarf domain in the HR diagram. Evidence for this hypothesis has been presented by Eggen (1970, 1971) based on six stars with apparently reliable luminosity determinations. Four of these stars (L870-2, R974, LTT6662, and W1037) have trigonometric parallax determinations greater than 0".05, and a fifth star (LTT6333) appears to have a common proper motion with β TrA (HD141891) which has a trigonometric parallax of about 0".08. The sixth star (HG7-138) has a position and a proper motion indicating a 90% probability of membership in the Hyades cluster (van Altena 1969).

A. G. Davis Philip and D. S. Hayes (eds.), The HR Diagram, 129-132.

Naval Observatory parallax determinations have been completed for the four stars north of -20° declination. The results are summarized in Table I below, where M_V (Eggen), the visual absolute magnitude adopted by Eggen, is compared with M_V(USNO), the corresponding quantity obtained from the Naval Observatory parallax. Our deduced status of the star is given in the last column; the data do not support the proposed subluminous sequence.

TABLE I

STAR	USNO	M_v(USNO)	M_v(Eggen)	STATUS (USNO)
L870-2	0".084±0".004	+12.5±0.1	+11.8	Normal degenerate
R974	0".012±0".004	+8.2±0.7	+12.1	Extreme subdwarf
HG7-138	0".003±0".003	+7.9±2.2	+12.7	Dwarf or subdwarf
W1037	0".023±0".004	+11.0±0.4	+12.8	Extreme subdwarf

Regarding the two southern stars, we note the following: 1) The relationship between LTT6333 and β TrA is suspect, since the two stars are widely separated ($\rho \simeq$ 157") and their proper motions differ by $\Delta\mu$= 0".05 yr^{-1} (Luyten 1957); and 2) The luminosity of M_V = +12^m.7 adopted by Eggen for LTT6662 is based on a reliable Yale parallax determination (0".099) but is consistent with an extreme subdwarf interpretation for this star.

3. FIELD SUBLUMINOUS STARS

Over 400 field stars have been suggested as being subluminous by Eggen (1971 and earlier papers). These suggestions were usually based on either 1) an observed ultraviolet excess larger than the maximum expected from abundance effects, or 2) photometry which, when considered in conjunction with a measured proper motion, indicated an excessive tangential velocity if the star were considered to be a normal dwarf or subdwarf. Over 50 of the 200+ candidates north of δ = -20° have been included on the USNO parallax program. Of the 42 stars with reliable solutions, 12 are normal degenerates. The remaining 30 stars have parallaxes which for the most part are small (<0".020). Considering only the 17 candidates for which the measured parallaxes exceed their mean errors, we examined the locations of the 12 stars with 0^m.6≤(B-V)≤1^m.3 in the M_V vs. (B-V) color magnitude diagram; the M_V vs $(V-I)_{KM}$ diagram was employed for the 5 later candidates. When compared with the average dwarf sequences defined by the nearby stars (π≥0".05), we find no stars convincingly subluminous in the sense of Eggen's definition. However, we do find considerable evidence for extreme subdwarfs up to 2.5 mag subluminous, although the large errors in absolute magnitudes associated with the small parallaxes involved limit the conclusiveness of this

evidence.

We can, however, supplement the above sample with 54 additional stars from the USNO Parallax Program which apparently have kinematic and photometric properties similar to Eggen's subluminous candidates. Examination of this larger body of data leads to the following preliminary conclusions: 1) stars as subluminous as 1.5 mag below the mean main sequence (defined by the nearby stars) seem well established for the color range of $0^m.6 \leq (B-V) \leq 1^m.3$ in the M_V vs. (B-V) diagram. Subluminosities as large as 2.0 mag are less well established but are highly likely, while subluminosities as large as 2.5 mag seem doubtful but still possible. No evidence is found for stars more subluminous than 2.7 mag. 2) Stars as subluminous as 2.0 mag below the mean main sequence (defined by the nearby stars) seem well established for the color range of $0^m.8 \leq (V-I)_{KM} \leq 1^m.9$ in the M_V vs. $(V-I)_{KM}$ diagram. Furthermore, subluminosity as large as 2.5 mag seems quite likely over this color range. No evidence is found within this sample of stars for subluminosities in excess of 2.9 mag. 3) Stars with tangential velocities as large as 250 km/sec are well established in this sample, while values as large as 300 to 350 km/sec are probable. No reliable evidence for tangential velocities above 400 km/sec is found.

4. CONCLUSIONS

Available Naval Observatory data do not support the presence of field stars between the normal degenerate sequence and the subdwarf region. However, our data do support the existence of subdwarfs as much as 2.0 to 2.5 magnitudes below the main sequence defined by the nearby stars in both the M_V vs. (B-V) and M_V vs. $(V-I)_{KM}$ diagrams.

REFERENCES

Chin, C. and Stothers, R. (1971). Astrophys. J. 163, 555.
Eggen, O. J. (1969). Astrophys. J. 157, 287.
Eggen, O. J. (1970). Astrophys. J. 159, 945.
Eggen, O. J. (1971). In IAU Symposium No. 42, White Dwarfs, W.J. Luyten, ed., D. Reidel, Dordrecht, p. 8.
Luyten, W. J. (1957). A Catalogue of 9867 Stars in the Southern Hemisphere with Proper Motions Exceeding 0"2 Annually, Lund Press, Minneapolis.
van Altena, W. F. (1969). Astron. J. 74, 2.

DISCUSSION

LIEBERT: What was the trigonometric M_V for G165-47?

DAHN: $M_V = +12\overset{m}{.}5 \pm 0\overset{m}{.}3$.

LIEBERT: I wonder if Dr. Greenstein might comment on his spectra of the low luminosity subdwarf or subluminous stars mentioned in his talk -- $M_V \sim +13^m$, I believe -- compared with the spectra shown for G165-47?

GREENSTEIN: I can't remember the details of the G165-47 spectrum, except that it has strong hydride bands. In any case, the colors indicate a luminosity $M_V \approx +10\overset{m}{.}8$, rather than $+12\overset{m}{.}5$ from the recent good parallaxes. The evidence is therefore that even my "subluminous" sequence is not sufficiently faint. The more depressing prospect is that any color-magnitude diagram will be still affected in the late M's by differential line-blocking. The blanketing vector is so large, from the model-atmosphere to the observed colors, that composition changes might make a high-velocity dM look like a sdK and a late sdM like an early dM, and make its predicted luminosity too bright and its space motion too large.

EMPIRICAL EVIDENCE FOR CRYSTALLIZATION SEQUENCES OF DA TYPE WHITE DWARFS

S. Tapia

Lunar and Planetary Laboratory
University of Arizona

The determination of the distribution of white dwarfs in the HR diagram is marred by two observational difficulties: The recent discoveries of degenerate stars with unusual spectral characteristics and the controversy over the existence of crystallization sequences. Observationally, the origin of the controversy can be traced to the single band displayed on the M_V vs. (b-y) diagram by several white dwarfs with accurate parallax (Graham 1972). The discrepancy between this single band and the two bands evident in the M_V vs. (U-V) diagram (Eggen and Greenstein 1965) is far too great to be attributed to observational errors (Eggen 1969). Theoretically, the number of sequences in the HR diagram is directly related to the cooling properties of degenerate dwarfs, for which crystallization is expected on rather general grounds. On one hand, Van Horn (1968) considered a discontinuous transition to the solid phase with the release of latent heat and the addition of it to the total internal energy. Since the amount of latent heat depends upon the chemical composition, bright and faint crystallization sequences should be observed according to the different stages of evolution reached by the parent stars. On the other hand, Mestel and Ruderman (1967) proposed a gradual transition to the solid phase, and consequently the formation of a single cooling sequence. The purpose of this contribution is to report observational results indicating the existence of two crystallization sequences of DA type white dwarfs.

An analysis of several photometric attempts to derive atmospheric parameters of degenerate dwarfs led Wickramasinghe and Strittmatter (1972) to suggest photometric bands designed to optimize determinations of gravity and effective temperature. For DA white dwarfs they propose three intermediate bands that measure the

A. G. Davis Philip and D. S. Hayes (eds.), The HR Diagram, 133-136.

continuum plus one band that measures the absorption of Hβ. With these spectral bands two reddening-free indices are formed. The first index, w, is directly related to the strength of Hβ while the second index, d, provides a broad baseline comparison between the Balmer and Paschen continua. For the hotter white dwarfs (T_{eff}> 12000 K) w is only sensitive to gravity and d is only sensitive to temperature. However, a combination of both indices can extend their use to T_{eff} = 7000 K.

A sample of 46 DA white dwarfs, most of them with accurate determinations of trigonometric parallax, was observed with this new photometric system employing a computer controlled photometer and the Steward Observatory 2.29 m telescope. The observed w and d indices were compared with synthetic w and d indices computed with the grids of model atmospheres published by Wickramasinghe (1972) and Wehrse (1976) to determine gravities and temperatures. In general, the results agree with a value log g = 8 for the majority of the white dwarfs, as determined in an earlier analysis (Weidemann 1975). Derived uncertainties for the effective temperatures are typically 700 K. In the mentioned sample of DA white dwarfs 25 stars have absolute magnitudes determined to $0^m.25$ or better. These stars have been included in the M_V vs. log T_{eff} diagram presented in Fig. 1. In this diagram nine stars for which the trigonometric parallax are larger than 0".06 plus seven Hyades members outline two bands of different slope. Aside from the smaller number of stars Fig. 1 bears strong resemblance to the M_V vs. (U-V) diagram obtained by Eggen and Greenstein (1965). Also, two bands with different slopes are obtained if the (u-b) colors published by Graham (1972), instead of the (b-y) colors, are plotted versus the absolute magnitude. It appears therefore, that the two bands have been observed in three different photometric systems.

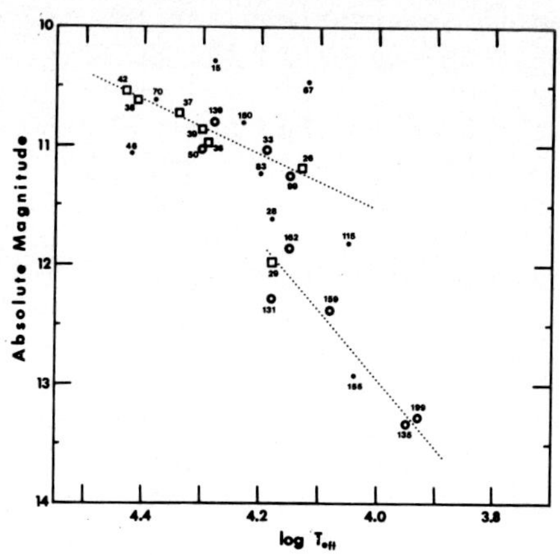

Fig. 1. The M_V vs. log T_{eff} diagram for 25 DA white dwarfs with uncertainty smaller than $0^m.26$ in the distance modulus. Squares represent Hyades members. Circles are used for stars with parallax larger than 0".06. Identification is made in terms of the EG numbers.

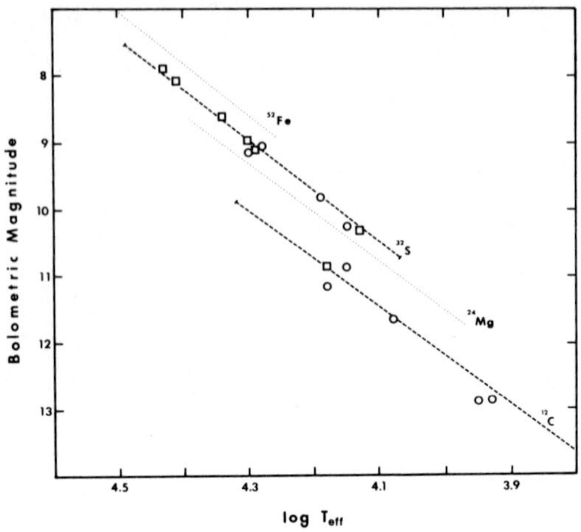

Fig. 2. The M_{bol} vs. log T_{eff} diagram for 16 DA white dwarfs with accurate distance determination. Symbols as in Fig. 1. Crystallization sequences computed by Van Horn are presented with dashed and dotted lines. The ^{32}S sequence is restricted here to mass in the range 0.88 to 0.40 solar units, with larger masses at the upper end.

Adding the bolometric corrections to the absolute magnitudes produces a homologous transformation of these bands into the M_{bol} vs. log Teff diagram. The results are presented in Fig. 2, where the 16 stars with better distance determination are included. The two bands now display the same slope of the crystallization sequences derived by Van Horn (1968). However, the interpretation is not straightforward: Current trends in nucleosynthesis research question the presence of elements much heavier than carbon in the core of degenerate dwarfs.

REFERENCES

Eggen, O.J. and Greenstein, J.L. (1965). Astrophys. J. 141, 83.
Eggen, O.J. (1969). Astrophys. J. 157, 287.
Graham, J.A. (1972). Astron. J. 77, 144.
Mestel, L. and Ruderman, M.A. (1967). Mon. Not. R. Astron. Soc. 136, 27.
Van Horn, H.M. (1967). Astrophys. J. 151, 227.
Wehrse, R. (1976). Astron. Astrophys. Suppl. 24, 95.
Weidemann, V. (1975). In Problems in Stellar Atmospheres and Evelopes, B. Baschek, W.H. Kegel and K. Traving, eds., Springer-Verlag, Berlin, p. 173.
Wickramasinghe, D.T. and Strittmatter, P.A. (1972). Mon. Not. R. Astron. Soc. 160, 421.
Wickramasinghe, D.T. (1972). Mem. Roy. Astron. Soc. 76, 129.

DISCUSSION

WEIDEMANN: There are three arguments against your hypothesis of crystallization sequences: first, the new model atmospheres are changing the locations in such a way that the average radius is constant from the highest to the lowest temperatures, as mentioned by Shipman and as shown in my HR diagram; second, it has been demonstrated that even if latent heat is released at crystallization the additional energy would be radiated away so smoothly - over a larger luminosity interval - that no sequence would be visible in the HR diagram (Koester, Astron. Astrophys. 16, 459, 1972), and third, according to more recent investigations by the van Horn group, the latent heat is probably freed over a larger range of state variables - which again tends to smooth out visible effects in the HR diagram.

TAPIA: I think that we should take a more open attitude on this subject. Using the best temperature indicators from three different photometric systems, in each instance we find two sequences of DA white dwarfs on the HR diagram. If there are doubts that the system described in this paper is free of contamination from the Balmer lines, let me say that the separation of DA white dwarfs is also detected when the near monochromatic color index (2.80) - (1.85) is used in the abscissa of the HR diagram (Greenstein, Astron. J. 81, 332, 1976). Better yet, I was pleased to see that in Fig. 1 of your paper the Hyades white dwarfs define a tight sequence with the same slope as the brighter sequence found with my data. Retaining only the high quality bolometric magnitudes in your Fig. 1, there are indications of two sequences also. It is useful to know which model atmospheres or color indices give only one sequence. However, that there are two sequences appear to be a persistant observational result. Even if there are some reasons to believe that crystallization has unobservable effects on the evolution of white dwarfs, the nature of these two sequences remains to be explained.

BIDELMAN: What are the sources of the absolute magnitudes of the stars of your upper sequence? Are they from trigometric parallaxes?

TAPIA: For the Hyades members the distance modulus $3\overset{m}{.}29 \pm 0\overset{m}{.}20$ was adopted. For the other DAs, trigometric parallaxes were used, mainly the new results from the Naval Observatory.

PART IV

HR DIAGRAMS, CLUSTERS

Robert P. Kraft and A. Renzini on stage.

SPECTROPHOTOMETRY IN STELLAR ASSOCIATIONS

A. Gutiérrez-Moreno and H. Moreno

Universidad de Chile

When analyzing the spectral energy distributions of early type stars as measured at Cerro Calán, we find that many of them have a feature which we have called ΔD (Gutiérrez-Moreno et al., 1968); the internal error of the measure of ΔD for a single star is ±0.02. An analysis of the spectrum scans and the energy distributions shows that this feature may be interpreted as an emission in the Balmer continuum. A few stars have values of $\Delta D<0$, and a comparison of their scans with those of normal stars ($\Delta D=0$) suggests an absorption larger than normal in the Balmer continuum.

In this paper we propose to analyze the behavior of ΔD in two stellar associations, the Orion aggregate (Moreno, 1974) and the Upper Scorpius group (Gutiérrez-Moreno and Moreno, 1977). Practically all the stars observed are located in the HR diagram in the region of pre-main sequence stars. We have preferred to obtain the value of ΔD from the absolute energy distributions, calling it ΔD_a. To avoid possible systematic errors depending on right ascension we reduced the two groups independently to the absolute Oke and Schild system (1970) by using secondary standards located as close as possible to the corresponding right ascension zone.

A comparison of these two stellar groups with field stars shows that positive ΔD values are more frequent in the associations than in the field stars. This fact would confirm the theory that pre-main sequence stars have circumstellar envelopes which produce emission in the Balmer continuum.

We found that ΔD_a is well correlated with c_1°, the dis-

A. G. Davis Philip and D. S. Hayes (eds.), The HR Diagram, 139–142.

tance of a star to an arbitrary "main sequence" in the c_1^o, $(b-y)_o$ plane, and that they have similar behavior when correlated with other photometric parameters. It is interesting to mention the correlation between ΔH and ΔD_a, where ΔH, the distance of any star to the mean (Hδ, Hβ) relation, is a measure of the emission in the lines. It is possible to appreciate a correlation between these two indices of shell strength, but there are two stars, HD36917 (V372 Ori) and HD142983 (48 Lib), which show emission in the lines, but a rather large negative ΔD_a. This fact suggests that the values of $\Delta D_a < 0$ correspond also to a shell effect.

We have also measured an ultraviolet gradient, ϕ_u ($2.831\ 1/\lambda < 3.030\ \mu^{-1}$). These gradients have been reduced to the Paris system (Chalonge et Divan, 1952), and corrected for reddening according to the usual procedures. The standard deviation of the transformation to the absolute system is ± 0.08, in good agreement with the probable error of ± 0.05 found for one gradient on the basis of the errors of the measured magnitudes. These errors are comparable with those quoted by Ardeberg (1969). Nevertheless, when comparing the absolute intrinsic gradients $\phi_{u,o}$ with the mean values corresponding to the spectral types, we find a scatter much larger than could be expected on the basis of the errors previously computed. Besides, it is evident that Orion stars have

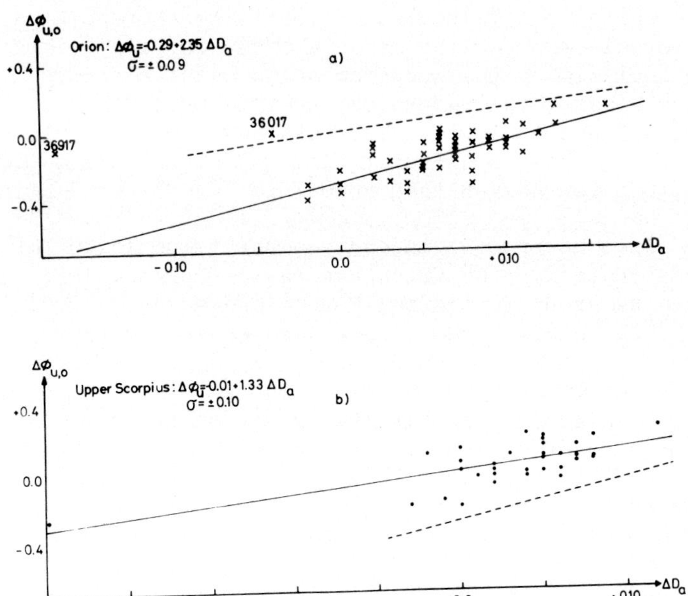

Fig. 1. Correlation between ΔD_a and $\phi_{u,o}$; a) the dashed line represents Upper Scorpius relation; b) the dashed line corresponds to Orion stars.

smaller $\Phi_{u,o}$ gradients than Upper Scorpius stars.

An analysis of the residuals found shows that they are functions of ΔD_a. The correlation (Fig. 1) implies that stars with emission in the Balmer continuum have gradients larger than normal, while those with an excess absorption have gradients smaller than normal. The effect for positive ΔD_a values confirms quantitatively the conclusion of the early work by Barbier and Chalonge (1941) in the sense that stars with emission in the Balmer continuum are redder (have larger gradients) than normal stars. We notice that Orion and Scorpius stars satisfy different relations, with different zero points. Thus, the gradients are smaller for Orion stars, which confirms the Schild and Chaffee (1971) conclusion that the energy distributions of several Orion stars suggest that continuum temperatures are higher than those inferred from the MK spectral types. Upper Scorpius stars satisfy a relation similar to that of the field stars.

The same type of correlation is found when comparing our results with uvby photometry. If we determine the difference between the observed gradients and the mean gradients corresponding to a given observed (u-b), we find that the residuals are also functions of ΔD_a. The stars HD36017 and HD36917 seem to behave more like Upper Scorpius stars. We are not sure if this fact has a physical meaning.

The fact that positive and negative ΔD_a values satisfy the same relation for a given group of stars suggests again that both have the same origin, that is to say, that both originate in circumstellar shells. The difference between the two types of phenomena would be due to different characteristics of the shell.

It is evident that the spectral characteristics analyzed here have influence in the photometric spectral classifications. For example, the S_Q classification is based in a measure of the Balmer jump D; but D appears too small for stars with $\Delta D_a > 0$, which would be classified earlier than their MK type; and it appears too large for stars with $\Delta D_a < 0$, which would be classified later.

REFERENCES

Ardeberg, A. (1969). Astron. Astrophys. **3**, 257.
Barbier, D. and Chalonge, D. (1941). Ann. d'Astrophys. **4**, 30.
Chalonge, D. and Divan, L. (1952). Ann. d'Astrophys. **15**, 201.
Gutiérrez-Moreno, A. and Moreno, H. (1977), in preparation.
Gutiérrez-Moreno, A., Moreno, H. and Stock, J. (1968). Publ. Dept. Astr. Univ. Chile **1**, 127.
Oke, J.B. and Schild, R.E. (1970). Astrophys. J. **161**, 1015.
Schild, R.E. and Chaffee, F. (1971). Astrophys. J. **169**, 529.

DISCUSSION

WARREN: Have you looked for and/or noticed any correlation between the parameter ΔD and spectral type?

GUTIÉRREZ-MORENO: No, we have never compared our ΔD_a with spectral types; we have compared them with differences between MK types and photometric types; and we have found that a slight correlation seems to exist.

A PHOTOMETRIC STUDY OF ARA OB1

William Herbst

Carnegie Institute, D.T.M.

Robert J. Havlen

University of Virginia

ABSTRACT

Photoelectric and photographic UBVRI photometery have been obtained for over 700 stars in the field of Ara OB1, including many in the central cluster NGC 6193. Association members were identified mainly on the basis of their reddenings. The association sequence is well populated blueward of $(B-V)_o = -0^m.10$, $(U-B)_o = -0^m.35$. Beyond this "turn-up" point stars lie only above the ZAMS. There are clear gaps in the color magnitude diagrams just redward of the turn-up points. The width of these gaps is 0.07 mag. in (U-B). They cannot be easily explained by pre-main sequence models involving only gravitational contraction or by those involving ^{3}He burning. The difficulty of identifying the faint pre-main sequence stars (if they exist) amidst the field stars is discussed and a limiting luminosity function for the association is derived.

A. G. Davis Philip and D. S. Hayes (eds.), The HR Diagram, 143-144.

DISCUSSION

BUSCOMBE: The determination of radial velocities for stars in this association would be particularly useful, both for internal dispersion and differential galactic rotation. The published data are as yet only for a few Radcliffe spectrograms because the Mount Stromlo observations were made in a season when the thermal stability of the spectrograph and reliability of the comparison source were unsatisfactory. A considerable number of binaries and possible spectrum variables deserve more attention from spectroscopic observers. High-resolution data in interstellar absorption lines would be of very great interest at this galactic longitude.

HAVLEN: I am presently working on velocities for a number of the early B stars in NGC 6193, but I agree that a much more extensive study is needed especially in order to understand the interaction of the stars with the enveloping dust and gas.

WARREN: Have you looked for evidence of the possible existence of subgroups in the Ara OB1 association by dividing the group into inner and outer regions and constructing separate diagrams?

HAVLEN: Our attempts along that line were unsuccessful.

FOUR-COLOR AND Hβ DIAGRAMS FOR OPEN CLUSTERS - A SURVEY

D. L. Crawford

Kitt Peak National Observatory*

Intermediate band photometric data for stars in at least 50 open clusters now exist. This paper reviews the status and desiderata for the uvbyβ subset. Similarities and differences between clusters are highlighted, and the relation of the photometric differences to physical differences is noted. Some of these are stellar age, chemical abundance, rotational velocity, and duplicity.

As the stars in the cluster are at the same distance and presumably have nearly the same age and chemical composition, a "separation-of-parameters" is possible and is a great advantage over analysis of non-cluster stars.

We derive interstellar reddening and absorption values (and the variation over the cluster field), the true distance modulus, the ultraviolet excess and δm_0 (empirical measures of chemical composition), and show color-magnitude and other diagrams for representative clusters. Some discordances with theory are noted.

(This full paper, with additional details, will be published shortly in the Astron. J.).

* Operated by the Association of Universities for Research in Astronomy Inc., under contract with the National Science Foundation.

A. G. Davis Philip and D. S. Hayes (eds.), The HR Diagram, 145–146.

DISCUSSION

COX: What is the effect on your indices of helium poor B stars? Can they affect the distance moduli?

CRAWFORD: We don't have data for many. Most do not fit the calibrations, but some do. There are not enough of them to adversely affect the calibration statistically.

ROMAN: The agreement of the distance moduli determined from various parameters is impressive. To what extent are the calibrations independent? What is the systematic accuracy of the calibrations?

CRAWFORD: They are independent in the sense that entirely different data were used. But, naturally, we all have used nearly the same clusters and the same stars. Only a few A and F-type stars have large parallaxes for example.

I feel that things are good in color to $\pm 0^m.01$, and in M_V to about $\pm 0^m.2$ for the late B to F stars. It's larger for earlier B stars, as the relation between M_V and B is so steep (± 0.4 or 0.5 perhaps).

MORGAN: How would a change in the modulus of the Hyades of $0^m.25$ affect your calibration?

CRAWFORD: The Hyades was not used at all in this calibration, so a change in the distance modulus has no effect whatsoever. The natural question, then, is what do we get for the Hyades? The problem is that we get this funny effect in the c_1 index for the Hyades F stars. If you assume that $\delta c_o = 0$ (which it isn't), then you get a distance modulus of $3^m.2$. If you assume that it is real, and due to some real luminosity effect, which I don't think it can be, then you get $3^m.4$. But I don't think that's a valid result. If you use the A-stars, where you don't see any Hyades anomaly, we get $3^m.2$.

MORGAN: You showed a couple of slides in which you have β vs. c_1, and there were gaps, which you mentioned. Could these gaps be due to your including both cluster and field stars in the distribution?

CRAWFORD: I think those slides only had field stars. There is no question that there are gaps and non-linearities. There are explanations for some and none at all for others. Demarque and I have discussed the one found in the late B's; it falls just where they expect the CNO opacity problem.

MORGAN: Have all of the Pleiades stars which you said were binaries been certified spectroscopically?

CRAWFORD: Most of them have been, but not all.

THE MAIN SEQUENCE OF NGC 6231 AND THE CALIBRATION OF ABSOLUTE MAGNITUDES

R. F. Garrison

David Dunlap Observatory, University of Toronto

It is fitting to mark the Centennial of Henry Norris Russell's birth with a discussion of a new approach to the calibration of absolute magnitudes for MK spectral types. With the addition of the NGC 6231 main sequence down to A0, the material for the cluster fitting method using very carefully determined MK types is complete. The data will be published elsewhere (Garrison and Schild 1978).

The use of NGC 6231 (nucleus of I Sco), the southern hemisphere's answer to the Perseus double cluster, allows access in the calibration scheme to numerous supergiants of all types, peculiar stars of early type, and O-type stars above the main sequence. If similar, high accuracy data could be obtained for the double cluster, the grid would be considerably strengthened.

Garrison (1973) discussed the use of the three well known moving clusters in a cluster-fitting procedure for determining the calibration of MK types. The Hyades and the Alpha Persei cluster diagrams overlap over the entire F-star range, giving good opportunity for comparison of rotations and metallicities in the two clusters. No differences were found. The Alpha Persei and the II Sco diagrams overlap over the entire A-star range, allowing comparison of rotation and abundances there. No significant differences were found. The new data for NGC 6231 (I Sco) overlap with the II Sco data over the entire B-star range. It is not possible to compare metallicities in B stars at classification dispersion in the blue-violet region of the spectrum because of the lack of sensitive lines, but it is unlikely that they differ significantly. The rotations are very similar.

A. G. Davis Philip and D. S. Hayes (eds.), The HR Diagram, 147–150.

To complete the cluster fitting, the composite main sequence was extended to 04 V by adding the diagrams of NGC 2244 and NGC 2264. The data are from Morgan et al. (1965), but R = 3 was used (Turner 1976). The clusters were assumed to be at different distances and thus were considered separately.

Fig. 1 was constructed by forming means for each spectral type for each cluster and drawing a line with only a small amount of smoothing. Spectroscopically peculiar stars, stars with luminosity classes of IV-V and brighter, and stars which are obvious two-line binaries at MK resolution were left out of the means.

The random errors, considerably reduced by the above procedures, are further reduced in the fitting by the long overlapping segments. It is obvious that a shift of even a few tenths of a magnitude in any of the cluster sequences would produce a considerably poorer fit. Thus, the new calibration should be good to a few tenths of a magnitude or better. Since the range of values for the Hyades distance modulus, which forms the foundation of this procedure, is of the same order of magnitude, it is unlikely that we will be able to do much better.

Support through the National Research Council of Canada is gratefully acknowledged.

REFERENCES

Garrison, R.F. (1973). In Problems of Calibration of Absolute Magnitudes and Temperatures of Stars, B. Hauck and B.E. Westerlund, eds., D. Reidel, Dordrecht, p. 124.

Garrison, R.F. and Schild, R.E. (1978). Astrophys. J. (submitted).

Morgan, W.W., Hiltner, W.A., Neff, J.S., Garrison, R.F. and Osterbrock, D.E. (1965). Astrophys. J. **142**, 974.

Turner, D.E. (1976). Astrophys. J. **210**, 65.

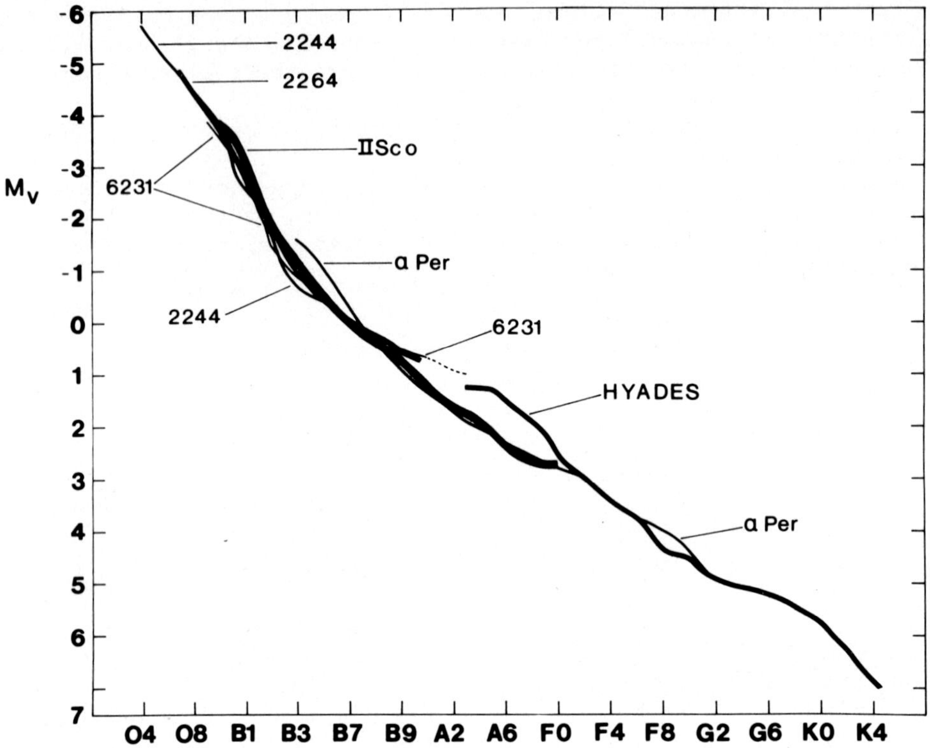

Fig. 1. Composite HR diagram for calibration of MK types.

DISCUSSION

MORGAN: In NGC 6231 you have a super-supergiant, ζ^1 Sco. In your experience, is this star unique, or are there other associations containing such an object?

GARRISON: I haven't taken any photographs or spectra of stars similar to ζ^1 Sco.

MORGAN: This points up our lack of knowledge about such cases. We need investigations of rich O-associations, in the greatest detail possible, because we see such a great variety in the brightest objects present.

THE HYADES CLUSTER — TOO DISTANT?

Phillip J. Flower

Joint Institute for Laboratory Astrophysics
University of Colorado and National Bureau of Standards

1. THEORETICAL AND OBSERVATIONAL ZERO AGE MAIN SEQUENCE

Theoretical zero age main sequences (TZAMS) from Flower (1976) are compared with two observed zero age main sequences (ZAMS) in Fig. 1.: ZAMS40, based on a distance of 40 pc (Sandage 1957) and ZAMS46 on a distance of 46 pc (see Flower 1977 for T_{eff}: (B-V): BC scales). Although ZAMS46 lies above any TZAMS, the TZAMS can be shifted in luminosity by changing the chemical composition. An equation giving the effects of Y (helium) and Z (metals) on the TZAMS was found by Flower (1976) to be

$$\Delta \log L/L_\odot = 0.432\ (\log Z_R - \log Z) - 0.838\ (Y_R - Y) \qquad (1)$$

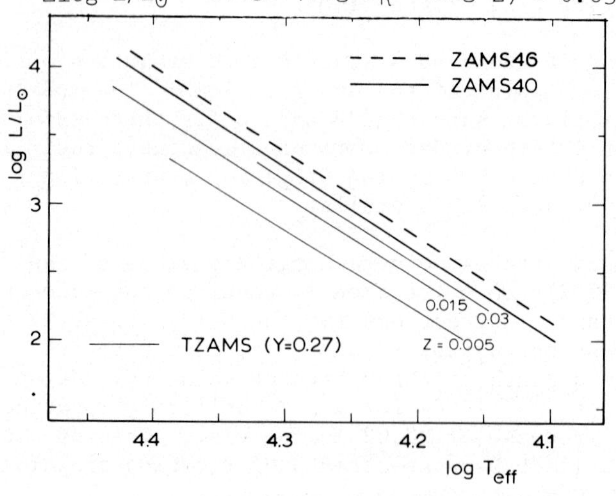

Fig. 1. Theoretical and Observational Zero Age Main Sequences.

A. G. Davis Philip and D. S. Hayes (eds.), The HR Diagram, 151–154 and 405.

where $\Delta \log L/L_\odot \equiv \log(L/L_\odot)_R - \log(L/L_\odot)$ and R indicates values for a reference composition. Using eq. (1), one finds that metal abundances in the range of $0.03 \leq Z \leq 0.05$ for Y between 0.22 and 0.32 give coincidence with ZAMS40. However, for ZAMS46 the required metal abundances are $0.06 \leq Z \leq 0.10$ for the same range in Y. Before assigning any significance to these high metallicities, the uncertainties inherent in such comparisons must first be evaluated; i.e., three other factors, besides variations in Y and Z, may have a significant influence on such comparisons: 1) uncertainties in stellar opacities; 2) evolutionary effects; and 3) effects of stellar rotation. These will now be evaluated.

2. THE "EVOLUTIONARY DISTANCE" TO THE HYADES CLUSTER

Although the luminosity of the ZAMS is quite sensitive to Z, a change in Y or Z has very little effect on the slope of the ZAMS (Flower 1976). Therefore, a ZAMS constructed by MS (main sequence) fitting, starting with the Hyades, represents MS stars with the same chemical composition as the Hyades stars. Consequently, the distance estimation from comparisons between the TZAMS and the upper part of the observed ZAMS refers to the distance to the Hyades cluster itself.

Fricke et al. (1971) found that the location of the TZAMS is very sensitive to changes in opacity at temperatures $\geq 10^6$ K. Using their results and the L-T_{eff} relation for upper MS models (Flower 1976), one finds that to eliminate the difference in luminosity between ZAMS46 and the TZAMS, opacities must be increased by factors of 3 and 2 for $Z = 0.03$ and $Z = 0.04$ respectively. However, at temperatures $\geq 10^6$ K opacities are not expected to increase in the near future by more than 20% (Mertz 1977, private communication). This increase in opacity corresponds to an increase of 0.042 in $\log L/L_\odot$ of the TZAMS.

Any brightening of the Hyades stars due to evolution off the MS will shift the zero point of the observed ZAMS. The models of Demarque and Gisler (1975) show that Hyades stars three magnitudes fainter than the brightest Hyades stars have increased their luminosities by no more than 0.008 in $\log L/L_\odot$; for stars fainter than this, the luminosity increase is smaller.

Since rotational effects do not change the slope of the MS (Meyer-Hofmeister 1972), no correction is required for moderately massive rotating stars. Corrections for the Hyades rotating stars will be small because the Hyades MS stars are slow rotators, rotating at less than a tenth of their breakup velocity. Using the models of Faulkner et al. (1968) for a 1 $M_\odot$ uniformly rotating star, one finds an increase of ≤ 0.02 in $\log L/L_\odot$. This probably represents an upper limit because effects of rotation are minimized

since the lower envelope of the MS band is emphasized when determining the ZAMS.

These factors give a 0.07 increase in log $L/L_{\odot}$ of the TZAMS. Adding this to the TZAMS for (Y,Z) = (0.25,0.04), log $L/L_{\odot}$ = 3.25 at log T_{eff} = 4.3, results in the TZAMS being 0.15 mag brighter than ZAMS40 or in a Hyades distance modulus of (m-M) = $3\overset{m}{.}18$. If the range in Z of the Hyades stars is between 1.5 and 2.5 times solar Z, then the minimum and maximum moduli are $3\overset{m}{.}05$ and $3\overset{m}{.}29$.

The modulus, (m-M) = $3\overset{m}{.}18$, agrees quite well with that from "secondary indicators," (van Altena 1974) and from the convergent-point results of Corbin et al. (1975), but disagrees with Hanson (1976) and Anthony-Twarog and Demarque (1977). However, McAlister (1977) has reanalyzed Hanson's solution for possible systematic magnitude effects and finds (m-M) = $3\overset{m}{.}18$. Unfortunately, the models used by Anthony-Twarog and Demarque were not calibrated to reproduce the T_{eff} of the Sun; this would require a larger mixing length. Because a larger mixing length would result in a smaller modulus and because the masses derived for the Hyades binaries depend on the adopted modulus, the consistency between the CM diagram and the ML relation for the Hyades binaries they found for a modulus of $3\overset{m}{.}34$ would be wiped out. Hence, if the mixing length chosen for the models must be the same for both the Sun and the Hyades stars, then a smaller modulus is required from comparisons with the Hyades CM diagram.

REFERENCES

Anthony-Twarog, B. J., and Demarque, P. (1977). Astron. Astrophys. **57**, 471.
Corbin, T. E., Smith, D. L., and Carpenter, M. S. (1975). Bull. American Astron. Soc. **7**, 337.
Demarque, P., and Gisler, G. R. (1975). Astron. Astrophys. Suppl. **20**, 237.
Faulkner, J., Roxburgh, I. W., and Strittmatter, P. A. (1968). Astrophys. J. **151**, 203.
Flower, P. J. (1976). Ph.D. Thesis, University of Washington.
Flower, P. J. (1977). Astron. Astrophys. **54**, 31.
Fricke, K., Stobie, R. S., and Strittmatter, P. A. (1971). Mon. Not. R. Astron. Soc. **154**, 23.
Hanson, R. B. (1976). Report to Commission 24, IAU, General Assembly 16, Grenoble.
McAlister, H. A. (1977). Astron. J. **82**, 487.
Meyer-Hofmeister, E. (1972). Astron. Astrophys. **16**, 282.
Sandage, A. (1957). Astrophys. J. **125**, 435.
van Altena, W. F. (1974). Publ. Astron. Soc. Pacific **86**, 217.

DISCUSSION

HANSON: The Hyades distance modulus $3^m.42 \pm 0^m.20$ (m.e.) (Hanson 1975 Astron. J. 80, 379) was determined from Lick absolute proper motions in the central 6° x 6° region of the cluster. In more recent work I have extended my proper motion measurements to 4 additional regions, North, South, East and West of the cluster center; the resulting 16° extent on the sky gives a much better hold in the convergent point. The resulting distance modulus is $3^m.31 \pm 0^m.06$ (m.e.). This value is in excellent agreement with recent Hyades trigonometric parallax determinations at Van Vleck and Lick observatories.

SEARS: Did you consider the mass-luminosity relation of the Hyades binaries?

FLOWER: It seems that I forgot to mention the mass range of my models. Since I have only computed models for moderately massive stars, $3 \leq M/M_{\odot} \leq 10$, it was impossible for me to consider the Hyades mass-luminosity relation.

SEARS: The preceding paper by Garrison assumed that clusters overlap, down to the Hyades, in the HR diagram. Perhaps your work indicates that this assumption may not be correct. Is that a possible interpretation of your results?

FLOWER: This is a very interesting point. If clusters are overlapped, starting with the Hyades to form a ZAMS, then, because the slope of the ZAMS is independent of chemical composition, the resulting ZAMS is just an extension of the Hyades main sequence. This means that comparisons between this observed ZAMS and main sequences of clusters will give a correct distance to the clusters only if the clusters have the same composition as the Hyades. If the composition differs from the Hyades, then the intrinsic luminosity of the cluster main sequence will be different from that of the observed ZAMS, and, hence, a wrong distance to the clusters will result.

ABT: I think that there is a fourth effect that you should consider. If I understand correctly, you are comparing models computed for single stars with real stars, many of which are double. Because of duplicity the real ZAMS should be about $0^m.2$ above the computed ZAMS.

FLOWER: It is true that unresolved double stars will be above the single stars of a cluster (as in the Hyades and Pleiades clusters), but the single stars should still define a lower envelope, which is the cluster main sequence unaffected by duplicity.

CAYREL de STROBEL: We used the main sequences calculated by Hejlesen for normal metal abundance ($Z = 0.02$), and three solar-type Hyades stars fitted perfectly.

FLOWER: If you reduce the metal abundance of the Hyades, you bring them closer. You could pick a metal abundance, and I can tell you which theoretical ZAMS will fit.

CAYREL de STROBEL: Why concentrate only on Z? What about Y?

FLOWER: Yes, if you change Y by 0.05, you change log L by

CONTINUED ON PAGE 405

THE OCCURRENCE OF PECULIAR STARS IN OPEN CLUSTERS

Helmut A. Abt

Kitt Peak National Observatory

Hugo Levato

La Plata Observatory

We have classified on the MK system a total of 455 stars in 12 open clusters and associations. The classification is based on wide (1.2 mm) spectra of two reciprocal dispersions (39, 128 Å mm^{-1}) obtained with the Kitt Peak 2.1 m and 90 cm reflectors, respectively. The higher dispersion is necessary to show the subtle peculiarities found in some stars. The clusters are the Orion Nebula cluster, Orion OB1 association, Lacerta OB1 association, IC 2602, IC 4665, Pleiades, M39, M34, NGC 2516, NGC 6633, NGC 6475, and Coma; two of these were done with Morgan.

When the frequency of 13 Ap (Si) and six Ap (Hg, Mn) stars (relative to the total numbers of stars within their appropriate ranges of absolute magnitudes) are plotted against age, we see that it takes about 10^6 yr to produce Ap stars and then the number increases steadily with age. The frequency for the older clusters (10^7-$5x10^8$ yr) agrees well with that for field stars.

The mean rotational velocities of Ap (Si) stars decrease inversely with the square root of their age. In the magnetic Ap stars this must be due to magnetic braking but the mechanism in the Ap (Hg, Mn) stars is unknown. The rotational velocities of Am stars decreases inversely with the quarter root of their age. In the oldest clusters (10^8 yr) apparently not all the slow rotators have become Ap or Am stars. We can only speculate that the magnetic stars are gradually braked and develop peculiar abundances by diffusion while the members of close binaries are braked slowly by tidal effects and become Am stars.

A. G. Davis Philip and D. S. Hayes (eds.), The HR Diagram, 155-156 and 405.

In the Orion Nebula cluster ($10^{5.7}$ yr) half of the stars have unusually broad hydrogen line wings while in the remaining Orion association ($10^{6.7}$ yr) only 2% of the stars show that effect. This broadening is not due to rotation but to some unexplained effect in stars right on the ZAMS, according to Morgan. The effect decreases inversely with age.

We found many stars with weak shells characterized by broad helium lines but sharp lines due to Ca II, Si II, C II, Fe II, Mg II, and often Balmer line cores. One-third of the B giants among field stars show the effect. When these stars are added to more extreme shell stars (e.g. Pleione, AD 23878), we find no age dependence for shell stars, or for Be stars either.

The Am stars show no age dependence either. They occur in the youngest clusters so they are produced quickly ($<10^6$ yr). The frequency of Am stars in clusters is the same as for field stars. But there are statistically significant differences from cluster to cluster, evidently reflecting differences in frequencies of binaries.

DISCUSSION

APPENZELLER: You showed us evidence for an increase of the percentage of peculiar stars with cluster age. Is there also a correlation between the type and strength of the spectral pecularities of the individual peculiar stars and age?

ABT: Yes, the peculiarities in the younger groups are less pronounced.

McNAMARA: Why do you associate such different ages to the various subgroups in the Orion OBI association? The observed age spreads seem to be within each sub-group, and not as separate well defined age regions within the association.

ABT: We took the ages from the literature. Wayne (Warren), would you care to reply to this?

WARREN: When one compares the individual subgroups of the association, many factors indicate that they display various ages. While a unique age may not be defined for the region immediately surrounding the Orion Nebula, it is clear that the stars in this region are in general younger than stars in the Belt and Northwest subgroups.

ROSENDHAL: Would you care to comment on the implications of the existence of rapidly rotating metallic line stars on the hypothesis that diffusion in a quiescent atmosphere is responsible for the existence of the metallic-line phenomenon?

ABT: Frankly, No! Diffusion is the only viable theory that we have. So if it fails, we are up the creek!

STENCEL: How do your results on Ap stars in clusters compare with the recent work by Hartoog (1976 Astrophys. J.)?

ABT: Our material is based on two dispersions, one of which

CONTINUED ON PAGE 405

THE CLUMP GIANTS: A SPECTROSCOPIC SURVEY OF G AND K GIANTS IN OPEN CLUSTERS

Catherine A. Pilachowski

University of Washington

Walter K. Bonsack

University of Hawaii

1. INTRODUCTION AND OBSERVATIONS

The clump giants in old and middle-aged open clusters provide a sample of stars whose evolutionary state can be clearly identified as core helium burning (Cannon 1970). The evolutionary state of the Ba II stars, on the other hand, is at present undetermined. The enrichment of s-process elements in the atmospheres of these peculiar stars suggests that the stars have at least passed through double shell burning, where the heavy elements may be produced and mixed to the surface. The Ba II stars are, however, too faint to be associated with this phase of stellar evolution; the absolute magnitudes and temperatures are consistent with the core helium burning phase. The barium stars should then occur in the clumps of Population I clusters in similar proportion to their numbers among the field giants.

A systematic spectroscopic survey of clump giants is in progress using the Cassegrain image tube spectrograph of the 2.2 m telescope of the Mauna Kea Observatory. Clusters have been selected for observation from among those with well-defined and well-populated clumps. A total of 82 giants in 14 open clusters, including NGC 752 and Praesepe, reported here, have been observed. Eleven MK standards of similar temperatures and luminosities, and eleven barium stars have been observed for calibration purposes. The earliest main sequence stars in each cluster are also being observed for rotation velocities.

The spectra cover the region from Ca II H and K through Hα at 50 Å mm^{-1}. The half width of the instrumental profile is 2 Å.

A. G. Davis Philip and D. S. Hayes (eds.), The HR Diagram, 157-160.

Intensity tracings of the plates have been produced with equipment at the University of Hawaii. Equivalent widths have been measured with respect to the local continuum for approximately 30 features on each of the tracings. Lines which are used in spectral classification in this part of the HR diagram were selected from those which are relatively free of blending, or are blended with lines of similar characteristics. Intensities at two points in the G band and at 4208 Å in the CN band were also measured.

2. ANALYSIS

From the line strength measurements a set of seven spectral indices was constructed: hydrogen line strength (Hα+Hβ+Hγ), calcium line strength (4226 Å), overall metal line strength (Cr I 4565, Fe I 4647, Fe I 4667 Å), barium (4554, 4934, 6141 Å), strontium (4077, 4215 Å), CN, and CH. Except for the overall metals, each index was normalized to the strength of nearby iron peak metal lines.

The ratio of hydrogen index to calcium index was chosen as a temperature parameter. The hydrogen lines are most sensitive to temperature through G type stars, while 4226 Å increases strongly with decreasing temperature through the K giants. The ratio of H/Ca correlates with (B-V) for normal field giants, but more observations are needed to calibrate the relationship. No simple luminosity indicator could be found. For this study of clump giants a luminosity indicator is not required since relative V magnitudes are available for stars in a cluster.

The barium index is clearly higher for the field barium stars, even the marginals, than for normal standards, including the two class II stars (Fig. 1). A stellar mass effect is also apparent from examination of the data for three groups of stars: the field stars, the Praesepe giants, and the NGC 752 giants. The mean Ba index for each group appears to correlate with cluster age. Praesepe is a younger cluster than NGC 752, and its giants are more massive than those in the older cluster. At the same L and T_{eff}, log g will be higher in Praesepe, so the Ba II lines will be weaker.

The G band index behaves similarly with H/Ca as with spectral type, remaining constant through the G type stars, and dropping sharply in strength in the early K's. NGC 752 and Praesepe giants both have normal CH, and some, but not all, of the Ba II stars have enhanced CH. CN appears enhanced in most of the Ba II and CN stars, and not in the bright giants. Praesepe giants appear to have slighly strong CN. Some, but not all of the NGC 752 giants have weak CN.

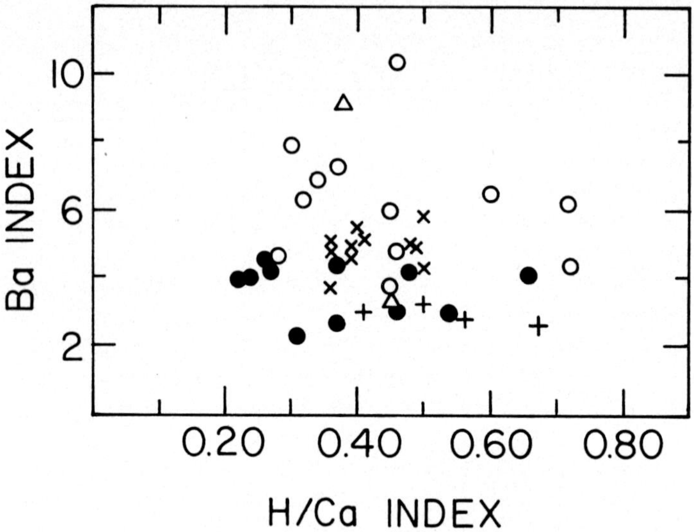

Fig. 1. Barium index vs. H/Ca temperature index. Field stars: normal giants, filled circles; barium stars, open circles. Cluster giants: Praesepe, +'s, NGC 752, X's; NGC 2420, triangles.

Preliminary data for two clump giants in NGC 2420 suggest that one of them (star S) may be enriched in s-process elements and carbon. NGC 2420 contains a previously known barium-carbon star (McClure et al. 1974) at approximately one magnitude brighter than the clump. The overall metal indices suggest that the stars are metal poor, but both stars appear to have enhanced CH. Star S is also enhanced in CN, Sr, and Ba relative to star Q and to the normal field stars. A high carbon abundance may be characteristic of this cluster; NGC 2420 seems prone to produce peculiar stars, but also contains some normal stars. Corroborative observations of star S and others in the clump will be obtained.

Some of these indices can be calibrated using either spectrum synthesis or observations of stars at high dispersion for detailed analysis. The quantitative classification indices will be useful for a variety of purposes besides identifying peculiar stars in clusters. Overall cluster abundances of CH, CN, and Fe group elements can be determined. The barium index is also useful for estimating the mean mass of a clump giant.

We acknowledge partial support of this work from the NSF.

REFERENCES

Cannon, R.D. (1970). Mon. Not. R. Astr. Soc. 150, 111.
McClure, R.D., Forrester, W.T. and Gibson, J. (1974). Astrophys. J. 189, 409.

K-LINE PHOTOMETRY: COMA AND NGC 6475

Richard C. Henry

The Johns Hopkins University

James E. Hesser

Cerro Tololo Inter-American Observatory

ABSTRACT

Photometry of the K line of calcium in A-type stars is useful in segregation of Am stars and in estimating the metal abundance of field stars and of stars in clusters. To date, K-line photometry data for six clusters have been analyzed using uvbyβ data: Hyades, Pleiades, IC 2391, and IC 2602 (Hesser and Henry 1971), Orion (Hesser, McClintock, and Henry 1977), and Praesepe (Henry, Anderson, and Hesser 1977). We present similar analyses for the Coma and NGC 6475 star clusters, using the uvbyβ data of Crawford and Barnes (1969) and of Snowden (1976) respectively. The Coma stars closely resemble field stars. NGC 6475, on the other hand, is most remarkable. Of 31 stars for which complete data are available, 9 are metal-rich, 3 are metal-poor, and 19 (about 2/3 of the stars!) are Am.

REFERENCES

Crawford, D. L., and Barnes, J. V. (1969). Astron. J. 74, 407.
Henry, R. C., Anderson, R., and Hesser, J. E. (1977). Astrophys. J. 214, 742.
Hesser, J. E., and Henry, R. C. (1971). Astrophys. J. Suppl. 202, 453.
Hesser, J. E., McClintock, W., and Henry, R. C. (1977). Astrophys. J. 213, 100.
Snowden, M. S. (1976). Publ. Astron. Soc. Pacific 88, 174.

A. G. Davis Philip and D. S. Hayes (eds.), The HR Diagram, 161–162.

DISCUSSION

SNOWDEN: I would like to comment on my four-color photometry of NGC 6475, which you used. These m_1 indices are probably more accurate than they are precise. In other words, they contain a fair amount of scatter. Also you should know that the observations were conducted differentially with respect to one or two cluster substandards. So, it would be easy to check my zero point simply by reobserving these two stars carefully.

ABT: NGC 6475 is a cluster with an unusually large number of Am stars of the early A type, like Sirius, that have primarily slightly weak K lines. Most of those stars are known spectroscopic binaries.

MEASUREMENTS OF ABUNDANCES AND AGES OF OLD DISK CLUSTERS

Alan Hirshfeld, Robert D. McClure and Bruce A. Twarog

Yale University Observatory

The old disk clusters represent a striking contrast to the globular clusters which will be discussed extensively at this meeting in that the disk clusters show few anomalies. This makes a discussion of their abundance measurements of particular interest at this meeting, and may shed some light on the cause of the peculiar behavior of the halo objects. It is also appropriate to discuss the ages of the clusters since the HR diagram is a major tool for obtaining the ages of stars. A complete sequence of stars of one age in a cluster can be fitted to the ZAMS of theoretical models, and accurate abundances averaged over numerous stars in the cluster can be estimated. We will discuss the ages and abundances of 11 disk clusters for which UBV and DDO photometry are available in the literature. This limits the sample to those for which good color-magnitude (C-M) diagrams exist. In addition we present the C-M diagrams of two clusters that we have studied and that are not yet published elsewhere. We will use these clusters to illustrate our procedures.

NGC 5822 is an intermediate age cluster whose C-M diagram is shown in Fig. 1. We have done iris photometry from 8 plates in each color, and believe that the scatter present near the main sequence is due to binary stars. Notice that the blue edge of the main sequence is well defined.

The color-color diagram for NGC 5822 is shown in Fig. 2 for stars near the principal sequences in the C-M diagram. Crosses represent stars slightly above the main sequence which could be binaries. The reddening, determined from DDO photometry in NGC 5822 and calibrations of Janes (1977a, b), is $E(B-V) = 0\overset{m}{.}12$ for giants, which is equivalent to $0\overset{m}{.}13$ for main sequence stars (see Hartwick and McClure 1972). After dereddening along appropriate

A. G. Davis Philip and D. S. Hayes (eds.), The HR Diagram, 163-166 and 405.

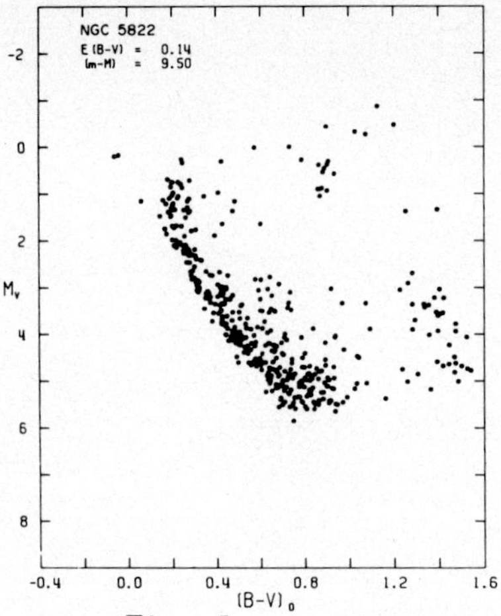

Fig. 1

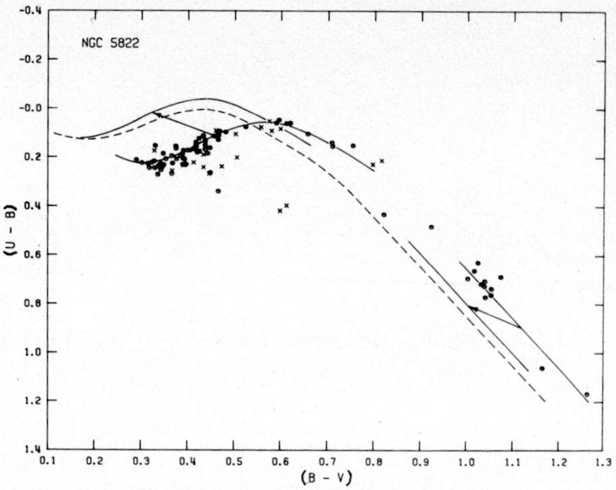

Fig. 2

trajectories (see Crawford and Mandwewala 1976), both giants and dwarfs show an ultraviolet excess of $\delta(U-B) \sim 0^m.05$ with respect to the Hyades sequences (indicated by the dashed curve).

The C-M diagram for NGC 2506 is shown in Fig. 3. We have superimposed a set of isochrones constructed by Ciardullo and Demarque (1977) and converted by them to the observational plane as described by Ciardullo and Demarque (1978). The Hyades main sequence fits models for Z =0.03, Y =0.30, and distance modulus $3^m.30$ (Hanson 1975, Anthony-Twarog and Demarque 1977). The distance modulus for NGC 2506 results, then, from fitting to the theoretical isochrones. The resulting ages for all the open clusters fitted in this way have almost no dependence on helium abundance.

The resultant data for clusters in our sample is shown in Table I. Values of [Fe/H] were obtained by consulting the $\delta(U-B)$ calibrations of Wallerstein (1962), Eggen (1964), Wallerstein and Helfer (1966) and Alexander (1973) and calibrations of CN strength by Janes (1975). For NGC 2243 a UV excess measurement, made by Norris and Hawarden (1977) from DDO photometry, has been averaged with the (U-B) abundance determination. Zero points of [Fe/H] were adjusted to give + 0.2 for the Hyades. The correlation of [Fe/H] values determined by the two methods is shown in Fig. 4, a correlation which is surprisingly tight. A similar correlation for globular clusters would be impossible because of the large fluctuations in CN strengths among stars of a given cluster. The contrast between halo and disk clusters has been pointed out by Hesser et al. (1976). Examination of their Fig. 1 shows large variations in CN strengths for metal-rich globular clusters such as 47 Tuc. Their Fig. 2, on the other hand, shows no scatter (other than observational) among stars in disk clusters, even for clusters like NGC 2420 which appears to be similar in metal abundance to 47 Tuc. (Demarque and McClure 1977).

Metal abundances are plotted versus age for these clusters from

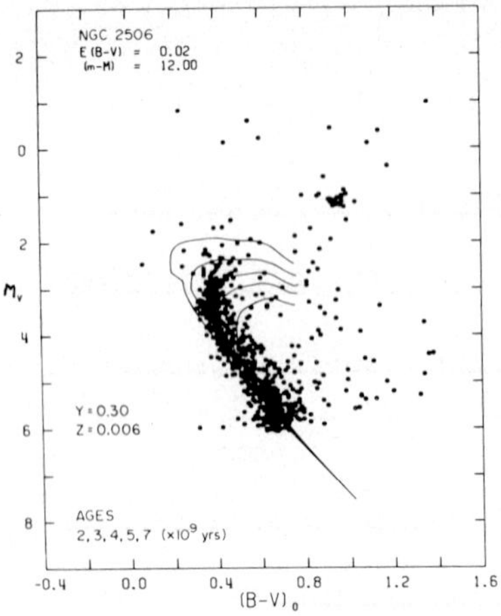

Fig. 3

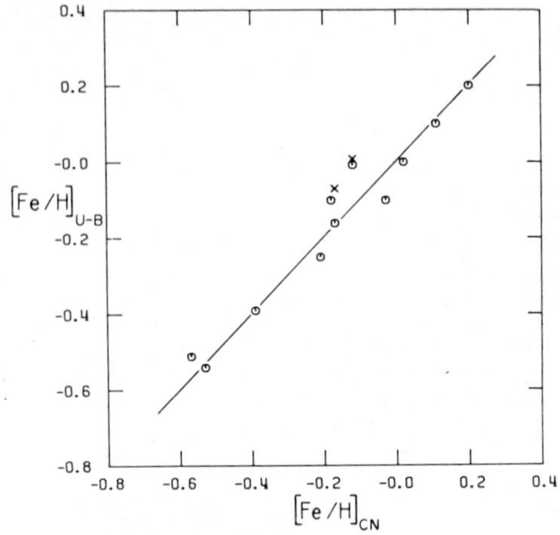

Fig. 4

similar data elsewhere by McClure and Twarog (1977). Differences from their data are small (new observations have been included here for NGC 5822 and NGC 2243).

As suggested by McClure and Twarog (1977) there appears to be no evidence for a variation of abundance with age for disk objects in the last 5-billion years. Rather, the most obvious trend is that between abundance and galactocentric distance (see Janes 1977b). The old galactic clusters which show a significant metal deficiency (NGC 2243, NGC 2420 and NGC 2506) are located 2-4 kpc in the anti-center direction. This trend of decreasing metal abundance with galactocentric distances has been discussed more fully by Janes (1977b) for a much larger sample of clusters. Another result is that at a given age there appears to be a significant spread in metal abundance of about ± 0.3 dex. It is necessary to conclude that material in the disk has not been well mixed.

TABLE I

Cluster	Age(yr)	δ(U-B)	δ CN	$[Fe/H]_{U-B}$	$[Fe/H]_{CN}$
Hyades	0.7×10^9	$0^m.00$	0.07	0.20	0.20
NGC 2477	0.7	0.02	0.05	0.10	0.11
NGC 5822	1.0	0.05	0.02	-0.10	-0.03
NGC 2360	1.3	0.04	-0.01	-0.07	-0.17
NGC 7789	1.6	0.08	-0.02	-0.25	-0.21
NGC 752	1.7	0.05	-0.01	-0.10	-0.18
NGC 3680	1.8	0.03	0.00	0.00	-0.12
M 67	3.2	0.03	0.03	0.00	0.02
NGC 6819	3.5	0.06	-0.01	-0.16	-0.17
NGC 2243	3.9	0.11	-0.10	-0.51	-0.57
NGC 2506	4.0	0.13	-0.09	-0.54	-0.53
NGC 2420	4.0	0.10	-0.06	-0.39	-0.39
NGC 188	5.0	0.03	0.00	0.00	-0.12

This work was supported by NSF Grant Number AST 76-22672.

REFERENCES

Alexander, J.B. and Branch, D. (1973). Mon. Not. R. Astr. Soc. 161, 409.
Anthony-Twarog, B. and Demarque, P. (1977). Astron. and Astrophys. 57, 471.
Ciardullo, R.B. and Demarque, P. (1977). Trans. Yale Univ. Obs. 33.
Ciardullo, R.B. and Demarque, P. (1978). IAU Symp. No. 80, The HR Diagram, A.G.D. Philip and D.S. Hayes, eds., Reidel, Dordrecht, p. 345.
Crawford, D.L. and Mandwewala, N. (1976). Publ. Astr. Soc. Pacific 88, 917.
Demarque, P. (1978). IAU Symposium No. 80, The HR Diagram, A.G.D. Philip and D.S. Hayes, eds., Reidel, Dordrecht.
Demarque, P. and McClure, R.D. (1977). Astrophys. J. 213, 716.
Eggen, O.J. (1964). Astron. J. 69, 570.
Hanson, R.B. (1975). Astron. J. 80, 379.
Hartwick, F.D.A. and McClure, R.D. (1972). Publ. Astr. Soc. Pacific 84, 288.
Hesser, J.E., Hartwick, F.D.A. and McClure, R.D. (1976). Astrophys. J. Letters 207, L113.
Janes, K.A. (1975). Astrophys. J. Suppl. 29, 161.
Janes, K.A. (1977a). Publ. Astr. Soc. Pacific 89, 576.
Janes, K.A. (1977b). Preprint.
McClure, R.D. and Twarog, B.A. (1977). IAU Coll. No. 45, Chemical and Dynamical Evolution in Our Galaxy, in preparation.
Norris, J. and Hawarden, T.G. (1977). Submitted to Astrophys. J.
Wallerstein, G. (1962). Astrophys. J. Suppl. 6, 407.
Wallerstein, G. and Helfer, H.L. (1966). Astron. J. 71, 350.

DISCUSSION

KRAFT: Do these three clusters share the low helium which you found in NGC 2420?

McCLURE: I can't say anything about NGC 2243, but NGC 2506 is almost identical to NGC 2420.

NORRIS: One might argue that the difference in CN properties between the open and globular clusters supports a primordial origin of the abundance anomalies in the latter. The globular clusters, being more massive, may be able to produce a second generation of stars possessing anomalies, while in the open clusters the ejecta from the first generation of supernovae may escape from the system.

SCHWARZSCHILD: Your concluding slide seems to me a beautiful contribution to the basic question of whether age or position in the Galaxy is the key factor determining the heavy element abundance of stars. Originally age was considered the main factor. More recently this one-sided view has been shown to be very incomplete. Now your conclusion indicates that for the last 5 billion yrs. position in the Galaxy is the dominant

CONTINUED ON PAGE 405

EVIDENCE FOR NON-HOMOGENEITY IN THE METAL ABUNDANCES OF STARS IN GLOBULAR CLUSTERS

Robert P. Kraft

Lick Observatory, University of California, Santa Cruz

Evidence for non-homogeneity in the metal abundances of giant stars in various globular clusters is reviewed. The non-homogeneity is of two kinds: (1) differences in the value of $[\frac{A}{H}] \equiv \log (\frac{A}{H})_* - \log (\frac{A}{H})_\odot$ from star to star in the same cluster, for particular values of A, and (2) non-constancy of the cluster mean value of $[\frac{A}{H}]$ as a function of A. Observational evidence for the existence of variations of both kinds is cited, and discussed for four "nucleosynthetic" groups of elements: viz., s-process, Fe-peak, α-process, and CNO. The rival hypotheses of core-envelope mixing and primordial abundance variations are examined in the light of the observations.

This paper will appear in an early volume of Annual Reviews of Astronomy and Astrophysics.

A. G. Davis Philip and D. S. Hayes (eds.), The HR Diagram, 167-168.

DISCUSSION

BELL: Rodgers' calibration of ΔS in terms of [Ca/H] contains some approximations which can be improved and a preliminary calibration by Manduca and me reduces the [Ca/H] value for a given ΔS by about -0.6.

KRAFT: That's good - I'm very happy to hear this.

LLOYD EVANS: Radial velocities in M22 show that stars with strong metal lines and stars lying to the blue of the giant branch (brighter than the Cepheid V 11) are field stars. Discussion of the photometry shows that the width of the giant branch is largely accounted for by observational error.

RENZINI: One brief comment on the CNO abundance determination by direct observations: C and O are primary elements while N is mostly secondary. Just for this reason one may expect quite large fluctuations of the nitrogen abundance. Therefore, the really important elements to observe for getting the CNO abundance are C and O, even if this is a difficult job.

RUBIN: When do you foresee that there will be enough data to examine the metal variation in globular clusters as a function of dynamical and geographical properties?

KRAFT: I suppose a few years - one really wants to know CNO as well as the Fe-peak, and this may take some time, especially if we want reliable values.

A COMPARISON BETWEEN THE RELATIVE HEAVY METAL ABUNDANCE OF THE EXTREME HALO FIELD STARS AND GIANT STARS OF GLOBULAR CLUSTERS

M. Spite and F. Spite

Paris-Meudon Observatory

Until very recently, estimates of the chemical composition at early epochs of the Galaxy were made through the analysis of halo field stars. Recently, however some medium dispersion studies of stars in globular clusters have appeared. It would be expected that both approaches would lead to identical results. However, a quick glance at the literature gives the opposite impression.

According to Dickens and Bell (1976) and Mallia (1977) the s process elements (barium, strontium) are, in some stars of some globular clusters, strongly enhanced (by a factor of more than thirty). This is attributed to mixing effects which occur at late stages of stellar evolution. Up to now, no barium line enhancement has been reported in Pop II field stars in spite of the fact that some of these stars have high luminosity (Fig. 1). In fact the enhancement of Ba has been observed in very few globular cluster stars, and it can be expected that the discrepancy will be removed when more Pop II field stars of high luminosity are observed.

A second point is that no over-deficiency of the same s elements has been reported in globular clusters. This over-deficiency, first discovered (but overestimated) by Wallerstein et al. (1963), has been reported several times in Pop II stars, but not in all of them (Koelbloed 1967). The barium over-deficiency, when it exists, is interpreted as a characteristic of the gas from which the star was made so it can give valuable information about the metal enrichment in the early stages of the Galaxy. We have investigated the occurrence of the barium over-deficiency among the Pop II stars by observing three new extreme halo field stars noted by Bond (1970) and Bidelman and MacConnell (1973). The analysis of these stars

A. G. Davis Philip and D. S. Hayes (eds.), The HR Diagram, 169–172. All Rights Reserved.

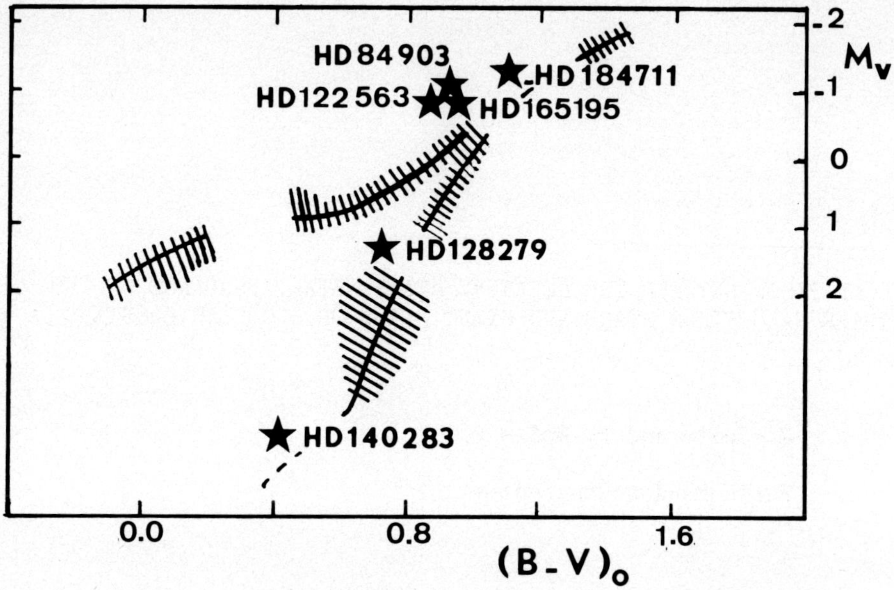

Fig. 1. Position of the analyzed field stars in the HR diagram, in relation with the NGC 6656 diagram.

shows that when the temperature is deduced from the (R-I) color index the overall metal abundance [M/H] is less than -2.3. In Fig. 2 the variation of the barium overdeficiency [Ba/Fe] is plotted as a function of [Fe/H] for these stars and for all the metal deficient stars selected in the literature with the condition [Fe/H] <-1.3. The overdeficiency of Ba appears only for [Fe/H] <-1.5, where [Ba/Fe] becomes a linear function of [Fe/H]. It is then easy to understand why the overdeficiency of Ba has not yet been observed in the globular clusters. The overall metal abundance of ω Cen is estimated as - 1.3 (Dickens and Bell 1976) or - 1.7 (Mallia 1977); in this case, an overdeficiency of Ba is unlikely. NGC 6397, with [M/H] = -2 (Mallia 1977) is the most deficient globular cluster studied and Fig. 2 predicts, in this case, an overdeficiency of Ba by a factor a little larger than two, which would imply a decrease of less than thirty percent of the equivalent width of the 4554 Å line of barium. Such a small decrease seems well below the observational errors.

It would be very interesting to observe in detail a few stars in globular clusters with very low iron content in order to check if the barium overdeficiency appears as would be expected if the halo field stars like the globular clusters, are remnants of the earliest phases of Galaxy formation.

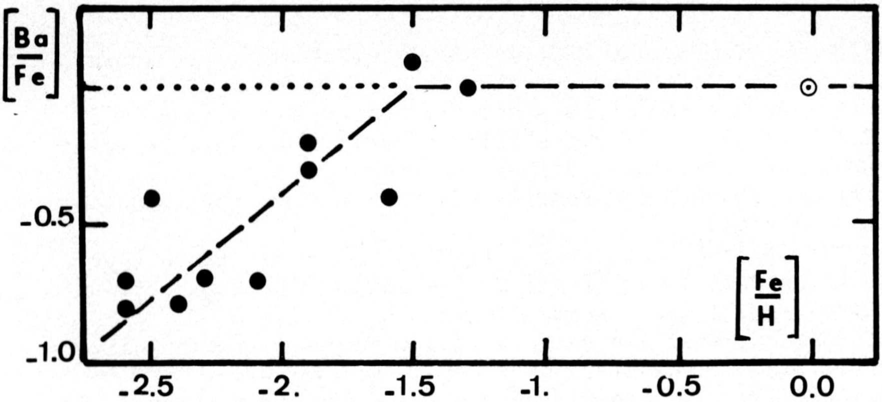

Fig. 2. The overdeficiency of Ba is a function of iron abundance and vanishes at about [Fe/H] = -1.5.

REFERENCES

Bidelman, W.P. and MacConnell, D.J. (1973). Astron. J. 78, 687.
Bond, H.E. (1970). Astrophys. J. Suppl. 117, 22.
Dickens, R.J. and Bell, R.A. (1976). Astrophys. J. 207, 506.
Koelbloed, D. (1967). Astrophys. J. 149, 299.
Mallia, E.A. (1977). Astron. Astrophys. 60, 195.
Spite, M. and Spite, F. (1977). submitted to Astron. Astrophys.
Wallerstein, G., Greenstein, J.L., Parker, R., Helfer, H.L. and Aller, L.H. (1963). Astrophys. J. 137, 280.

DISCUSSION

CAYREL de STROBEL (TO KRAFT): How many clusters are known having a metal deficiency greater than [Fe/H] = -2.5?

KRAFT: As far as I know, there is nothing worse than -2.2.

SPITE: At -2.2, it is possible to measure the deficiency of barium if we have a high dispersion spectrum.

CAYREL de STROBEL (TO KRAFT): What is the distance modulus of that cluster?

KRAFT: It is M92, for which (m-M) = $14^{m}.2$. Are there any cases known with lower [Fe/H] in the southern hemisphere?

NORRIS: No, I don't know of any.

DICKENS: Several years ago we did a curve-of-growth analysis of one of the bright stars in NGC 6397 and got -2.0.

WALLERSTEIN: If I recall correctly Helfer, Greenstein, and I in 1959 found a substantial (about 0.5 in the log) deficiency of barium and other heavy s-process elements in a star in M92 for which we found [Fe/H] to be about -2.3. This was based on a curve-of-growth analysis of 18 Å/mm blue spectrograms.

SPITE: This is a reason to think that the barium abundance in globular clusters has the same behavior that in the field stars. More observations would be needed to reach a definite conclusion.

GREENSTEIN: I do not believe that the older analyses are capable of determining a metal deficiency to a factor of two. The modern techniques of model atmospheres and synthetic spectra are needed. For resonance lines of an ion the boundary temperature is decisive.

SPITE: The abundances presented here for all the field stars are determined through the use of model atmospheres. More details will be found in the paper submitted to Astronomy and Astrophysics.

KRAFT: There's also this question of star counts. ω Cen is a very rich cluster. If you take the ratio of barium stars to the total number of giants in ω Cen and apply it to a metal-poor cluster, would you expect to see any barium stars?

DICKENS: I don't know. Perhaps Norris could comment.

NORRIS: I wouldn't like to comment at this point. Zinn and I are completing a survey of CN and CH stars in ω Cen. The Ba II anomaly appears to exist in many stars, but I can't give a percentage.

OBSERVATIONS OF CO AND OI IN STARS IN GLOBULAR CLUSTERS

George Wallerstein and Catherine Pilachowski

University of Washington, Seattle

During the past decade it has been evident that the morphologies of globular cluster color-magnitude diagrams cannot be described by only two parameters, age and metal abundance (Sandage and Wildey 1967, Hartwick 1968). The third parameter is unlikely to be helium because a normal helium abundance is needed to explain the pulsation of RR Lyrae stars (Deupree 1977) and because the helium abundance of the planetary nebula in the metal-poor cluster M15 is normal (O'Dell et al. 1964).

Since studies at classification dispersion and early analyses of high dispersion spectra have yielded little quantitative data on the abundances of C, N, and O in globular clusters we have been endeavoring to establish their abundances in stars in several clusters. We have approached the problem in two ways, by observing the 2.3 micron CO bands and the 6300 Å [OI] line in individual stars in globular clusters.

Observations of CO bands have been carried out by one of us (C.P.) using a photometric system very similar to that of Baldwin et al. (1973). Data have been obtained on the KPNO 2.1 and 4 meter telescopes for 30 stars in 9 clusters with emphasis on M71, M10, M13, M5 and M3. These clusters were selected because M71 has a high metal content while the others show contrasting color-magnitude diagrams. M10 and M13 have extremely blue horizontal branches while M5 and M3 have a mixture of red and blue horizontal-branch stars. The four clusters in question all have [Fe/H] values near -1.4. The results are shown in Fig. 1 where we plot the CO index against (V-K). The continuous line shows the index for standard

A. G. Davis Philip and D. S. Hayes (eds.), The HR Diagram, 173-176.

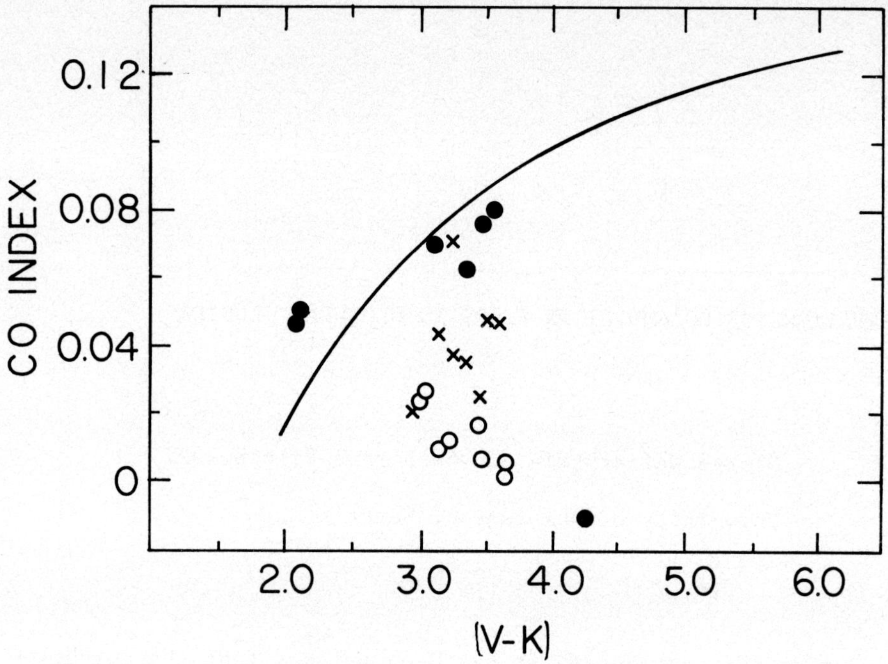

Fig. 1. CO index *versus* (V-K) color index.

---

stars. The stars in the cluster M71 appear to lie on the line for standard stars. There is a marked difference between stars in M3 and M5 as compared to stars in M10 and M13. The former deviate from the standard line by 0.035 mag while the latter deviate by 0.070 mag. This indicates a higher CO concentration and hence a higher abundance of carbon or oxygen (or both) in M3 and M5 as compared with M10 and M13. To be quantitative we must first determine the abundance of free oxygen in these stars.

From 4m. echelle spectra of one star in M22 and two stars each in M5 and M13 we have derived the ratio of OI to the metals by a curve-of-growth analysis. Our results are of a very preliminary nature for the following reasons. The analysis is by a curve-of-growth for OI as compared with singly ionized iron, titanium, and scandium. Since oxygen is almost entirely neutral and the other three elements are almost entirely ionized, no ionization corrections need be made. No account of association of oxygen in CO is taken since carbon is normally 1/3 of oxygen and hence CO does not greatly deplete the supply of OI. The temperatures were obtained from the colors. Improved temperatures from line intensities will be obtained later and our final analysis will be done by model atmospheres.

TABLE I

DERIVED PARAMETERS FOR THREE GLOBULAR CLUSTERS

Cluster	Star	$\Delta\theta_{eff}$	$-\log W/\lambda$ (OI)	[O/m]
M5	I-68	0.18	4.83	+0.8
	IV-81	0.24	5.12	+0.25
M13	II-67	0.24	5.30	-0.1
	III-56	0.20	5.11	+0.2
M22	IV-102	0.16	5.23	+0.55

Our comparison has been done with the Hyades as standard stars (Helfer and Wallerstein 1968). The results are shown in Table I.

From the table we find that [O/m] is +0.5 for M5 and virtually zero for M13. Hence we see that OI as well as CO is enhanced in M5. The enhancement of CO indicates that carbon is also enhanced since adding O without C will not enhance CO very much. We plan to carry out the necessary model atmospheres, etc. to predict CO enhancement as a function of C and O abundance and the other relevant parameters; but we can conclude from the data available that oxygen and probably carbon are enhanced in M5 relative to M13.

It therefore appears that somehow M5 has retained a higher ratio of helium burning products to metals than has M13. Furthermore the increase in oxygen shows that helium burning occurred in a high temperature, low density environment so as to produce oxygen as well as carbon. This means that the helium burning must have taken place in very massive stars. The lifetimes of stars in the range of 10-100 $M_\odot$ are so short that mass lost from their surfaces during their evolution could easily have been collected by the stars near 0.8 $M_\odot$ that are presently red giants while they were still contracting and hence had a large geometrical cross section to capture intracluster gas.

If the metals and CNO now present in the red giants of globular clusters were formed prior to star formation in the cluster, there is almost no way to speculate as to their mode of formation.

The observations reported upon in this paper were made at the Kitt Peak National Observatory, which is operated under contract with the National Science Foundation. This research was supported by grants from the National Science Foundation. We thank Mr. Bernard Bates for measuring and plotting the equivalent widths and curves-of-growth.

REFERENCES

Baldwin, J.R., Frogel, J.A. and Persson, S.E. (1973). Astrophys. J. 184, 427.
Deupree, R.G. (1977). Astrophys. J. 214, 502.
Hartwick, F.D.A. (1968). Astrophys. J. 154, 475.
Helfer, H.L. and Wallerstein, G. (1968). Astrophys. J. Suppl. 16, 1.
O'Dell, C.R., Peimbert, M. and Kinman, T.D. (1964). Astrophys. J. 140, 119.
Sandage, A.R. and Wildey, R.L. (1967). Astrophys. J. 150, 469.

DISCUSSION

BUTLER: For all of the clusters studied, what does the plot of [Fe/H] vs. [O/H] look like?

WALLERSTEIN: We really don't have enough data to say anything about this.

BUTLER: I just want to comment that a group of us are in the process of analyzing oxygen-line (7774 Å) strengths in a sample of field RR Lyrae stars known to have a large range in [Fe/H]. We do not find the scatter which appears to be present in the [Fe/H], [O/H] diagrams for clusters.

WALLERSTEIN: In M13 the horizontal branch is so blue that there are only 2 or 3 RR Lyrae stars. It may be that in the general field as well as in M13 there are so few RR Lyrae stars with low [CNO/Fe] that you haven't observed any.

ZINN: I would like to remark that Leonard Searle and I have made a survey of 19 globular clusters for metal abundance. We find that M5 is more metal-rich than M13 by approximately 0.4 dex. Perhaps this difference can explain the disagreement between the results discussed by Wallerstein and Butler.

CASTELLANI: I agree that nobody, at present, believes in globular cluster stars without helium. Nevertheless, I wish to point out that a variation in the original helium as low as $\Delta Y \sim 0.03 - 0.06$ can produce strong consequences in the expected morphology of HR diagrams. So, even if we accept "regular" helium in globular clusters ($Y \gtrsim 0.22$), one cannot exclude the occurrence of such kind of variations in determining the observed "peculiarities".

KRAFT: Did you find any variation in [O/Fe] for the stars in a given cluster?

WALLERSTEIN: Not a significant difference. We have only two stars in each cluster.

THE CHEMICAL ABUNDANCES OF THE HALO CLUSTERS OF THE GALAXY

R. Canterna and R. A. Schommer

University of Washington

1. INTRODUCTION

Photometric metal abundances of individual red giants in eight extremely distant halo globular clusters and the Draco and Ursa Minor dwarf spheroidal galaxies have been obtained using the Washington broad-band system, C, M, T_1, T_2 (Canterna 1976). Observations were made at the KPNO 2.1-m and CTIO 1.5-m telescopes. In Table I we list for each system the mean value of [Fe/H], the number of stars observed in each system, n, the Galactocentric distance, R_{GC}, the intrinsic color of the giant branch at the level of the horizontal branch (HB), $(B-V)_{o,g}$, and the fraction of HB stars bluer than the RR Lyrae gap, f_B. Sources for unpublished color-magnitude diagram (CMD) data are: Pal 11 (Canterna and Schommer), Pal 12 (Canterna and Harris), and Ursa Minor (Schommer, Olszewski and Kunkel).

TABLE I

System	[Fe/H]	n	R_{GC}	$(B-V)_{o,g}$	f_B
Draco	-2.3	3	65 kpc	$0^m.95$	0.07
Ursa Minor	-2.1	3	70	0.70	0.75
Pal 3	-2.3	2	100		
Pal 4	-2.4	3	95	0.81	0.00
Pal 11	-0.7	4	7		
Pal 12	-1.6	3	18	0.93	0.00
Pal 13	-1.9	3	25	0.90	0.00
NGC 2419	-2.2	4	100	0.70	0.75
NGC 5053	-2.4	4	15	0.67	0.70
NGC 7006	-2.0	4	32	0.72	0.16

A. G. Davis Philip and D. S. Hayes (eds.), The HR Diagram, 177–182.

2. THE HEAVY-ELEMENT ABUNDANCE OF THE GALACTIC HALO

In Fig. 1 we present a plot of [Fe/H] vs. R_{GC} for each system listed in Table I and M92 and M71. No appreciable [Fe/H] gradient beyond 20 kpc is shown in Fig. 1. A mean value of [Fe/H] = -2.2 ± 0.3 for the outer halo satellites is found. The most distant galactic halo systems have nearly the same heavy-element abundances as the nearby globular clusters M92 and M15. This result is consistent with the low-dispersion spectroscopic survey of red giants in the outer halo systems by Cowley, Hartwick, and Sargent (1977; hereafter CHS), who find no gross differences between the spectral features of these giants and those in M92 and M15; none were found to be more metal-deficient that the M15 giants.

A few field halo-population stars are known that are somewhat more metal-deficient than the stars in the halo clusters (Wallerstein et al. 1963; Sneden 1974; Bond 1977), but such stars are extremely rare. No galactic stars are known with [Fe/H] $\lesssim -3.0$.

We believe that these observations imply one of the following conclusions: (a) the primordial material from which the first stars formed in the proto-Galaxy had [Fe/H] ≈ -2.2 to -3.0, in contradiction to current theories of big-bang nucleosynthesis; (b) the primordial material had [Fe/H] $<< -3.0$, but old stars, especially those in clusters, no longer reflect the primordial composition because of mixing, accretion, or other processes that alter the stellar surface composition; or (c) low-mass stars cannot form from primoridal material because a certain minimum heavy-element content is required.

3. THE ANOMALOUS COLOR-MAGNITUDE DIAGRAMS OF THE OUTER HALO SYSTEMS

Since the heavy-element abundances of the outer halo systems are similar to those in M92 and M15, it is instructive to compare correlations of [Fe/H] with established CMD parameters, such as HB type, S, ΔV, and $(B-V)_{o,g}$, between the nearby globular clusters and the outer halo systems. With the exception of NGC 2419 (Racine and Harris 1975) and Ursa Minor (Schommer et al.) the outer halo satellites have one common property, a highly populated red HB that is usually seen in metal-rich clusters. This anomaly was first pointed out by Sandage and Wildey (1967) from their investigation of NGC 7006. Rood (1973) has shown that the anomaly can be explained by a lower helium abundance, a higher CNO abundance, or a younger age for the anomalous clusters.

In Fig. 2 we present a second related CMD anomaly for these systems. Butler (1975) has shown that $(B-V)_{o,g}$ is strongly correlated with [Fe/H] for the nearby globular clusters. In Fig. 2

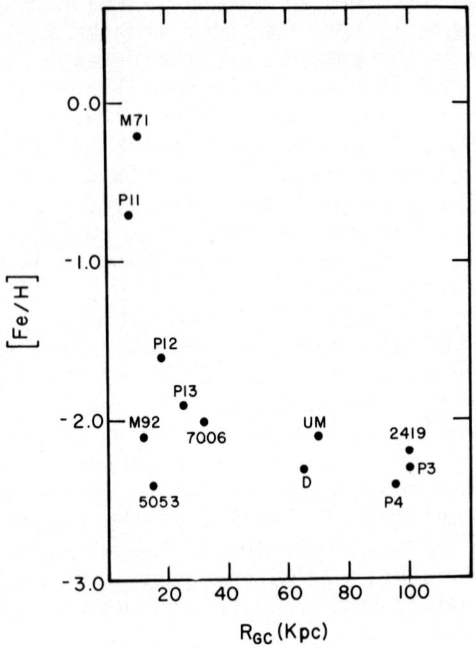

Fig. 1. [Fe/H] versus R_{GC}

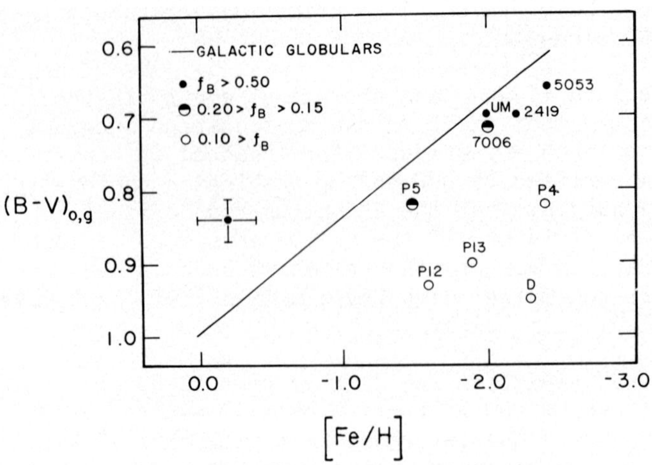

Fig. 2. $(B-V)_{o,g}$ versus [Fe/H]

the solid line shows this relation for 14 well studied galactic globular clusters with R_{GC} < 15 kpc. We also present the data for the extremely distant halo systems, separating each system according to the fraction of HB stars bluer than the RR Lyrae gap. The results for Pal 5 (Sandage and Hartwick 1977), have also been included in Fig. 2. The subgiant branch colors of all extremely distant halo systems with no blue HB stars of $f_B \leq 0.07$ (Draco) are too red for their respective heavy-element abundance by $0\overset{m}{.}10$ or more in (B-V). This anomaly cannot be explained by large errors in the reddening estimates since these systems all lie at high galactic latitudes.

Knowledge of the He or CNO abundance parameters and the age of each system may provide an insight into which of these parameters is responsible for the CMD anomalies. It has generally been assumed when interpreting observations of globular cluster giants that any photometric or spectroscopic evidence for strong CN or CH bands is due to an enhancement of CNO/Fe in the entire system (Hartwick and McClure 1972). In view of the recent evidence for strong CN bands due to mixing in the atmosphere of giants in ω Cen (Dickens and Bell 1976) one should observe caution with this interpretation until high-dispersion spectroscopic data are available.

NGC 7006 does not show the red subgiant branch anomaly. Although Hartwick and McClure (1972) have shown evidence for enhanced CN bands in some giants in NGC 7006, which may explain the anomalous red HB, the CHS spectra and Washington photometry (Canterna and Schommer 1977) reveal that not all giants in 7006 exhibit strong CN bands. The strong CN stars in 7006 may be due to mixing in the atmosphere and not to a primoridal abundance difference in the CNO elements.

Hartwick and McClure (1974) have shown that giants in the Draco galaxy, the system that exhibits the most extreme subgiant branch anomaly, do not show any evidence for enhanced CN bands. This result has been verified by CHS and by Washington photometry. Hartwick and McClure suggest that the anomalous red horizontal branch of Draco can be explained if the helium abundance is lower than normal or if Draco is younger than clusters such as M92 or M15; both results are consistent with Castellani's (1975) analysis of the RR Lyrae variables in Draco.

On the other hand, giants in Pal 12 have been shown to possess extremely strong CH and CN bands from the results of CHS and Washington photometry. The intermediate metallicity of this cluster, [Fe/H] = -1.6, has been verified from the δ(U-B) excess of the two brightest giants. Canterna and Harris (Priv. Comm.) have concluded that a primordial abundance difference in the CNO elements is the only explanation for the anomalous red HB and subgiant branch of Pal 12. The age of Pal 12 is 14 ± 3 billion

yrs., similar to that of M92 and M3.

There appears to be no one, common parameter that can completely explain the anomalous red HB and subgiant branch anomalies in the outer halo systems. We conclude that each halo system must be treated independently with regard to their age, He and CNO abundances. Separation of the progenitors of the Fe-peak, He, and CNO elements in the evolution of the proto-Galaxy and the collapse-enrichment scenarios are more complex than a simple free-fall collapse model. Although the observational evidence regarding the [Fe/H] abundances of the galactic halo can be explained by a one zone, free-fall model (Hartwick 1976), this simple enrichment-collapse scenario may not be coupled with the progenitors of the CNO or He elements, and cannot account for different epochs of star formation. Enrichment in isolated fragments, as proposed by Searle (1977), may account for these local variations.

REFERENCES

Bond, H.E. (1977). Private communication.
Butler, D. (1975). Publ. Astr. Soc. Pacific **87**, 559.
Canterna, R. (1976). Astron. J. **81**, 228.
Canterna, R. and Schommer, R.A. (1977). Astrophys. J. in press.
Castellani, V. (1975). Mon. Not. R. Astron. Soc. **172**, 59.
Cowley, A.P., Hartwick, F.D.A. and Sargent, W.L.W. (1977). preprint.
Dickens, R.J. and Bell, R.A. (1976). Astrophys. J. **207**, 506.
Harris, W.E. (1976). Astron. J. **81**, 1095.
Hartwick, F.D.A. (1976). Astrophys. J. **209**, 418.
Hartwick, F.D.A. and McClure, R.D. (1972). Astrophys J. **176**, L57.
Hartwick, F.D.A. and McClure, R.D. (1974). Astrophys. J. **193**, 321.
Racine, R. and Harris, W.E. (1975). Astrophys. J. **196**, 413.
Rood, R. (1973). Astrophys. J. **184**, 841.
Sandage, A.R. and Wildey, R. (1967). Astrophys. J. **150**, 469.
Searle, L. (1977). In The Evolution of Galaxies and Stellar Populations. B.M. Tinsley and R.D. Larson, eds., Yale Univ. Obs., New Haven, p. 219.
Sneden, C. (1974). Astrophys. J. **189**, 493.
Wallerstein, G., Greenstein, J.L., Parker, R., Helfer, H.L. and Aller, L.H. (1963). Astrophys. J. **137**, 280.

DISCUSSION

NORRIS: I would like to note that Bessell has found a southern giant, CD-38°425, in which [Ca/H]∿-3.5. Secondly I'd suggest that no present photometric system can handle abundances below [Fe/H]∿-2.5. So even if systems having such abundances existed in your sample you wouldn't necessarily know about it.

CANTERNA: I agree. The Washington system breaks down around -3, and we really can't differentiate between -3 and -4. However, in the Cowley, Hartwick and Sargent spectra, none of the giants had H and K or iron lines that were weaker than the M15 giants. So, it appears that none of the systems they've observed are more metal-poor than M15 or M92 which is in agreement with our results. I agree that if there are more metal-poor objects which we haven't observed, spectra must be used to distinguish between -3 and -3.5.

NORRIS: There is, however, growing evidence that the dwarf spheroidal galaxies contain a large range in abundance. In the Fornax system, in particular, there is apparently a range in the abundances of the globular clusters as shown by van den Bergh. According to Danziger et al. this system also contains a planetary nebula with a horrendously high metal abundance.

CANTERNA: I agree. Washington photometry of globular clusters verifies van den Bergh's result.

KRAFT: I suggest that you have to watch out for membership in discussing the metal abundance of individual stars in dwarf ellipticals.

ZINN: I would like to make another remark about the globular cluster survey made by Leonard Searle and myself. We have surveyed 19 clusters and the most distant of which is NGC 7006. We find that there is no evidence for a metal abundance gradient from 8 to 20 kpc from the galactic center. This agrees with your results. However, we find that the upper envelope of the plot of [Fe/H] vs. galactic radius is approximately [Fe/H] = -1.2, which is a signficantly higher metal abundance than your results indicate.

CANTERNA: I may agree with that, but our results pertain to systems beyond 20 - 30 kpc. We also get some clusters with a high metallicity around 15 - 20 kpc, such as Pal 12.

ZINN: Our results for NGC 7006, for example, are considerably different from yours.

CANTERNA: That may be true, but our results are in agreement with the spectra of giants in NGC 7006 from Cowley, Hartwick and Sargent.

A COMBINED PHOTOMETRIC/SPECTROSCOPIC STUDY OF RR LYRAE STARS IN THE GLOBULAR CLUSTER OMEGA CENTAURI

Dennis Butler* and Roger A. Bell

University of Maryland

R. J. Dickens and Elizabeth Epps

Royal Greenwich Observatory

1. INTRODUCTION

Most CM diagrams of well observed galactic globular clusters have giant branch widths which are understandable in terms of realistic estimates of random (and systematic) error associated with data acquisition and reduction. Not Omega Centauri. The scatter in (B-V) color among Omega Centauri giant branch stars is considerably greater than that allowed by observational error, differential reddening, or star-to-star mass and/or helium abundance variation (Demarque and Geisler 1963; Woolley 1966; Dickens and Woolley 1967; Rood 1973; Cannon and Stobie 1973; Cannon and Kontizas 1974; Iben 1974; Norris and Bessell 1975; Hesser et al. 1976; 1977). This problem is an old one, but one which has recently received a flurry of attention. Indeed, our next speakers have investigated possible causal relationships between interior mixing and Omega Centauri giant (or subgiant) color. I see from the program that they intend to discuss such things today, so we shall try to avoid those topics as much as possible.

Instead, we wish to concentrate on an analysis of spectroscopic and photometric data for the cluster's RR Lyrae stars. Working within the formalism defined by some of the latest stellar atmosphere, stellar evolution, and pulsation calculations, we find that we are able to make several interesting deductions and

---

*Visiting Astronomer, Cerro Tololo Inter-American Observatory, supported by the National Science Foundation under contract No. NSF-C 866. Currently at Yale University.

A. G. Davis Philip and D. S. Hayes (eds.), The HR Diagram, 183-194. All Rights Reserved.

inferences concerning (1) the chemical composition dependence of RR Lyrae absolute magnitudes, and (2) the chemical composition dependence of total mass loss during earlier evolutionary states. Almost all the material we shall report and discuss today is new, in the sense that it is drawn from papers (either in press, or in preparation) by various permutations of Butler, Bell, Dickens and Epps.

1.1 Rationale and Spectroscopic Method

In Fig. 1 we superimpose a series of giant branch loci for the well observed globular clusters M92 (Sandage and Walker 1966), M5 (Arp 1962), and 47 Tuc (Hartwick and Hesser 1974) with a schematic representation of the Dickens and Woolley (1967) Omega Centauri CM diagram. One major parameter which is unquestionably different for each of the first three clusters is [Fe/H] = log (Fe/H)* - log $(Fe/H)_{\odot}$. According to Hesser et al. (1977), [Fe/H] = -2.2, -1.0 and -0.4, respectively, for M92, M5 and 47 Tuc. If we suppose that either [Fe/H] or something predictable from it dominates giant-branch color, then we can deduce there might well exist a large range ($-0.8 \gtrsim$ [Fe/H] $\gtrsim -1.7$) in chemical composition among members of Omega Centauri, including the RR Lyraes.

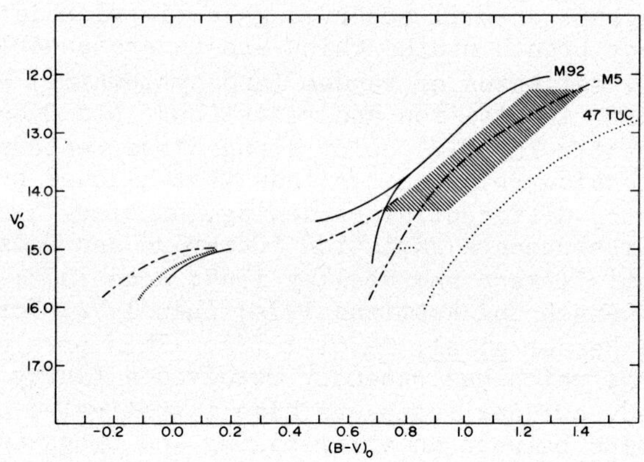

Fig. 1. Superposition of mean giant-branch loci for the well observed globular clusters M92 ([Fe/H] = -2.2), M5 ([Fe/H] = -1.0) and 47 Tuc ([Fe/H] = -0.4) with a schematic representation of the Omega Cenaturi giant branch. Intrinsic giant-branch width was obtained from a quadratic combination of observed width (0.24) and an estimate of observational scatter of $\pm$ 0$^m_.$07 in (B-V). It is important to warn the reader that different investigators find quite different values of giant-branch width. For example, Cannon and Stobie (1973) give a lower limit of 0.3, while Bessell and Norris (1976) prefer 0.4.

Although Omega Centauri is not very rich in RR Lyrae stars when things are normalized in terms of the number of red-giant progenitors, or in terms of total cluster luminosity, it nevertheless contains nearly 150 of them (with the cluster's great mass partly responsible for such a large absolute number).

In the visual region, low dispersion spectra of RR Lyrae stars are dominated by strong absorption lines of hydrogen (Balmer series) and the H and K-lines of singly ionized calcium. In the temperature domain occupied by RR Lyrae stars, hydrogen-line strength is extremely (and almost entirely) temperature dependent whereas K-line strength is sensitive to both temperature and calcium abundance. One can use measurements of hydrogen-line strength to allow for the K-line's temperature dependence, thus clearing the way for an abundance determination. This is the essence of the Preston ΔS method (Preston 1959; Butler 1975). In practice, one assigns to each RR Lyrae star a spectral classification based on hydrogen-line strength Sp(H), and one based on K-line strength Sp(K). ΔS is simply the difference in tenths of a spectral class, between Sp(H) and Sp(K). It is a parameter having values ranging from ~ 0 to 10, and internally calibrated through high dispersion (8-16 Å/mm) curve-of-growth abundance analyses of ~ 40 spectra of 13 bright, field RR Lyrae stars. The fundamental calibration is reproduced in Fig. 2. The great power, and hence great appeal of the technique, is obvious from the figure (see caption).

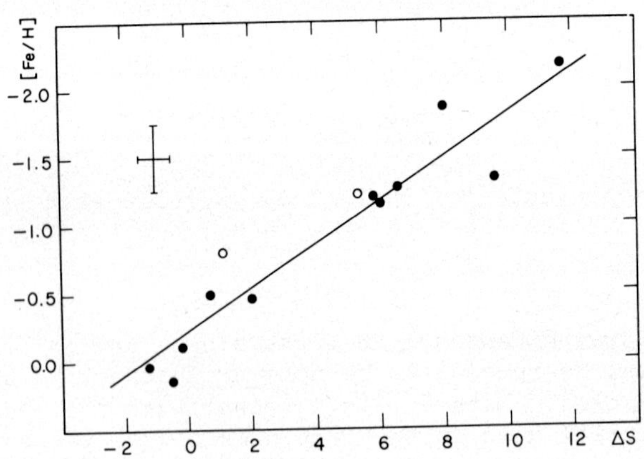

Fig. 2. A single ΔS measurement of precision ±1 leads to an [Fe/H] value of precision comparable to that obtainable from a high dispersion curve-of-growth analysis. Filled circles, ab-type RR Lyraes; open circles, c-types.

2. DATA AND DISCUSSION

Several years ago, we began a major observing program in which (1) ΔS measurements (and hence [Fe/H] determinations) would be made for a large number of RR Lyrae stars in Omega Centauri, and (2) new

photographic photometry would be obtained for the purpose of analyzing pulsation characteristics. Today, leaving all details regarding observing technique, data reduction procedure, etc., to the publications already mentioned, we report on what has come of it.

Our spectroscopic data are displayed in Fig. 3. Many data points represent mean values from 3 or 4 observations, with individual observations having a maximum weight W=5. The two underlined symbols represent variables which can be called "BL Her stars", i.e., variables having P ~ 1-3 days, and magnitudes ~ 1 mag brighter than the RR Lyraes. A striking characteristic of the figure is that the scatter in [Fe/H] is considerably greater than that allowed by error estimates. Also, the c-types as a group are systematically more metal deficient than the ab-types. Freeman and Rodgers (1975) have used a slight variation of the ΔS method in determining calcium

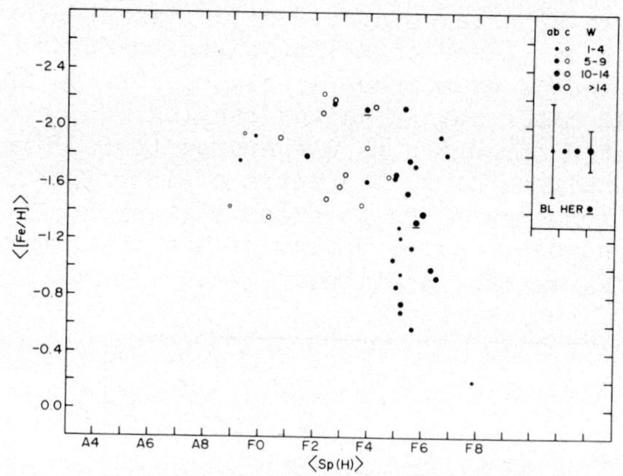

Fig. 3. Spectroscopic data for individual variable stars in Omega Centauri: filled circles, ab-type RR Lyraes; open circles, c-types; underlined circles, BL Her stars.

abundances for 25 Omega Centauri RR Lyrae stars, and in Fig. 4, we present their data, re-reduced to be more or less on the ΔS system, and combined with ours. Data in Fig. 4 constitute the fundamental spectroscopic material used in the remainder of the present report.

In Fig. 5, we present a montage of spectra showing the extreme star-to-star variation in K-line strength. All stars were observed at a phase near minimum light, so they all have roughly similar temperatures, as is obvious from the hydrogen-line strengths. These spectra were obtained just a few months ago, and have not been included in Figs. 3 or 4. Significantly, when the spectra are ordered by K-line strength, they are also ordered by [Fe/H] from previously obtained data. From data presented in Figs. 3, 4 and 5, and from the lack of scatter in Fig. 2, we are confident that there exists a large range in [Fe/H] among individual RR Lyrae stars in

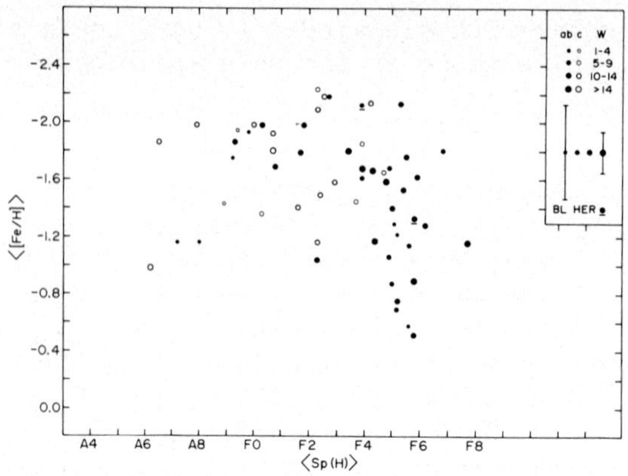

Fig. 4. Freeman and Rodgers' (1975) spectroscopic data combined with ours.

Omega Centauri. If the giants share the same range in [Fe/H] as the RR Lyraes, namely Δ [Fe/H] ~ 1, then much, if not all, of the great giant branch width is explained quite naturally.

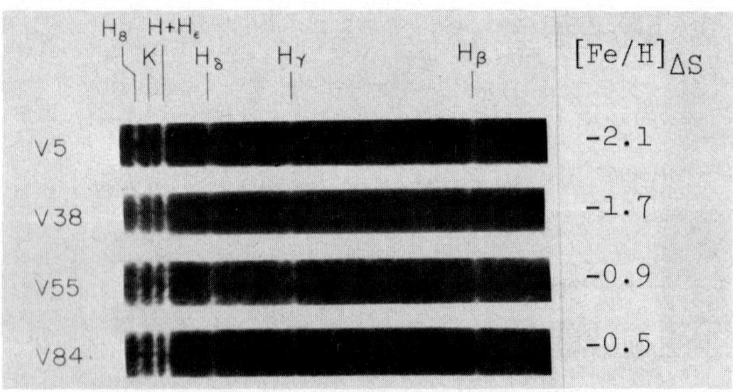

Fig. 5. Reproduction of spectra of 4 Omega Centauri RR Lyrae stars showing an enormous range in K-line strength. The spectra are widened to ~ twice the value typical for an observation of weight 5.

2.1 Oosterhoff Characteristics

There exists yet another line of evidence (although less direct than the spectroscopy) supporting a large abundance range. If we sort data in Fig. 4 into two bins: one for stars having [Fe/H] > -1, and the other, for [Fe/H] < -1 (hereafter, "metal-rich", and "metal-poor" bins, respectively), then we can compare Oosterhoff characteristics of stars in each bin with those of stars in other globular clusters. We made the comparison in Table I, in which a summary of Oosterhoff characteristics for other globular clusters has been taken from van Albada and Baker (1972). Note

that stars in the metal-rich bin appear to have Oosterhoff group I characteristics, while those in the metal-poor bin, group II.

TABLE I

OMEGA CENTAURI CHARACTERISTICS

Bin:	Metal-Rich	Metal-Poor
n	7	49
<P(ab)>	0.57	0.49
<P(c)>	0.31	0.37
n(c/ab)	0.17	0.63

GLOBULAR CLUSTER CHARACTERISTICS

Group:	I	II
<P(ab)>	0.55	0.65
<P(c)>	0.32	0.37
n(c/ab)	0.2	0.8
metals	weak	very weak

The period-frequency diagram for cluster RR Lyraes (Fig. 6) shows the same thing; stars in the metal-rich bin favor an M5-like ([Fe/H] = -1) distribution, while those in the metal-poor group M15, or M92 ([Fe/H]= -2).

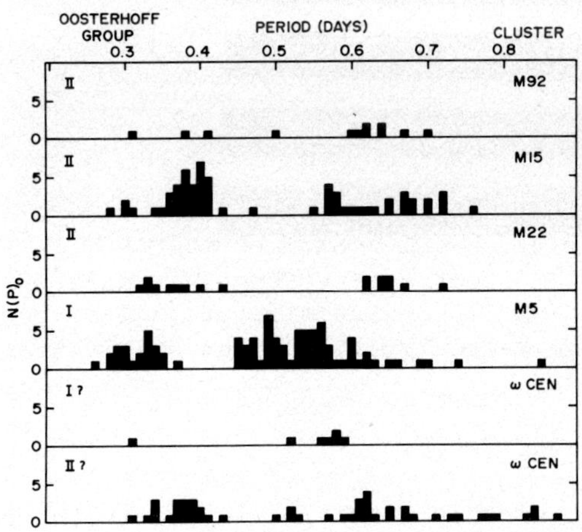

Fig. 6. Period-frequency distributions for Omega Centauri RR Lyrae stars belonging to the metal-rich (I?), and metal-poor (II?) bins, compared with distributions for other clusters.

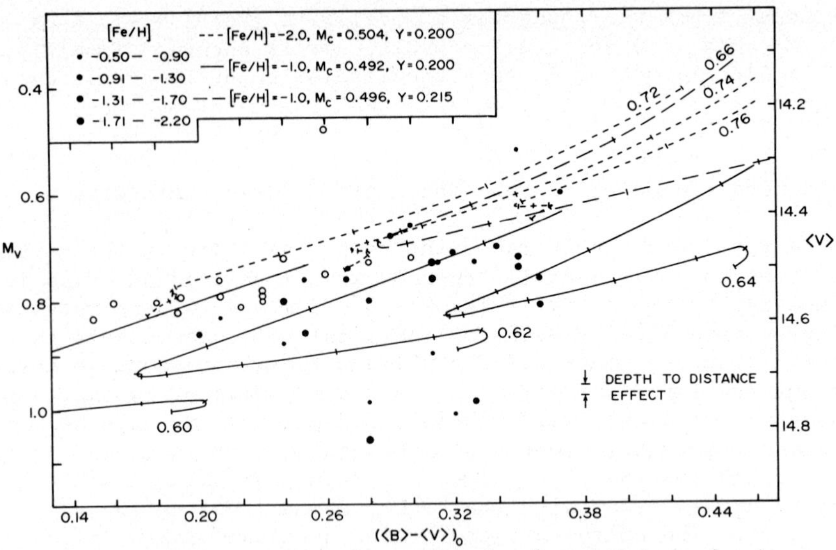

Fig. 7. Evolutionary tracks for model horizontal-branch stars superimposed on the CM diagram of Omega Centauri RR Lyrae stars. Filled circles, ab-types; open circles, c-types. Tick marks delineate intervals of 10^7 years. Track labels give values of total mass. See text for discussion.

2.2 CM Diagram for the RR Lyrae Stars

In Fig. 7 we give our new CM diagram for Omega Centauri RR Lyrae stars. Mean apparent visual magnitude <V> is plotted against <B> - <V>, corrected for an adopted color excess of E(B-V) = 0.11. Evolutionary tracks result from interpolation in the grids of Sweigart and Gross (1976; 1977); theoretical quantities were related to observational ones <u>via</u> a new grid of color-temperature-chemical composition relations appropriate for RR Lyrae stars.

Under the assumptions (1) that $\Delta \log Z = \Delta$[Fe/H] and (2) a star with log Z = -3 has [Fe/H] = -1.4, we began by considering the main-sequence to helium-flash evolution of two stars having [Fe/H] values (-1 and -2) characteristic of the observed range for RR Lyrae stars in Omega Centauri; the stars are otherwise identical, having $M/M_\odot = 0.9$. The metal-rich star takes 1.6×10^9 years longer to get from main-sequence to helium flash. In order for both stars to reach helium-flash at the same time, the metal-rich one must have evolved from a slightly more massive (~0.02 $M_\odot$) main-sequence progenitor. At the time of helium flash, the core masses are significantly different, with the metal-rich star having $M_c = 0.492\ M_\odot$, and the metal-poor one, 0.504 $M_\odot$. For the horizontal-branch phase of evolution, we retrieved several sets of sequences: one for a star with [Fe/H] = -1, $M_c = 0.492\ M_\odot$, and $M/M_\odot = 0.72$, 0.74, and 0.76, and another for [Fe/H] = -2, $M_c = 0.504\ M_\odot$, and $M/M_\odot = 0.60$, 0.62, and 0.64. Both sets are for Y = 0.20. An

additional track was found for a model with [Fe/H] = -1, M_c = 0.496 $M_\odot$, $M/M_\odot$ = 0.66, and Y = 0.215. It is apparent from Fig. 7 that if Δ[Fe/H] = Δlog Z, then the metal-rich RR Lyraes in Omega Centauri are $\sim$ 0.1 $M_\odot$ less massive than the metal-poor ones.

2.3 Chemical Composition Dependence of RR Lyrae Luminosity

We note with considerable interest that stars in the metal-rich group have very nearly the same mean magnitude as those in the metal-poor group: $\langle V \rangle = 14.^{m}62 \pm 0.^{m}07$ (m.e.) for the metal-rich variables, and $14.^{m}53 \pm 0.^{m}02$ for the metal-poor ones. This is not expected from the tracks of Fig. 7, and that is because we have left out one important effect. Red giant models of Sweigart and Gross (1977) indicate differential, Z-dependent envelope helium enrichment through convective mixing as the star is ascending the giant-branch for the first time. It turns out that Δlog Z = 1 leads to ΔY_e = 0.005, with the higher Z model having the greater enrichment. The extra envelope helium abundance makes up for much of the luminosity difference which arises from a large star-to-star range in Z. We should not be surprised, then, that magnitudes of Omega Centauri RR Lyrae stars have such a slight Z-dependence.

It is of interest to compare (in Table II) our results with those from statistical (Hemenway 1975) and photometric (McDonald 1977) parallaxes for field RR Lyrae stars, even though conclusions drawn from such a comparison are fraught with numerous implicit assumptions regarding the stars' backgrounds and evolutionary histories.

TABLE II

RR LYRAE ABSOLUTE MAGNITUDES

$[Fe/H]_{\Delta S}$	$\langle M_V \rangle^{stat}_{H}$	$\langle M_V \rangle^{phot}_{Mc}$	$\langle M_V \rangle^{evol}_{\omega\ Cen}$
>-0.7	$+0.^{m}21 \pm 0.54$	$+0.^{m}82 \pm 0.20$	----
-0.7 to -1.2	$+0.44 \pm 0.43$	$+0.58 \pm 0.20$	$\langle M_V \rangle \pm 0.^{m}07$
<-1.2	$+0.44 \pm 0.46$	$+0.55 \pm 0.20$	$\langle M_V \rangle - 0.^{m}09 \pm 0.^{m}02$

If we adopt McDonald's (1977) value of $\langle M_V \rangle = +0.^{m}55$, then we find Y = 0.23, and $M/M_\odot$ = 0.68 and 0.77 for our metal-rich, and metal-poor groups, respectively. On the other hand, if we compare the Omega Centauri RR Lyrae gap width with widths from calculations of Deupree (1977), we find).23 $\lesssim$ Y $\lesssim$ 0.28, with a "best" value which must await the completion of a formal analysis of scatter in $\langle B \rangle - \langle V \rangle$. No matter what value of Y we finally settle on, the 0.1 $M_\odot$ mass difference between the metal-rich and metal-poor stars remains, as does the very small luminosity difference. Our mass result is strongly coupled to the theoretical formalism employed,

but the luminosity result is purely empirical.

3. CONCLUDING REMARKS AND PROSPECTS FOR FUTURE WORK

I. Under the assumption that we understand the basic features of stellar evolution for low-mass stars, we conclude that a large-scale spatial chemical composition inhomogeneity existed in the proto-cluster cosmic cloud which was to become Omega Centauri. Several hundred solar masses of heavy elements must have been deposited in various regions of the cloud prior to, or during collapse. New high-dispersion spectroscopic equipment expected to be operational soon at CTIO raises the truly fascinating possibility of studying explosive nucleosynthesis elemental yield for a system which has been isolated from the galactic chemical genetic pool for nearly 15 billion years.

II. Considering the large range ($0\overset{m}{.}24$ - $0\overset{m}{.}40$) in Omega Centauri giant-branch width adopted by different investigators, it is clear that great need exists for a new Omega Centauri CM diagram -- a diagram obtained not for the purpose of simply locating major sequences, but one obtained for the purpose of determining precise values of cosmic scatter in the various sequences.

III. Finally, we note the intriguing possibilities of (1) checking the stellar evolution theory results concerning mass differences between metal-rich and metal-poor Omega Centauri RR Lyrae stars, and (2) checking the possibility that the supernovae which contaminated the proto-cluster cloud might have built up "primordial" overabundances of one or more of the CNO group elements. One can use results from linear, nonadiabatic pulsation calculations (van Albada and Baker 1971; Iben 1971) to determine mass values for individual stars. In the [Fe/H] - $M/M_\odot$ diagram, our Fig. 7 predicts a slope of -9.5 for the case in which $\Delta\log Z = \Delta$[Fe/H].

We wish to express our gratitude to Drs. P. Demarque, J. Mengel, and A. Sweigart for many useful discussions during the course of this investigation. We also wish to thank the CTIO staff members for their able and cheerful support, and Mike Wolff for helping with some of the spectroscopic data reduction. This research has been supported in part by NASA under grant NGL-033, and by NSF under grants MPS 74-00790, and AST 76-83413; this support is gratefully acknowledged.

REFERENCES

Arp, H.C. (1962). Astrophys. J. **135**, 311.
Butler, D. (1975). Astrophys. J. **200**, 68.
Cannon, R.D. and Stobie, R.S. (1973). Mon. Not. R. Astron. Soc. **162**, 207.

Demarque, P. and Geisler, J. (1963). Astrophys. J. 137, 1102.
Deupree, R.G. (1977). Astrophys. J. 214, 509.
Dickens, R.J. and Woolley, R. (1967). R. Obs. Bull. 128, E255.
Freeman, K.C. and Rodgers, A.W. (1975). Astrophys. J. (Letters) 201, L71.
Hartwick, F.D.A. and Hesser, J.E. (1974). Astrophys. J. (Letters) 194, L129.
Hemenway, M.K. (1975). Astron. J. 80, 199.
Hesser, J.E., Hartwick, F.D.A. and McClure, R.D. (1976). Astrophys. J. (Letters) 207, L113.
Hesser, J.E., Hartwick, F.D.A. and McClure, R.D. (1977). Astrophys. J. (Suppl.) 33, 471.
Iben, I., Jr. (1971). Publ. Astron. Soc. Pacific 83, 697.
Iben, I., Jr. (1974). In Ann. Rev. Astron. Astrophysics., Vol. 12, Annual Reviews, Palo Alto, p. 215.
McDonald, L.H. (1977). Ph.D. Thesis, University of California, Santa Cruz.
Norris, J. and Bessell, M.S. (1975). Astrophys. J. (Letters) 201, L75.
Preston, G.W. (1959). Astrophys. J. 130, 507.
Rood, R.T. (1972). Astrophys. J. 177, 681.
Sandage, A. and Walker, M.F. (1966). Astrophys. J. 143, 313.
Sweigart, A.V. and Gross, P.G. (1976). Astrophys. J. (Suppl.) 32, 367.
Sweigart, A.V. and Gross, P.G. (1977). preprint.
van Albada, T.S. and Baker, N. (1971). Astrophys. J. 169, 312.
van Albada, T.S. and Baker, N. (1972). In The Evolution of Population II Stars, A.G.D. Philip, ed., Dudley Obs. Rept. No. 4, p. 193.
Woolley, R. (1966). R. Obs. Annals, No. 2.

DISCUSSION

COX: There is a possibility of a wind from the RR Lyrae variable which is capable of enriching the metal content of the surface convection layers. The solar wind actually has a mass loss of hydrogen much more than for the heavier elements. Only 10^{-4} or 10^{-5} of the mass needs to be lost. Instead of trying to evoke evolutionary arguments to give different mass tracks with different helium contents, could you have a small mass range as traditionally used, with fixed helium, and have all anomalies due to an RR Lyrae wind?

BUTLER: Although we have not made calculations to examine this effect, I find it hard to believe that the large local helium overabundance which would result would not lead to unusual hydrogen-line strengths. The observed H-line strengths are normal. I agree, however, that detailed calculations would be valuable.

NORRIS: Is the range in calcium abundance observed in ω Cen unique? If Dr. Cox's suggestion is correct, one would expect to see a similar phenomenon in other clusters.

BUTLER: So far, it is unique in this respect.

WEIDEMANN: You had to invoke differential mass loss, dependent on abundance differences, in order to exaplin the position of both groups of stars. How large are these mass loss differences?

BUTLER: The exact difference depends on the adopted main-sequence mass and helium abundance of the progenitor. With plausible values of these parameters, one can deduce that a "metal-rich" RR Lyrae progenitor lost ∿ 0.2 $M/M_{\odot}$, and a "metal-poor" one ∿ 0.1 $M/M_{\odot}$.

FLOWER: Böhm-Vitense at the University of Washington has computed T_{eff}: (B-V): BC scales for a variety of metal abundances. How do your scales compare with hers and how sensitive are your results to the transformation scales?

BUTLER: I am not familiar with her color temperature or BC scales for RR Lyrae stars. In any case, it is necessary to calculate them for the correct microturbulent velocity parameter, and it is unlikely that we can make a direct comparison since our relations were computed for DBV = 3.6 km/sec, the value resulting from previous curve-of-growth determinations. The [Fe/H] spread and magnitude range results do not depend on the color temperature or BC relations. The <u>differential</u> mass result is <u>slightly</u> sensitive, while an <u>absolute</u> mass would be <u>extremely</u> sensitive to the relations adopted.

ZINN: I would like to remark that your results concerning the RR Lyrae stars in ω Cen may be inconsistent with stellar pulsation theory. If I understand your results correctly, you find that metal-rich variables have the same luminosities and effective temperatures as the metal-poor variables. Then you say that metal-rich variables must be less massive. If all these things are true, then the metal-rich variables have smaller surface gravities. Well, pulsation theory would then say that these metal-rich variables must have longer periods than the metal-poor variables. This is, of course, the opposite from the Oosterhoff effect. Have I missed something?

BUTLER: The temperatures (and luminosity) are only <u>approximately</u> equal. The average temperatures for an RR Lyrae in the metal-rich group is ∿ 300 K hotter than the corresponding quantity for an average star in the metal-poor group. If the "metal-rich" stars are ∿ 0.1 $M/M_{\odot}$ less massive than the "metal-poor" ones, then linear, non-adiabatic pulsation theory predicts a luminosity difference amounting to $\Delta V \sim 0.14$, with the most metal-rich star being the faintest. Since we found $\langle V \rangle = 14^{m}.62 \pm 0^{m}.07$ (m.e.) for the metal-rich group and $\langle V \rangle = 14^{m}.53 \pm 0^{m}.02$ for the metal-poor group, the size of the discrepancy is ∿ $0^{m}.05 \pm \sim 0^{m}.1$ (estimated error). Considering all of the uncertainties involved, and the assumptions ($\Delta \log Z = \Delta$[Fe/H]) central to the analysis, I don't think the "discrepancy" is a serious one. Note that if Δ[Fe/H] $\neq \Delta \log Z$, and if, in particular, the range in Z_{CNO} is less than the range in Z, then the mass differences could be reduced or eliminated completely.

KRAFT: Neglecting, for now, the question of the enrichment of the envelope, isn't it true that you can't change Y from your metal-rich to your metal-poor groups and have the ZAHB in the same place? Doesn't that set a limit on the range of Y between the two groups?

BUTLER: Yes, but I want to make clear that by "metal-rich" I don't mean solar abundance. The metal-rich stars are only moderately metal rich.

KRAFT: You say that the magnitudes of the two groups differ by no more than $0\overset{m}{.}1$, on the average. To what change in Y does this correspond?

BUTLER: Oh, to a ΔY of about 0.02 or 0.03.

THE ROLE OF MIXING AND PRIMORDIAL ABUNDANCE VARIATIONS IN 47 TUCANAE AND ω CENTAURI

John Norris

Mount Stromlo and Siding Spring Observatory,
Australian National University

1. INTRODUCTION

It has become clear in recent years that widespread and complex abundance variations exist among the stars of several globular clusters. ω Cen is the prime example, containing CH stars, weak-G-band stars, CN strong stars, TiO stars, and (not least) RR Lyrae stars possessing an anomalous range in calcium line strengths. (Harding 1962; Freeman and Rodgers 1975; Norris and Bessell 1975, 1977; Dickens and Bell 1976; Bessell and Norris 1976; Lloyd Evans 1977) The extent to which these phenomena result from primordial abundance variations or from the mixing to the surface of material processed by nuclear reactions is currently the subject of some debate. The present paper presents a short account of investigations of the giant and horizontal branches of 47 Tuc and of several giants in ω Cen. The results were obtained in collaboration with M. S. Bessell and K. C. Freeman, and will be fully reported in papers being prepared for submission to Astrophysical Journal. Most emphasis will be placed on 47 Tuc, where we have sought to address four problems. (i) How widespread are the CN anomalies reported by McClure and Osborn (1974), Bell, Dickens, and Gustafsson (1975), and Hesser, Hartwick, and McClure (1977) - both throughout the HR diagram and in relative frequency? (ii) Is there any correlation between the relative frequency of the CN anomaly with radial position in the cluster, which might correlate with the color gradients observed by Gascoigne and Burr (1956) and Chun (1976)? (iii) Is there any evidence for G-band variations similar to those found generally in the metal poor ([Fe/H]< -1) globular clusters? (cf. Zinn 1973, 1977; Mallia 1975; Norris and Zinn 1977) (iv) Are the CN variations accompanied by variations in the abundance of the heavy elements?

A. G. Davis Philip and D. S. Hayes (eds.), The HR Diagram, 195–204 and 406.

Finally we address the question of the relationship between the TiO stars in 47 Tuc and ω Cen.

2. 47 TUCANAE

Our investigation of 47 Tuc is based principally on 50Å mm^{-1} spectra obtained with the 1.9 m telescope at Mount Stromlo, together with several 33Å mm^{-1} spectra obtained with the 4 m telescope at Siding Spring Mountain. Two series of spectra were obtained at Mount Stromlo. The first consists of stars in the outer regions of the cluster (r > 10 arc min), and comprises 16 red giant branch, 12 asymptotic giant branch, and 14 horizontal-branch stars. The second comprises 16 red giant and 2 asymptotic giant branch stars in the inner regions (60 arc sec < r < 120 arc sec). These observations were supplemented by UBV and DDO intermediate band photometry obtained with the 1 m telescope on Siding Spring Mountain, together with the data of McClure and Osborn (1974), Cannon (1974), Lee (1977), and Hesser et al. (1977).

In Fig. 1 we show the blue spectral region in selected pairs of stars of similar magnitude and color on the red giant branch, the asymptotic giant branch, and the horizontal branch. Inspection

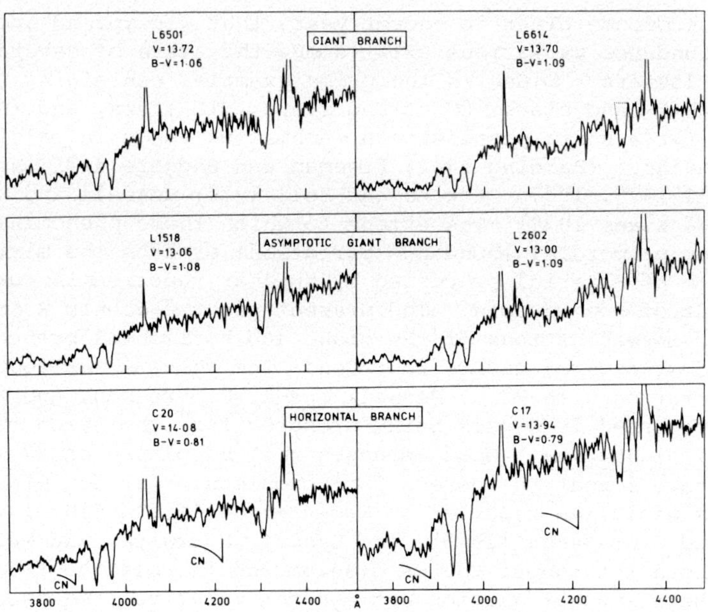

Fig. 1. Spectra of pairs of stars of similar magnitude and color on the red giant, the asymptotic giant, and horizontal branches. The CN strong stars are shown on the right. (Nomenclature from Lee [1977] and Cannon [1974]).

of the CN bands at 4216 and 3883 Å shows that the CN anomaly is widespread throughout the HR diagram, occurring on all three branches.

2.1 The Red Giant And Asymptotic Giant Branches

On each spectrum of the red giant and asymptotic giant branch stars we have computed a CN index defined by S(4142) = -2.5 log $[\int_{4120}^{4216} F_\lambda d\lambda / \int_{4216}^{4290} F_\lambda d\lambda]$. The existence of DDO photometry for 12 stars in our sample allowed us to compare S(4142) with the corresponding DDO index C(4142). A close correlation exists and was used to transform the C(4142) values for a further 15 stars, for which this index is available, onto the S(4142) scale. This body of data is shown in the [S(4142),(B-V)]- plane in Fig. 2. Adopting the lower envelope of the data as the lower limit to the CN index, the CN anomaly δS(4142) was measured for each star as the distance the star lay above the lower envelope at its observed (B-V). (For completeness it should be noted that the asymptotic giant branch stars appear to be systematically CN stronger by 0.05 mag than the red giants. To allow for this effect a lower envelope displaced upwards by this amount was adopted for this group.)

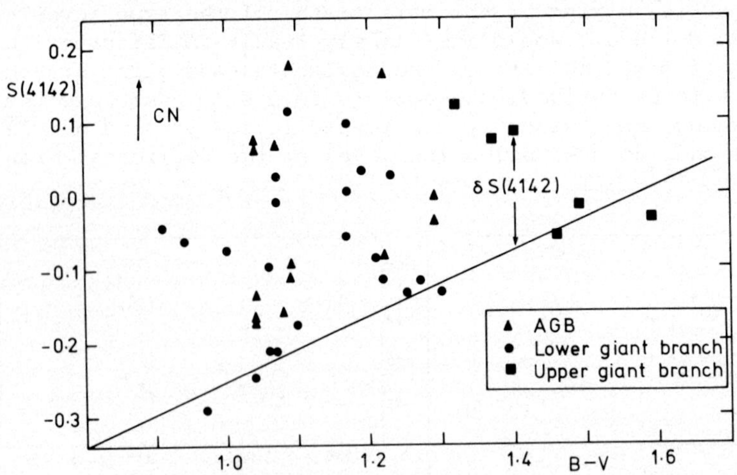

Fig. 2. 47 Tuc giants in the S(4142), (B-V) - plane. The direction of increasing CN is shown.

The histograms of δS(4142) represent the principal results of our investigation and are shown in Fig. 3 for the outer and inner regions of 47 Tuc. Two conclusions may be derived from these data. First, in the outer regions there appear to be two populations in approximately equal numbers. In this region this conclusion is approximately true for the red giant branch stars and for the

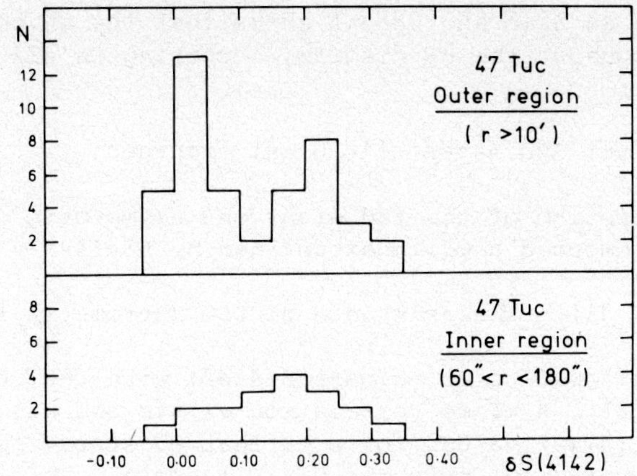

Fig. 3. Histogram of δS(4142) for stars in the inner and outer regions of 47 Tuc.

asymptotic giant branch stars, and as we shall see in what follows the same pertains to the horizontal-branch sample. These results are extremely difficult to understand in terms of mixing at the helium core flash (or the helium shell flashes). Any scenario involving partial mixing at the core flash, of the type investigated by Rood (1970), would most likely result in different proportions of mixed objects on the horizontal and giant branches. If mixing is to be invoked, one appears forced to postulate that it occurs below the luminosity considered in the present sample. This corresponds to just below the level of the horizontal branch.

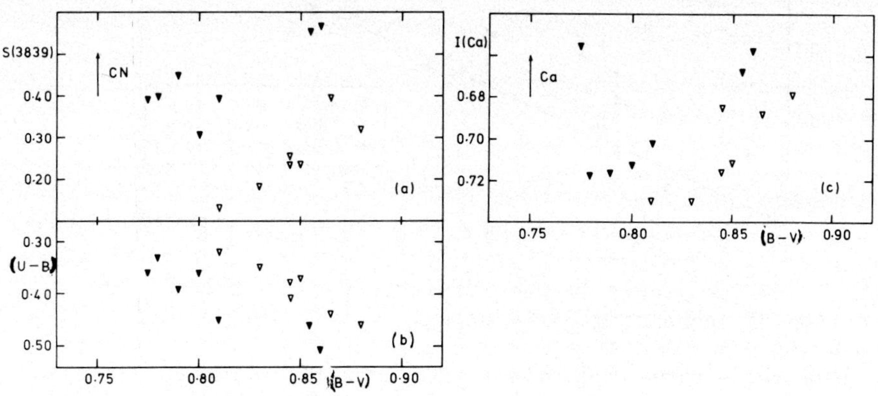

Fig. 4. Horizontal-branch stars in the (a) [S(3839), (B-V)] - plane, (b) [(U-B),(B-V)]- plane, and (c) [I(Ca),(B-V)]- plane. Filled and open symbols refer to strong and weak CN stars as defined in Fig.4a.

The second conclusion one may derive from Fig. 3 is that in contrast to the outer region, the inner region is CN strong. It seems reasonable to conclude that this effect is closely related to the color gradients observed in this cluster. (cf. Chun 1976)

2.2 The Horizontal Branch

As demonstrated in Fig. 1, CN anomalies exist also on the horizontal branch. From our spectra we have measured the CN index $S(3839) = -2.5 \log \left[\int_{3846}^{3883} F_\lambda d\lambda / \int_{3883}^{3916} F_\lambda d\lambda \right]$, and show our results in the [S(3839),(B-V)]- plane in Fig. 4a. (We note that the CN band at λ4216 has greatly weakened at the higher temperature of the horizontal branch). Here too there appear to be two classes of CN strengths. In Fig. 4b we show our sample in the [(U-B), (B-V)]- plane where the open and filled symbols represent weak and strong CN stars as defined in Fig. 4a. It is clear that the CN strong stars have the smaller ultraviolet excesses. Finally in Fig. 4c we plot a measure of the CaII H and K line strengths defined by $I(Ca) = \int_{3916}^{3985} F_\lambda d\lambda / \left[\int_{3883}^{3916} F_\lambda d\lambda + \int_{3985}^{4016} F_\lambda d\lambda \right]$. Here we see an apparent correlation of Ca with CN, in the sense that at a given (B-V) the CaII lines are stronger in the CN strong stars.

2.3 G-Band And MgI Measurements

Our G-band (CH molecule) measurements for all stars in the outer region are shown in the [W(G),(B-V)]- plane in Fig. 5,

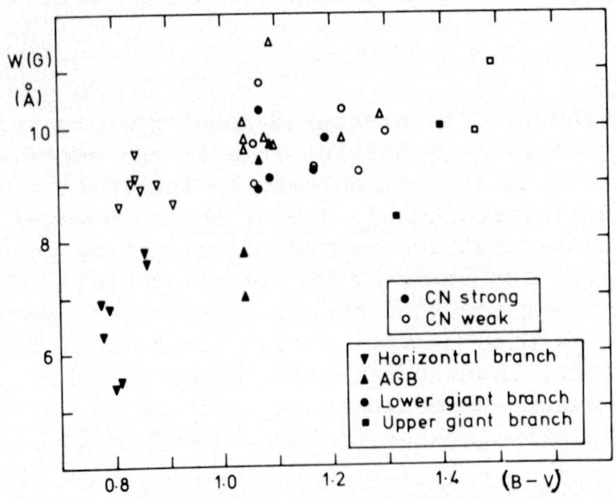

Fig. 5. The red giant branch, asymptotic giant branch, and horizontal-branch stars in the [W(G), (B-V)] - plane.

where W(G) is the equivalent width as illustrated in the upper right panel of Fig. 1. There is a clear anticorrelation, with CN strengthening accompanied by G-band weakening. The decrease in the effect at lower temperature may result from the effect of saturation as the G-band increases in strength. The simplest explanation of the anticorrelation is that the nitrogen enhancement inferred for the CN strong stars (Bell et al. 1975) is accompanied by a decrease in carbon, as might be expected from CNO processing.

To supplement our Ca line strength measurements we have obtained 33Å mm^{-1} spectra in the region of the MgI'b' lines for two CN strong and one CN weak red giants of similar magnitude and color, which are shown in Fig. 6. There is no obvious difference between the spectra, a result somewhat at odds with our Ca measurements.

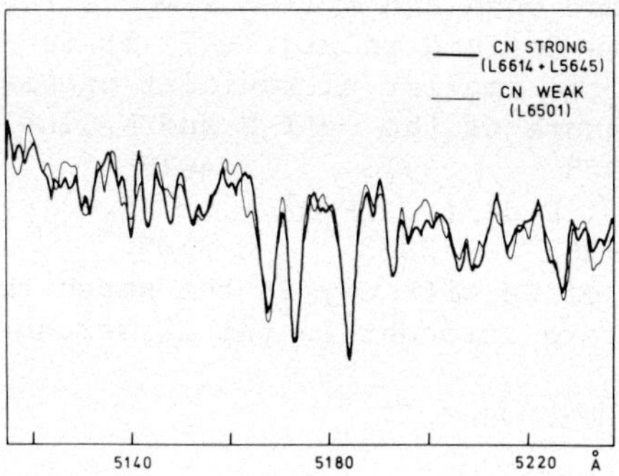

Fig. 6. Averaged spectrum of two CN strong stars compared with that of a CN weak star in the region of the MgI'b' lines. The three stars have very similar magnitude and color.

2.4 Discussion of the 47 Tuc Results

The CN and G-band results presented above seem to indicate that CNO processing plays an important role in the anomalies. The results for Ca and Mg are not so easy to interpret. Reference to the results of Gustafsson et al. (1975) shows, however, that interpretation of these features is made difficult by changes induced in atmospheric structure by CNO peculiarities. They find that CO and CN cause surface cooling and backwarming, respectively. In the CN strong, weak-G-band type of star found in 47 Tuc these effects will reinforce, leading to hotter outer regions. Further work is clearly necessary to evaluate the effects on CaII, MgI, and (B-V) before the effects seen in Figs. 4 and 5 can be fully understood.

We have argued above that if mixing is responsible for the observed anomalies, one is forced to postulate that it occurs below

the level of the horizontal branch. While very mild CNO mixing is predicted at the base of the giant branch by stellar evolution computations (cf. Faulkner and Iben 1967) a more drastic effect is needed to explain nitrogen enhancements of a factor 10, as reported by Bell et al. (1975). Somewhat radical modifications to the above mixing process have been briefly considered by Norris and Zinn (1977) in the context of G-band weakening; following Dearborn, Bolton, and Eggleton (1975), they suggest that greater CNO mixing might occur as the result of thermal instability on the lower giant branch, together with substantial mass loss. The alternative, of course, is that the variations are primordial to the cluster, resulting from two (or more) epochs of star formation. The attractiveness of this hypothesis is that it requires no radical departures from reasonably well understood astrophysical processes.

3. THE TIO STARS IN ω CEN AND 47 TUC

It is not clear to what extent conclusions similar to those presented above apply to ω Cen. According to Freeman and Rodgers (1975) the abundances of some of the RR Lyrae stars in ω Cen are 47 Tuc-like ([Fe/H] ~ -0.5). Norris and Bessell (1977) found no counterparts of these stars in a sample of 60 red giants, while Glass and Feast (1973) noted that the TiO stars possess ultraviolet excesses unlike those in their 47 Tuc counterparts. The recent report by Lloyd Evans (1977) of further anomalously red members of ω Cen which appear to lie on the 47 Tuc locus in the [I,(V-I)]-plane allows us to further investigate this problem.

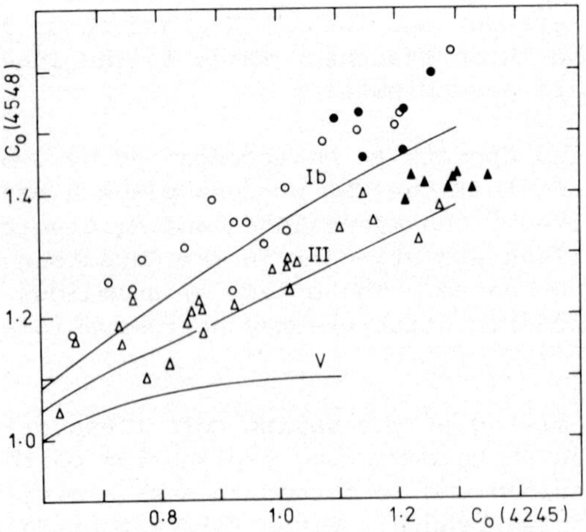

Fig. 7. ω Cen (circles) and 47 Tuc (triangles) red giants in the [C_o(4548), C_o(4245)]- plane. For ω Cen the filled circles represent the red stars of Lloyd Evans (1977); for 47 Tuc the filled triangles represent stars showing TiO bands.

We have obtained DDO photometry using the 1 m telescope at Siding Spring Mountain of several of the red ω Cen stars lying on the 47 Tuc locus (Lloyd Evans 1977, Fig. 1), together with several of the redder 47 Tuc stars. (Our spectra show TiO in both groups, though not in all ω Cen stars.) Our results, corrected for reddenings of E(B-V) = $0^m.11$ and $0^m.03$ for ω Cen and 47 Tuc, respectively, are shown in the [$C_o(4548)$, $C_o(4245)$]- plane in Fig. 7, together with the results of Bessell and Norris (1975) for ω Cen, and McClure and Osborn (1974), Hesser et al. (1977), and Bessell and Norris (1977) for 47 Tuc. The Population I sequences of Osborn (1971) are shown also for comparison. In this plane we see a clear separation of the ω Cen red stars from those in 47 Tuc, lending support to the results of Glass and Feast (1973), and Norris and Bessell (1977) that few, if any, of the giants in ω Cen are like those in 47 Tuc.

These results do not necessarily argue against primordial abundance variations in ω Cen. One possible explanation is that large primordial CNO variations are responsible for most of the peculiarities observed in globular clusters and that in ω Cen these CNO variations are coupled with large heavy element variations (as a result perhaps of a very different mass function for its first generation of stars). With such peculiar abundance ratios it would be surprising if the ω Cen giants were 47 Tuc-like.

4. SUMMARY

CN variations have been found on the giant branch, the asymptotic giant branch, and the horizontal branch of 47 Tuc. In the outer regions 'strong' and 'weak' CN stars occur in roughly equal numbers on the three branches, while in the inner regions the 'strong' CN stars predominate.

CN strengthening appears to be accompanied by G-band weakening, suggesting that the CNO process plays a major role in the observed anomalies. The elements Ca and Mg give contradictory results; the CaII lines are stronger in the CN strong stars, while the MgI lines are unchanged. The effect of anomalous C, N, and O abundances on atmospheric structure may be responsible for this behavior.

We argue that mixing at the helium core flash or the helium shell flash provides an unconvincing explanation of these effects. If mixing is responsible one is forced to seek a site below the level of the horizontal branch. Given this situation a primordial origin of the peculiarities becomes more attractive.

Finally we show that the anomalous red stars in ω Cen (Lloyd Evans 1977) do not appear to be like the giants in 47 Tuc,

as might have been expected from the abundances derived for the RR Lyrae stars in ω Cen. This places important constraints on any primordial origin of the peculiarities in this cluster.

REFERENCES

Bell, R. A., Dickens, R. J., and Gustafsson, B. (1975). Bull. Amer. Astron. Soc. 7, 535.
Bessell, M. S., and Norris, J. (1976). Astrophys. J. 208, 369.
________________. (1977). unpublished.
Cannon, R. D. (1974). Mon. Not. R. Astron. Soc. 167, 551.
Chun, M.-S. (1976). Ph.D. Thesis, Australian National University.
Dearborn, D. S. P., Bolton, A. J. C., and Eggleton, P. P. (1975). Mon. Not. R. Astron. Soc. 170, 7P.
Dickens, R. J., and Bell, R. A. (1976). Astrophys. J. 207, 506.
Faulkner, J., and Iben, I. Jr. (1967). Nature 215, 44.
Freeman, K. C., and Rodgers, A. W. (1975). Astrophys. J. (Letters) 201, L71.
Gascoigne, S. C. B., and Burr, E. J. (1956). Mon. Not. R. Astron. Soc. 116, 570.
Glass, I. S., and Feast, M. W. (1973). Mon. Not. R. Astron. Soc. 163, 245.
Gustafsson, B., Bell, R. A., Eriksson, K., and Nordlund, Å (1975). Astron. Astrophys. 42, 407.
Harding, G. A. (1962). Observatory 82, 205.
Hesser, J. E., Hartwick, F. D. A., and McClure, R. D. (1977). Astrophys. J. Supp. 33, 471.
Lee, S.-W. (1977). Astron. Astrophys. Supp. 27, 381.
Lloyd Evans, T. (1977). Mon. Not. R. Astron. Soc. 178, 345.
Mallia, E. A. (1975). Mon. Not. R. Astron. Soc. 170, 57P.
McClure, R. D., and Osborn, W. (1974), Astrophys. J. 189, 405.
Norris, J., and Bessell, M. S. (1975). Astrophys. J. (Letters) 201, L75.
________________ (1977). Astrophys. J. (Letters) 211, L91.
Norris, J., and Zinn, R. (1977). Astrophys. J. 215, 74.
Osborn, W. (1971). Astrophys. J. 186, 725.
Rood, R. T. (1970). Astrophys. J. 162, 939.
Zinn, R. (1973). Astrophys. J. 182, 183.
_______. (1977). Astrophys. J. in press.

DISCUSSION

DEMING: I have computed synthetic spectra in the region of the 4150 Å CN absorption under the assumption of various types of CNO processing. It turns out to be difficult to produce a CN-strong star if one mixes only the products of hydrogen-burning. With conversion of carbon into nitrogen the CN bands remain constant in strength until more than about 60% of the carbon is processed into nitrogen. If larger fractions of carbon are processed into nitrogen the CN gets weaker due to the carbon depletion. It is possible to strengthen CN by also processing oxygen into nitrogen, but this requires high temperatures and densities. But even with the processing of oxygen into nitrogen some initial overabundance of oxygen is needed to produce the large nitrogen enhancements which the CN-strong stars require. If one admits that some carbon from helium burning may be present then it is much easier to make a CN-strong star.

NORRIS: The need for large nitrogen enhancement is also clear from the work of Bell, Dickens, and Gustafsson. Cottrell and I are performing computations similar to yours to obtain more information on the amount of nitrogen required in 47 Tuc.

KRAFT: That's quite a general problem - there's always too much nitrogen required. The nitrogen features are just too strong for the amount of processing from carbon you would expect, even if you process all of the carbon to nitrogen. In this mass range, I quess, you don't expect to get much O processed into N, and perhaps none at all.

TAYLER: I wish to suggest that we must learn something about primoridal abundance variations versus mixing in ω Cen from the fact that ω Cen has a very broad giant branch whereas other generally similar clusters do not. The star in the broadened giant branch must know that it is in ω Cen rather than another cluster. If it has a given mass and chemical composition, it will be equally affected by any mixing processes that occur, whichever cluster it is in. I therefore believe that we must ask what makes ω Cen different from other clusters. One possible solution, which has been suggested, is that ω Cen is sufficiently massive to retain significant quantities of heavy elements produced in a first generation of stars and this could then account for the variations in metal content. In any case we should be looking for some gross property of ω Cen which distinguishes it from other clusters.

NORRIS: ω Cen is distinguished also by two other properties. First, it has a large ellipticity. Second it has low concentration for its mass. McClure and I have recently noted that M22 shares these properties and (alone among the clusters investigated at that time) possesses CH, CN, and Ba II stars as found in ω Cen.

KRAFT: I would add that the spread also depends on the choice of color system you use and the type of "metal" causing the spread.

CONTINUED ON PAGE 406

THE ANOMALOUS GIANT BRANCH OF ω CENTAURI

T. Lloyd Evans

South African Astronomical Observatory

ABSTRACT

The giant branch of ω Cen is much narrower in the I, (V-I) diagram than in the V, (B-V) diagram but at least 24 stars which lie on the red side of the giant branch are radial velocity members (Lloyd Evans 1977), (Lloyd Evans, to be published). Spectra of these stars show, with only one exception, stronger metal lines than stars of similar (V-I) color on the main giant branch (comparable to stars in 47 Tuc), strong CN bands where the stellar temperature allows and strong Ba II 4554 Å. The Ba II star RGO 371 (Dickens and Bell 1976) is a typical member of the group. The Cepheid V29 also shows strong CN bands and strong Ba II. A start has been made on a spectroscopic search for stars on the main giant branch showing these anomalies and the sample of stars with VI photometry is being enlarged. An explanation of the properties of these stars would appear to require both a large primordial metal content and mixing of processed material to the surface.

REFERENCES

Dickens, R.J. and Bell, R.A. (1976). Astrophys. J. 207, 506.
Lloyd Evans, T. (1977). Mon. Not. R. Astron. Soc. 178, 345.

A. G. Davis Philip and D. S. Hayes (eds.), The HR Diagram, 205–206.

DISCUSSION

KRAFT: You find some stars with over-all metal abundance (average Fe-group) looking about the same in 47 Tuc and ω Cen?

LLOYD EVANS: That's correct.

KRAFT: But you find stars with evidence of mixing in ω Cen and not 47 Tuc.

LLOYD EVANS: That's correct. Essentially all of the stars below the giant branch in ω Cen have rather strong CN and strong Ba II.

NORRIS: I don't think that Lloyd Evans' results are necessarily inconsistent with primordial abundance variations. If one has different mass functions for the first generation of stars one might envisage a situation in which C, N and O and the heavy elements are present in considerably different proportions in ω Cen than in 47 Tuc.

LLOYD EVANS: One might need to consider a local event, perhaps involving supernova ejecta retained by ω Cen, as the source of enriched material. Presumably Ba might be one of the elements involved as it is correlated with metals and CN in these stars.

BIDELMAN: There is a rather unlikely but perhaps not impossible way to explain non-evolutionary abundance differences among stars of the same cluster; the differential accretion of interstellar matter by the different stars during passage through the galactic plane.

LLOYD EVANS: The high velocity of the cluster relative to interstellar clouds in the galactic plane makes accretion unlikely.

PART V

HR DIAGRAMS, HORIZONTAL BRANCH

Jesse Greenstein and Cecilia Payne-Gaposchkin.

HORIZONTAL-BRANCH STARS IN THE LOG T_{eff}, LOG G DIAGRAM

A. G. Davis Philip*

Dudley Observatory and Union College

ABSTRACT

A review of the available Strömgren four-color data concerning horizontal-branch (HB) stars is presented. Several observers have studied seven globular clusters (with [Fe/H] values from -2.2 to -1.3) and more than 100 HB stars in fields at high galactic latitudes over the last ten years at Kitt Peak, Cerro Tololo, Mt. Wilson and Steward Observatories. The predictions of model atmospheres (Kurucz, 1976) allow one to calculate atmospheric parameters such as θ_{eff}, log g, and M_V (Philip, Miller and Relyea, 1976). These parameters then can be compared with the predictions of evolutionary models (Sweigart and Gross, 1976).

A comparison of the observed data with evolutionary tracks indicates that within the rms error of observation and the computational error of the models the data and the tracks agree quite well. This matches the case for Population I stars, where a similar analysis shows a good match between observational data and evolutionary tracks (Philip, et al. 1977).

1. INTRODUCTION

Humason and Zwicky (1947) were pioneers in the study of faint blue stars at high galactic latitudes. In the years since many additional surveys have been made, some of them colorimetric,

*Visiting Astronomer, Kitt Peak National Observatory and Cerro Tololo Inter-American Observatory, supported by the National Science Foundation under contract No. NSF-C866.

A. G. Davis Philip and D. S. Hayes (eds.), The HR Diagram, 209-220 and 406.

others spectroscopic. Some of the colorimetric surveys are those made by Luyten (A Search for Faint Blue Stars) in a long series of papers, Iriarte and Chavira (1957), Chavira (1958), Iriarte (1959) (The last three papers define the Tonantzintla stars.), Feige (1959), and Haro and Luyten (1962) (The PHL Stars.). Some of the spectroscopic surveys of early-type stars are those made by Slettebak, Bahner and Stock (1961) at the North Galactic Pole; Philip and Sanduleak (1968), Sanduleak and Philip (1968), and Slettebak and Brundage (1971) at the South Galactic Pole. Some spectrographic surveys of stars of all spectral types in various regions have been made by Upgren (1962) at the NGP and Philip (1966, 1968), Philip and Drilling (1970), Drilling and Philip (1970), and Philip and Relyea (1971) in various areas at high galactic latitudes.

As a result of all these surveys and other surveys a large number of faint blue stars have been identified. Four-color measures of such stars in the magnitude range $10<V<14^m$ have indicated that they are about equally divided between Pop I and Pop II (see Philip and Tifft (1971), Philip (1974), and Rodgers (1971)). Two important studies of blue halo stars have been made in recent years. Greenstein and Sargent (1974) made a spectroscopic study and Newell (1973) made a photometric study in another photometric system. In each paper log T_{eff}, log g diagrams are plotted showing the distribution of various types of halo stars. This paper will be concerned only with horizontal-branch stars.

The method of identifying early-type Pop II stars will be described, then four-color and Hβ measures of field and globular cluster horizontal-branch stars will be discussed. The method by which four-color measures are transformed to the astrophysical parameters log T_{eff} and log g is described in a paper in the poster session, Relyea and Philip (1978). These parameters are then compared with the predictions of evolutionary theories in Section 4.

2. FOUR-COLOR MEASURES OF HORIZONTAL-BRANCH STARS

A number of papers have been written concerning photometric classification in the Strömgren four-color system. Among these are papers by Graham (1970), Philip (1973), and Kilkenny and Hill (1975). A figure taken from the last mentioned paper and reproduced here as Fig. 1 illustrates the method. In the spectral range A0-F8 horizontal-branch stars can be recognized by their high c_1 indices and low m_1 indices. Philip (1972) has shown that the Hβ index of A-type HB stars is smaller than that for main sequence stars of Pop I of similar spectral type.

To identify Pop II A-type stars one first obtains four-color and Hβ measures of the candidate star. The candidate stars are selected from Schmidt spectral plates taken with a 4 degree or thin prism (dispersion 280 or 1360 Å/mm at Hγ). Since the four-color

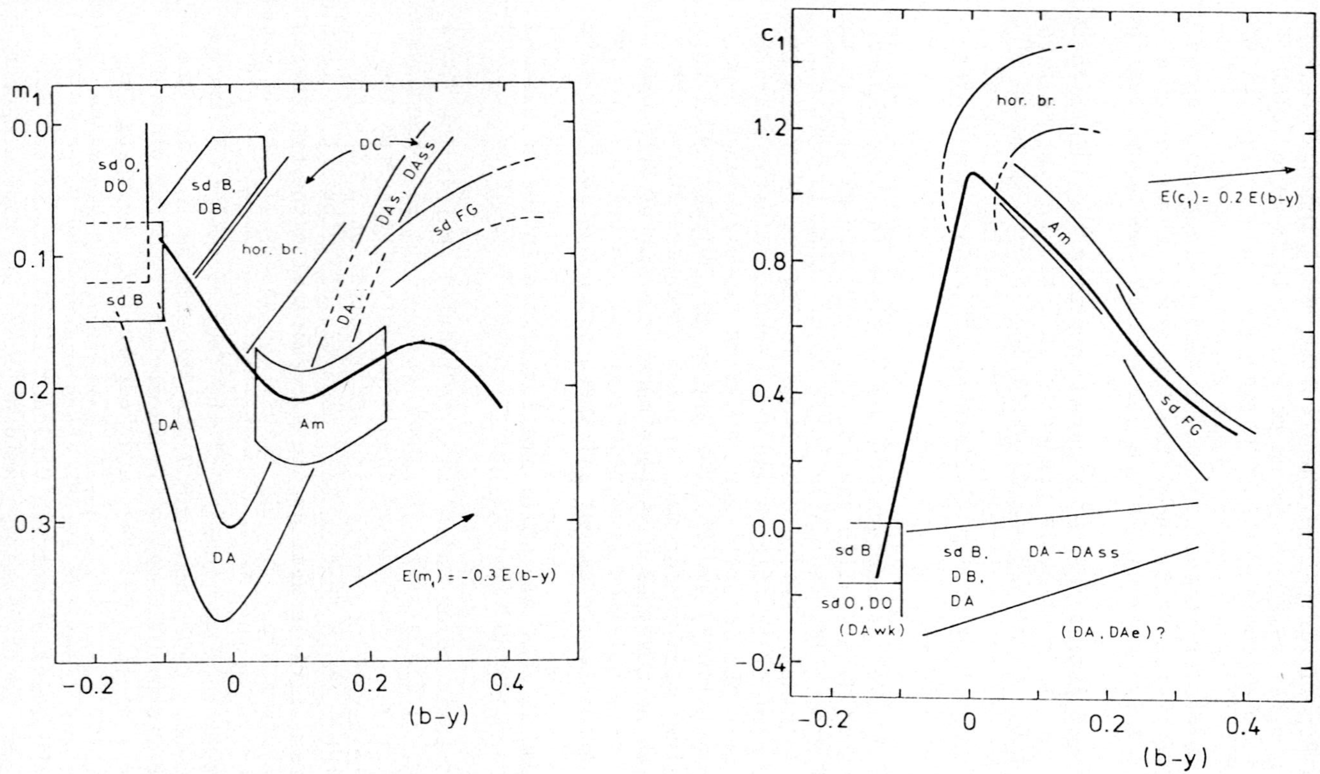

Fig. 1. m_1 versus (b-y) and c_1 versus (b-y). The location of various types of stars are indicated in the diagram as well as the ZAMS (heavy line). From Mon. Not. R. Astron. Soc. 173, 625.

indices are affected by interstellar reddening, the indices must be corrected. At the present time there is no formula by which one can deredden Pop II early-type stars so the reddening must be determined in other ways. A common method is to measure faint Pop I stars in the general field and then use the formulae derived by Crawford (1975) to derive the color excess in the region. The instrinsic four-color indices can be plotted in diagrams such as Fig. 1 and the photometric classification made. If Hβ measures are available this index can be used to confirm the classification.

On the next page a list of papers concerning the identification and measurement of HB stars is presented in Table I. In the first part of the table those FHB stars identified in part by means of four-color photometry are listed. Papers concerning the measurement of BHB stars in globular clusters and FHB stars are listed in parts two and three.

All the four-color measures of the HB stars in these papers plus unpublished data (Philip and Philip, 1978) are plotted in the following figures. The intrinsic $(c_1)_o$ color index is plotted against the intrinsic $(b-y)_o$ color index in Fig. 2, for field stars in the upper part of the figure and for globular cluster stars in the lower part. The interstellar reddening was calculated from tables in Philip (1973) and Philip, Miller and Relyea (1976) in which the values of E_{b-y} for many special areas are listed. For areas in which no measures have been made, a value of E_{b-y} was estimated by interpolation from nearby areas. In regions of high galactic latitude (where almost all these stars are located) this procedure works well and leads to errors of only a few hundredths of a magnitude in E_{b-y} and still smaller errors in the dereddened color indices $(c_1)_o$ and $(b-y)_o$.

The intrinsic color index $(m_1)_o$ is plotted against the $(b-y)_o$ index in Fig. 3; field stars are shown on the left side and cluster stars on the right. In each of these two sets of diagrams the ZAMS is indicated by the heavy line and the indices of the HB stars by points. Note that in each case the distribution of points in the diagram pairs is similar, in agreement with the hypothesis that the FHB stars are correctly classified as Pop II stars. Other evidence confirming this hypothesis has been noted in Philip (1973). This distribution of the indices of the four proto-type FHB stars (HD 2857, 86986, 109995 and 161817) is the same as that for the group of FHB stars. The velocity dispersion of 33 FHB stars is $\pm$ 113 km/sec (Philip 1969, 1970 and Rodgers 1971) matching the velocity dispersion of the globular clusters. The equivalent widths of the calcium K lines for 15 FHB stars measured by Rodgers (1971) fall below the relation between W(K) and $(B-V)_o$ for normal main sequence stars. More recently Danford (1976) has made a spectroscopic study of many FHB stars from the lists in Table I, Part 1 and he concludes that these stars are correctly classified as Pop II stars.

TABLE I

PAPERS CONCERNING DISCOVERY AND MEASUREMENT OF FIELD HB STARS AND MEASURES OF GLOBULAR CLUSTER BHB STARS

Part 1

Field Horizontal-Branch Stars

Identified by a combination of Schmidt Spectra and/or 4-Color Photometry

MacConnell, D.J., Frye, R.L., Bidelman, W.P. and Bond, H.E. (1971). Publ. Astron. Soc. Pacific 83, 98.

Philip, A.G.D. (1967). Astrophys. J. 148, L143. (List I)

_______ (1968). Astrophys. J. 152, 1107. (List II)

_______ (1969). Astron. J. 74, 209. (List III)

_______ (1969). Astron. J. 74, 812. (List IV)

_______ (1970). Astron. J. 75, 957. (List V)

_______ (1973). Publ. Astron. Soc. Pacific 85, 68. (List VI)

Part 2

Globular Cluster Blue Horizontal-Branch Stars

Measured in the 4-Color Photometric System

Graham, J.A. and Doremus, C. (1968). Astron. J. 73, 226. (NGC 6397).

Philip, A.G.D. (1972). In The Evolution of Population II Stars, A.G.D. Philip, ed., Dudley Obs. Rept. No. 4, 35.

_______ (1973). Astrophys. J. 182, 517. (M4).

_______ (1978). in preparation.

Part 3

Field Horizontal-Branch Stars

Measured in the 4-Color Photometric System

Bond, H.E. and Philip, A.G.D. (1973). Publ. Astron. Soc. Pacific 85, 332.

Drilling, J.S. and Pesch, P. (1973). Astron. J. 78, 47.

Graham, J.A. (1966). Publ. Astron. Soc. Pacific 78, 433.

_______ (1970). Publ. Astron. Soc. Pacific 82, 1305.

Hilditch, R.W., Hill, G. and Barnes, J.V. (1976). Memoirs Roy. Astron. Soc. 82, 95.

Kilkenny, D., Hill, P.W. and Brown, A. (1977). Mon. Not. R. Astron. Soc. 178, 123.

Kilkenny, D. and Hill, P.W. (1975). Mon. Not. R. Astron. Soc. 173, 625.

Newell, B. and Graham, J.A. (1976). Astrophys. J. 104, 804.

Philip, A.G.D. (1973). Dudley Obs. Rept. No. 8.

_______ (1973). IAU Symposium No. 50, Spectral Classification and Multicolor Photometry, Ch. Fehrenbach and B. Westerlund, eds. D. Reidel, Dordrecht, p. 230.

Philip, A.G.D. and Tifft, L.E. (1971). Astron. J. 76, 567.

Philip, A.G.D. and Philip, K.D. (1973). Astrophys. J. 179, 855.

Slettebak, G., Wright, R.R. and Graham, J.A. (1968). Astron. J. 73, 152.

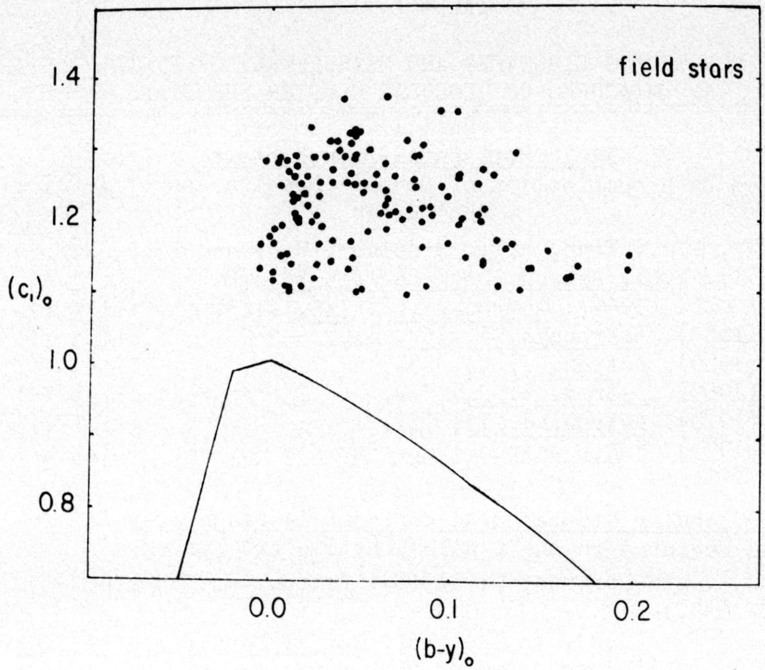

Fig. 2. $(c_1)_o$ <u>versus</u> $(b-y)_o$ for field (above) and cluster stars (below).

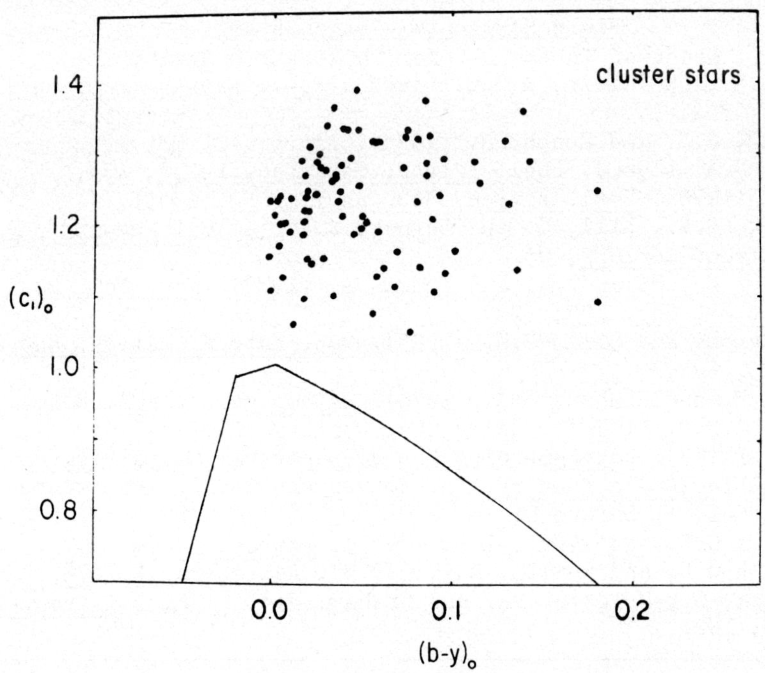

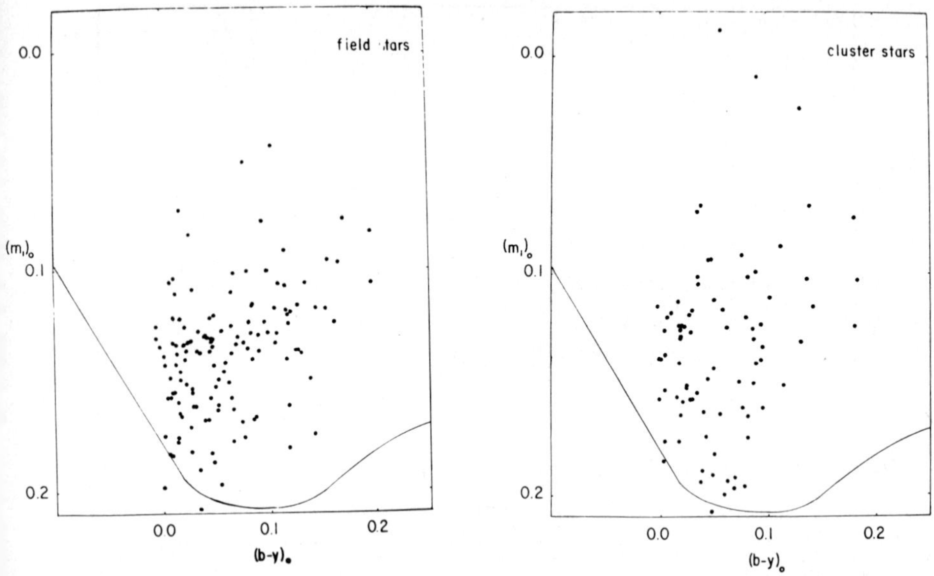

Fig. 3. $(m_1)_o$ <u>versus</u> $(b-y)_o$ for field (left) and cluster stars (right).

3. THE TRANSFORMATION OF PHOTOMETRIC INDICES TO LOG T_{eff} AND LOG G

The method of transforming the $(c_1)_o$, $(b-y)_o$ indices to log T_{eff} and log g is described fully in a paper by Philip and Relyea (1978) and outlined in a paper in the poster session, Relyea and Philip (1978). There are three grids which relate the color indices $(c_1)_o$, $(b-y)_o$ to log T_{eff}, log g for metal abundances equal to the solar metal abundance, 1/10 and 1/100 solar abundance. One plots the indices in the appropriate grid and then reads off log T_{eff} and log g. As a check that the method works it has been shown by Philip, <u>et al.</u> (1977) that the transformed indices of early-type stars in seven open clusters fall along isochrones calculated by Ciardullo and Demarque (1977). As an example of the fit the distribution of points in the log T_{eff}, log g diagram for the Hyades is reproduced from Philip <u>et al.</u> (1977) in Fig. 4. The isochrone lines represent ages of 0.5, 1.0, and 25 x 10^9 years. This agreement gives one confidence that the transformation is valid for early-type Pop I stars.

To check the transformation for Pop II stars the four-color indices of the four proto-type FHB stars were transformed to log T_{eff} and log g and the resulting values compared with those predicted by Kodaira (1975) and Oke <u>et al.</u> (1966). The log g values agreed well; no systematic trend was found. In Θ_{eff} however, a correction formula was derived to correct the temperatures calculated from the grid. The equation used is:

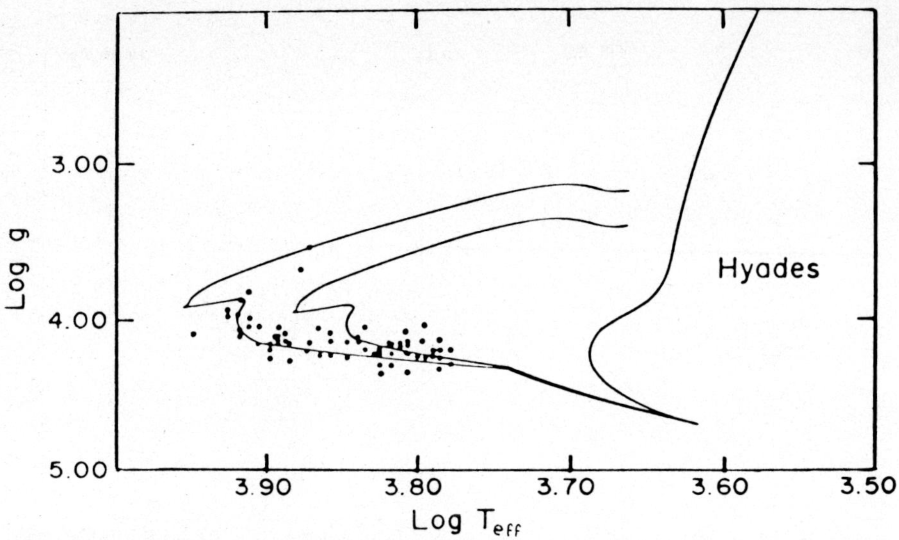

Fig. 4. Observations and theory compared for Pop I stars (From Publ. Astron. Soc. Pacific 89, 534).

$$\Theta_{eff} = 1.713\ \Theta_{grid} - 0.457. \quad (0.57 < \Theta_{eff} < 0.68)$$

4. COMPARISON WITH EVOLUTIONARY MODELS

The log T_{eff}, log g values are plotted in Fig. 5, field stars on the upper left and cluster stars on the upper right. Some of the theoretical horizontal-branch evolutionary tracks (Sweigart and Gross, 1976) are shown in the bottom part of Fig. 5. All the models shown are for core masses of 0.475 $M_{\odot}$ but the total mass, helium and metal abundances vary as shown in the diagram.

Because there are so many parameters that can vary one can not pick a unique set and determine masses and abundances for HB stars from these diagrams. Additional information will be needed to do this. But it is interesting that the models all have a similar slope in the log T_{eff}, log g diagram and that this slope is matched by the distribution of the transformed observational points in the diagrams for field and cluster stars. Extreme values of the physical parameters can be ruled out.

As in the case for Pop I stars, theory and observation seem to be agreeing within the rms errors determined from the theoretical and observational side.

5. CONCLUSIONS

Photometric indices of Pop I and Pop II stars of spectral

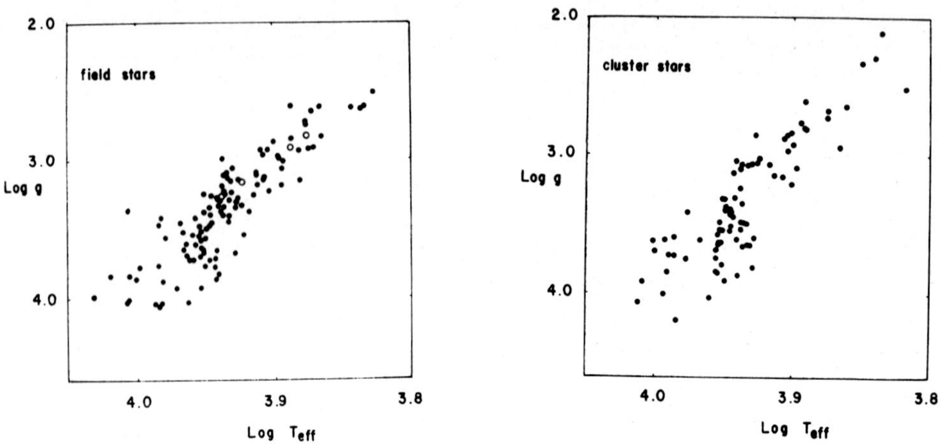

Fig. 5. Log g versus log T_{eff} for field stars and cluster stars. Evolutionary models are shown to the right.

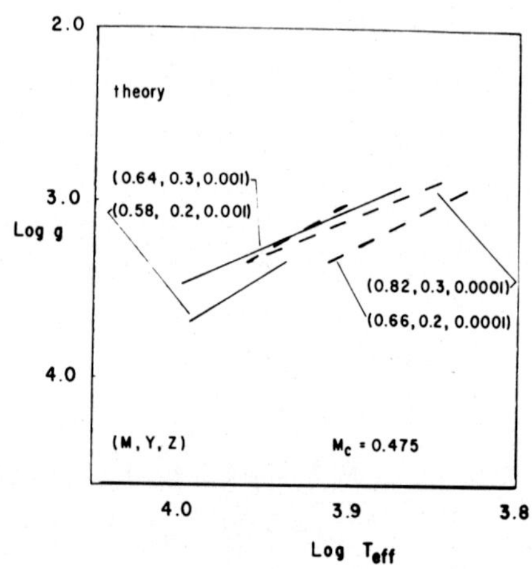

types A - F, luminosity classes V - III can be transformed successfully into a form of the theoretical HR diagram, the log g, log T_{eff}, diagram. For both Pop I and Pop II stars the distributions in such diagrams agree with the distributions predicted by current theory. The distributions also agree, in the spectral range A - F, with distributions determined by Greenstein (1974) and Newell (1973).

A temperature correction was needed for Pop II A - F stars. Since there were only four Pop II early-type stars for which log T_{eff} and log g are known the temperature correction used is not well determined. Additional bright FHB stars should be measured by a photoelectric scanning technique and fundamental values of log T_{eff} and log g derived. A correction formula to the $(c_1)_o$, $(b-y)_o$ grid derived from 20 such sets of observations would be a great improvement over the present situation.

BHB stars in many globular clusters of varying [Fe/H] values can be measured and transformed to the log T_{eff}, log g diagram. Since these stars are faint many measures will be needed with telescopes of aperture 2.5m or greater. If accurate data for many stars in different globular clusters could be obtained one could investigate to see if the scatter found in the giant branches of some globular clusters is also found in the horizontal branch. Such data would help to determine if the scatter is caused by inhomogeneity in the proto-cluster or if the scatter is caused by mixing in some later evolutionary stage.

REFERENCES

Chavira, E. (1958). Bol. Obs. Tonantzintla y Tacubaya No. **17**, 15.
Ciardullo, R.B. and Demarque, P. (1977). Yale Transactions Vol. **33**.
Crawford, D.L. (1975). In Multicolor Photometry and the Theoretical HR Diagram, A.G.D. Philip and D.S. Hayes, eds., Dudley Obs. Rept. No. 9, p. 17.
Danford, S.C. (1976). Thesis, Yale Univ., Order No. 76-29190
Drilling, J.S. and Philip, A.G.D. (1970). Bol. Obs. Tonantzintla y Tacubaya **5**, 307.
Feige, J. (1958). Astrophys. J. **128**, 267.
Graham, J.A. (1970). Publ. Astron. Soc. Pacific **82**, 1305.
Greenstein, J.L. and Sargent, A.I. (1974). Astrophys. J. Suppl. **28**, 157.
Haro, G. and Luyten, W.J. (1962). Bol. Obs. Tonantzintla y Tacubaya **3**, 117.
Humason, M.L. and Zwicky, F. (1947). Astrophys. J. **105**, 85.
Iriarte, B. (1959). Lowell Obs. Bull. **4**, 130.
Iriarte, B. and Chavira, E. (1957). Bol. Obs. Tonantzintla y Tacubaya No. **16**.
Kilkenny, D. and Hill, P.W. (1975). Mon. Not. R. Astron. Soc. **173**, 625.
Kodaira, K. (1975). Problems in Stellar Atmospheres, B. Bascheck, ed., Springer Verlag, N.Y., p. 149.
Newell, E.B. (1973). Astrophys. J. Suppl. **26**, 37.
Oke, J.B., Greenstein, J.L. and Gunn, J. (1966). In Stellar Evolution, R.E. Stein and A.G.W. Cameron, eds., Plenum Press, N.Y. p. 399.
Philip, A.G.D. (1966). Astrophys. J. Suppl. **12**, 394.
Philip, A.G.D. (1968). Bol. Obs. Tonantzintla y Tacubaya **5**, 297.
Philip, A.G.D. (1969). Astrophys. J. Letters **158**, 113.
Philip, A.G.D. (1970). Astron. J. **75**, 246.
Philip, A.G.D. (1972). In The Evolution of Population II Stars, A.G.D. Philip, ed., Dudley Obs. Rept. No. 4, p. 35.
Philip, A.G.D. (1973). In IAU Symposium No. 52, Interstellar Dust and Related Topics, J.M. Greenberg and H.C. van de Hulst, eds., D. Reidel, Dordrecht, p. 263.
Philip, A.G.D., Demarque, P., Sweigart, A.V. and Ciardullo, R.B. (1977). Publ. Astron. Soc. Pacific **89**, 534.

Philip, A.G.D. and Drilling, J.S. (1970). Bol. Obs. Tonantzintla y Tacubaya 5, 297.
Philip, A.G.D., Miller, T.M. and Relyea, L.J. (1976). Dudley Obs. Rept. No. 12.
Philip, A.G.D. and Philip, K.D. (1978). in preparation.
Philip, A.G.D. and Relyea, L.J. (1971). Bol. Obs. Tonantzintla y Tacubaya No. 36, 69.
Philip, A.G.D. and Relyea, L.J. (1978). in preparation.
Philip, A.G.D. and Sanduleak, N. (1968). Bol. Obs. Tonantzintla y Tacubaya No. 30, 253.
Philip, A.G.D. and Stock, J. (1972). Bol. Obs. Tonantzintla y Tacubaya No. 38, 201.
Philip, A.G.D. and Tifft, L.E. (1971). Astron. J. 76, 567.
Relyea, L.J. and Philip, A.G.D. (1978). In IAU Symp. 80, The HR Diagram, A.G.D. Philip and D.S. Hayes, eds., Reidel, Dordrecht, p. 443.
Rodgers, A.W. (1971). Astrophys. J. 165, 581.
Sanduleak, N. and Philip, A.G.D. (1968). Publ. Astron. Soc. Pacific 80, 437.
Slettebak, A., Bahner, K. and Stock, J. (1961). Astrophys. J. 134, 195.
Slettebak, A. and Brundage, R.K. (1971). Astron. J. 76, 338.
Slettebak, A. and Stock, J. (1959). In A Finding List of Stars of Spectral Type F2 and Earlier in a North Galactic Pole Region, Hamburg-Bergedorf.
Sewigart, A.V. and Gross, P.G. (1976). Astrophys. J. Suppl. 32, 367.
Upgren, A.R. (1962). Astron. J. 67, 37.

DISCUSSION

BUTLER: Suppose there is a population of metal-rich ([Fe/H]$\sim$0) blue horizontal-branch stars in the solar neighborhood. Is it possible to use your techniques in discovering any of these (as yet hypothetical) stars?

PHILIP: It would depend on the spectral type of the star being investigated. If the star is of spectral type B or A0 it is very difficult, but as the spectral type becomes later, A3, A5 to A7 a high metal abundance would show up in the metal index (the metal index, m_1, would become larger than that for a metal poor star of the same color).

McNAMARA: How does your $(c_1)_o$ versus $(b-y)_o$ diagram compare with Breger's and could you comment on the so called c_1 anomaly in the Hyades?

PHILIP: Dave Crawford, who is in the audience, would be the best person to comment on the c_1 anomaly in the Hyades. As for the comparison of our $(c_1)_o$ $(b-y)_o$ diagram with that of Breger's, Breger used the atmospheric models of Osmer and Peterson; Philip and Matlock used the 1966 atmospheric models of Mihalas. The various $(c_1)_o$, $(b-y)_o$ diagrams were compared in the paper in Multicolor Photometry and the Theoretical HR Diagram, Dudley Observatory Report No. 9 and all were found to agree within the rms errors.

CRAWFORD: Being naturally conservative, I have no suggestion about the Hyades solution. It could be due to blanketing or structure differences, due to metal abundance, or it could be helium differences, which haven't been talked about much here yet, or to other things. It is certainly there (the δc_1 anomaly for F-type stars), and I'd be glad to hear people's idea about the causes.

McNAMARA: I just worry about using the Hyades as a calibration for the diagram.

PHILIP: I am not using the Hyades as a calibration. It was just one cluster out of seven which were plotted to check out the systems as applied to Pop I stars. Kurucz's atmospheric models were used and Vega was used to make the zero point corrections. Since I don't like calibrations based on only one star I went through the literature and found 28 A-F type stars for which spectroscopic determinations of log T_{eff} and log g had been made. We then calculated the residuals (published - observed) and found that the rms errors were ± 0.015 in θ_{eff} and ± 0.2 in log g. These values are about the same as obtained from spectroscopic determinations.

LUB: Considering your plot of log T_{eff}, log g it seems to me that you have in there several stars which appear to be in the RR Lyrae instability strip; have you checked any of these stars for variability?

PHILIP: No, I have not made a systematic check for variability. On low dispersion spectral plates the differences would be hard to detect and in most cases I have only a few plates for each region. The four proto-type stars have been measured about 100 times each and I detect no signs of variability. The other program stars have been measured from 2 to 20 times each, and while I have no firm indications of variations, these data are not sufficient to decide on variability.

LUB: HD161817 is just 100 K beyond the blue edge of the RR Lyrae instability strip and you have some stars that would be in the instability strip.

PHILIP: That could be. It just means that someone has to obtain additional observations of those stars.

COX: Can I ask you a technical question about the calibration of your c_1 and m_1 indices? This morning we heard that Y can vary from less than 0.2 to over 0.35. What is the difference in your calibrations for these varying Y's?

PHILIP: The calculations are based on the Kurucz models which used a conventional Pop I helium abundance (0.9 hydrogen and 0.1 helium by number).

GREENSTEIN: Much theoretical attention has been devoted to horizontal-branch stars and it seems to me Dr. Philip's work provides an extension to what I and others have done. We worked on the hotter, you have worked on the cooler group. The plot of g and θ is almost certainly an extension of plots valid for the field horizontal-branch hot stars, with a

CONTINUED ON PAGE 406

THE EXPECTED MORPHOLOGY OF ZERO AGE HORIZONTAL BRANCHES

Vittorio Castellani

Istituto di Astronomia, Universita di Roma

On the basis of available theoretical evaluations of the H-burning evolutionary phases of Population II stars, it is at present quite simple to predict the expected location of Zero-Age Horizontal-Branch (ZAHB) stars for assumed values of the age and original chemical composition. A result of such a computation is disclosed in Fig. 1, for the labelled values of the assumed evolutionary parameters.

From a qualitative point of view, the HR diagram behavior is the one well known from the numerous ZAHB's computed since the first work by Rood (1973) appeared. Here it might be worth discussing some quantitative details.

The effect of the increase in surface helium, connected with the sinking down of the external convection during the red giant phase is shown in Fig. 2. If one choses Y=0.20, the ZAHB luminosity at the RR Lyrae gap increases, roughly as $\Delta \log L \sim 0.04$. Such a difference is obviously not a dramatic one. Nevertheless one has to realize that, if interpreted in terms of variation in the original He-content, a similar discrepancy would imply $\Delta Y \sim 0.03$ which can be a not negligeable value, when quantitative interpretations are attempted. As a matter of fact, increasing the original helium abundance decreases the size of the helium core at the beginning of H.B. evolution which counteracts the increased efficiency of the H-burning shell. As a result the ZAHB location is less sensitive to variations in the original helium abundance than in the "external helium". In this way it is easy to understand the above quoted results.

A. G. Davis Philip and D. S. Hayes (eds.), The HR Diagram, 221-224.

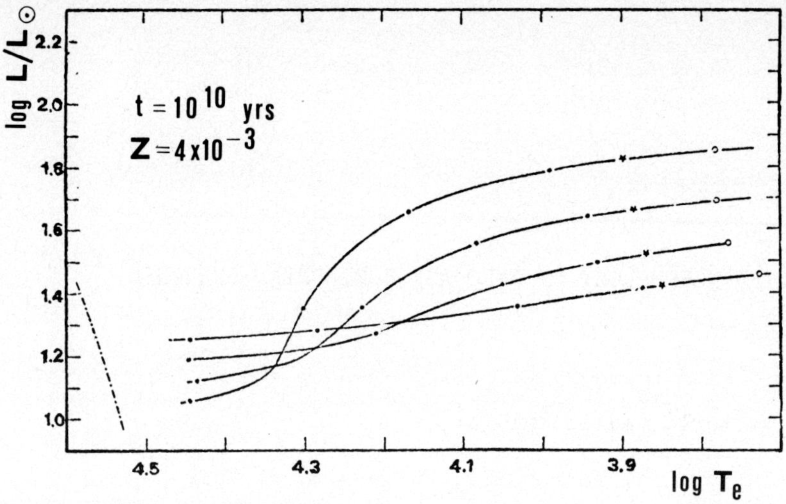

Fig. 1. ZAHB loci for $Z=4x10^{-3}$ and original Y= .40, 0.30, 0.20 and 0.10 (top to bottom in the red region). Stars indicate on each sequence the blue boundary for pulsational instability. (From Caloi et al. 1977.)

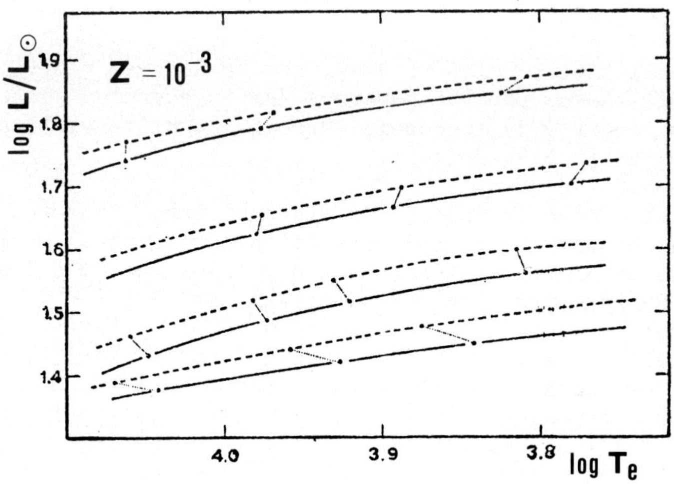

Fig. 2. ZAHB sequences showing the effect of the evolutionary increase in surface helium. Full lines: original helium, dashed lines: evolutionary helium.

A few further details may be worth discussing. The first one is the expected location of hot HB stars in the HR diagram. The occurrence of a very blue tail of HB stars has been observed (or suggested) in some "M 13 like" galactic globular clusters. From data reported in Fig. 1 we obtained the V, (B-V) loci reported in Fig. 3, adopting B.C.'s and C-T_{eff} relations as given by L. Rossi (1977) on the basis of detailed atmospheric computations. As a result, we find that the location of hot HB stars in the HR diagram is largely independent of the original chemical composition. Thus, the assumption made by Cannon (1974) in comparing the HR diagrams of globular clusters by superimposing the blue HB stars is supported. The observed differences in the location of red giant branches in the HR diagram remains (quantitatively) an open question.

A comparison with observations, as given in the same Fig. 3, seems to be in good agreement with the theoretical results, supporting the previous conclusions.

Of course, ZAHB structures depend on the assumption on the relative abundance of CNO elements. Fortunately, theoretical computations (Castellani and Tornambe 1977) indicate that varying CNO doesn't sensitively change the ZAHB locus, and the pulsational characteristics of RR Lyrae variables are expected to remain essentially correlated with He abundance, even if a "reasonable" fluctuation in CNO abundances is assumed. In this context, it might be worth noticing that in Population II systems there is not (yet) any clear indication about a discrepancy between "pulsational" and "evolutionary" masses, in the sense that "evolutionary" HB models can fit very well the observed period-frequency histograms of RR Lyrae pulsators, when suitable values of the evolutionary parameters are assumed.

In conclusion, one finds that "realistic" HB stars largely support the belief in a quantitative understanding of HB evolution. At the same time one is facing some new problems. Among these, I wish to mention the case of the Draco dwarf galaxy: if we believe for this system a value of Z as low as $Z \approx 3 \times 10^{-5}$ (Canterna 1975), evolutionary HB stars cannot reach the red side of the instability strip if $Y \lesssim 0.20$. On the contrary, Draco is a well known red-HB cluster. Why?

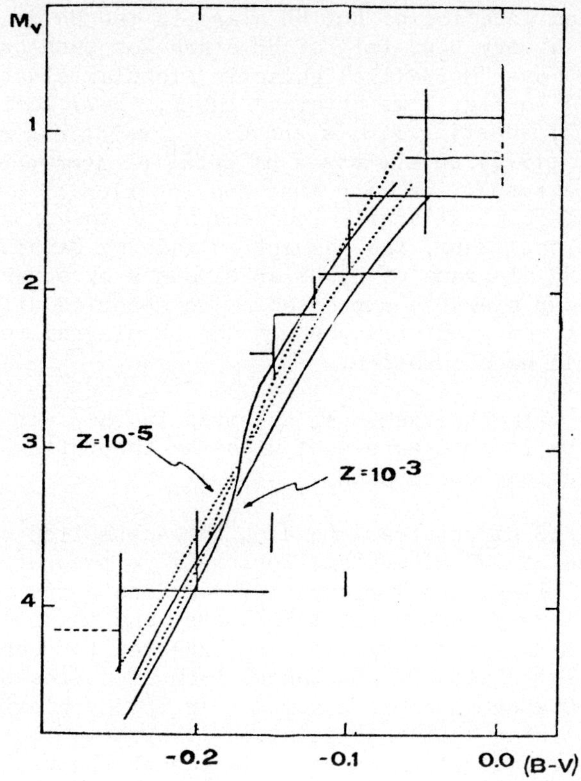

Fig. 3. The expected location in the C-M diagram of hot ZAHB stars for two different assumptions on Z and for Y=0.30 or 0.20 (top to bottom at the lower temperatures). Bars roughly indicate the locations of hot HB stars observed in NGC 6752 (Cannon and Lee 1973).

REFERENCES

Caloi, V., Castellani, V. and Tornambe, A. (1977). Astron. Astrophys. Suppl. (submitted to).
Cannon, R.D. (1974). Mon. Not. R. Astron. Soc. **167**, 551.
Cannon, R.D. and Lee, S.W. (1973). unpublished.
Canterna, R. (1975). Astrophys. J. **200**, L63.
Castellani, V. and Tornambe, A. (1977). Astron. Astrophys. (in press).
Rood, R.T. (1973). Astrophys. J. **184**, 815.
Rossi, L. (1977). Internal Report (Frascati).

EVOLUTION OF GLOBULAR CLUSTER STARS WITH MASS LOSS AND CORE ROTATION

F. Fusi-Pecci and A. Renzini

Osservatorio Astronomico, Italy

During the last decade theoretical and observational research on globular cluster stars has fairly conclusively demonstrated that the mass of horizontal-branch (HB) stars is substantially smaller (by $\sim 0.2\ M_\odot$) than the mass of their main-sequence progenitors (cf. Rood 1973, Fusi-Pecci and Renzini 1975, hereafter MPS V and FPR respectively, and the literature quoted therein). Also well established is the fact that HB stars in a single cluster do not follow a unique evolutionary path but that some dispersion in at least one HB parameter (e.g. total mass, core mass, or composition) is required. Further, a growing body of astrophysical evidence shows that the HB morphology is poorly correlated with the abundance of low-ionization potential elements (ALIPE) such as Mg, Ca, Si, and Fe. This has raised the well known problem of the "second parameter" in the sense that something else besides ALIPE has to vary from cluster to cluster, the most popular candidates being age, t, helium, Y, and CNO abundance, Z_{CNO}. Therefore, the main problems which research on globular cluster stars is presently facing are i) a quantitative knowledge of the mass loss process, ii) the identification of the origin of the HB dispersion in a single cluster, and iii) the identification of the second parameter(s). The solution of the last point has profound implications for the theory of the early chemical and dynamical evolution the Galaxy.

As far as point i) is concerned, extant mass loss rate (MLR) expressions for red giant stars (FPR, Reimers 1975, Fusi-Pecci and Renzini 1976) can be used to predict the total mass

A. G. Davis Philip and D. S. Hayes (eds.), The HR Diagram, 225-228 and 406.

in pre-HB phases. In both the FPR and Reimers MLR expressions a proportionality factor - respectively η_{FPR} and η_R - is present. Such coefficients can be determined by imposing the condition that the observed position of the HB in the HR diagram be reproduced. We find $\eta_{FPR} \simeq (4.4 \pm 0.2) \times 10^{-4}$ and $\eta_R \simeq 0.38 \pm 0.02$ (for details see Renzini 1977). The emphasis here is on the narrow range of the allowed MLR of red giants stars if observed HB's have to be reproduced. Conceivably, neither the theory of stellar winds nor direct observations in the near future will be able to provide an independent estimate of the mass loss rate of Population II red giants at this accuracy level ($\sim$ 4%). In other words, the only way to treat mass loss will continue to be through a "free parameter" (η) to be determined a posteriori. A second crucial question concerns the Z-dependence of the MLR. Assuming M proportional to Z^{α} (for Z larger than, say, 10^{-4}), one has $\alpha = -0.04$ for the FPR - MLR and $\alpha \simeq + 0.08$ for the Reimers-MLR. It is worth emphasizing that the arguments leading to the evaluation of α are very weak both in the FPR and in the Reimers approach. However, differences in the HB morphology using the FPR and Reimers MLR expressions are quite small. The fact that "metal rich" clusters have a stubby red HB sets a severe upper limit on α, i.e., $\alpha < \sim + 0.2$. This is an important constraint for the theory of stellar winds. Concerning point ii) we have explored the possibility that the HB spread is due to different initial rotation of the stars in one cluster. Following Mengel and Gross (1976) both the core mass and the luminosity at the helium flash (L_{fl}) increases for increasing initial stellar angular velocity (ω). Since total mass loss is a strong increasing function of L_{fl}, the mass of the HB stars decreases with increasing ω. (It is worth emphasizing that rotation affects mass loss <u>only</u> through its influence on L_{fl}, the MLR itself being unaffected by rotation in this approach.) Therefore, fast rotators will arrive on the HB with a larger core mass and a smaller total mass than slow rotators. As a consequence, the larger is ω the larger will be the initial effective temperature of HB stars. With an assumed distribution function for ω of the type

$$P(\omega) = \text{const.} \times \omega(\omega_c - \omega) \exp\left[-(\omega - \omega_o)/\sigma^2\right]$$

where ω_c is the main-sequence break-up angular velocity, we find that the required HB spread is well reproduced when $\omega_o = (1.8 \pm 0.1) \times 10^{-4}\ \text{sec}^{-1}$ and $\sigma/\omega_o = 0.4 \pm 0.1$. This value of ω_o is very close to the value expected for Population I solar mass stars ($\sim 2 \times 10^{-4}$) when the observed trend between A and F type stars is extrapolated to the lower main sequence (<u>cf</u>. Renzini 1977). The conclusion is that dispersion in the original rotational velocities provides a plausible explanation of the HB spread provided the cores of evolving Population II stars retain most of their original angular momentum. Langer (1977) has shown that a dispersion of CNO abundance $\Delta\log Z_{CNO} \simeq 0.3$ would

also produce the desired effect and Carbon et al. (1978) have actually observed this range of variation in the nitrogen abundance in M92. However, nitrogen is (mostly) a secondary element and its abundance is expected to suffer large fluctuations in comparison with carbon and oxygen which are primary elements (Trimble, 1975). In solar proportions (C:N:O ≈ 20:5:43) a variation by a factor of two in nitrogen will produce a variation of only 7% in Z_{CNO}. Therefore the observation of the CN band strength could be of little help for the determination of Z_{CNO}. What actually matters is carbon and, in particular, oxygen. A possible observational check for disentangling the two suggestions (rotational vs. compositional spread) comes from the fact that the bluer HB stars obtained following the rotational hypothesis are significantly brighter (by more than $0^m.5$) than those obtained in the framework of the composition hypothesis. We find some support for the rotational hypothesis from the photoelectric photometry of pre-gap blue HB stars of NGC 6752 (Newell and Sadler 1977). Further effort in this direction would be conclusive. Finally, concerning point iii), from our "rotationally" spread HB's we find the required ranges for the candidate "second parameters", namely $\Delta t \simeq 1$ billion years and $\Delta Y \simeq 0.03$ (in agreement with MPS V). A range of [CNO/Fe] = ±0.5 would also produce the desired effect (here for Fe we mean the whole ALIPE). As there is presently no hope of "measuring" such "small" age and helium variations, the only really crucial observations will be those attempting a determination of C and O abundances in globular clusters.

An extended paper covering these topics is in preparation.

REFERENCES

Carbon, D., Kraft, R.P. and Langer, G.E. (1978). Astron. J., in press.

Fusi-Pecci, F. and Renzini, A. (1975). Astron. Astrophys. **39**, 413.

Fusi-Pecci, F. and Renzini, A. (1976). Astron. Astrophys. **46**, 447.

Langer, G.E. (1977). preprint.

Mengel, J.G. and Gross, P.G. (1976). Astrophys. Space Sci. **41**, 407.

Newell, E.B. and Sadler, N.J. (1977). preprint.

Reimers, D. (1975). Mem. Soc. Roy. Sci. Liege 6th ser., **6**, 369.

Renzini, A. (1977). In Advanced Stages in Stellar Evolution, P. Bouvier and A. Maeder, eds., p. 149.

Rood, R.T. (1973). Astrophys. J. **184**, 815.

Trimble, V. (1975). Rev. Mod. Phys. **47**, 877.

DISCUSSION

WEIDEMANN: You might be interested to learn that Reimers (together with Kudritzki at Kiel) has recently recalibrated the mass loss formula by an investigation of α Her and Antares B and now finds $\eta_R = 0.3$. Which law of rotation has been assumed for the first giant branch? The core mass increase which you need in order to get the higher luminosities is probably obtained for differential rotation. However there are indications that on the first giant branch angular momentum must also be transported outward effectively (Weidemann, Astron. Astrophys. 59, 411, 1977).

RENZINI: What matters here is that the inner $\sim 0.3\ M_{\odot}$ of the stellar core retains its original angular momentum up to the core helium flash, and we are suggesting an observational test to see if this is actually the case: the blue end of the HB should be $\sim 0\overset{m}{.}5$ brighter than in the non-rotating case - Dr. J. Mengel could briefly comment on how rotating models were computed.

MENGEL: Our calculation was very crude. We assumed solid-body rotation on the main sequence followed by conservation of angular momentum in spherical shells in radiative zones. The convective envelope was assumed to be non-rotating. No transport of angular momentum was included.

KRAFT: I thought you said that the total mass loss rate was changed by rotation. Maybe you meant that the total mass loss, itself, was what was changed.

RENZINI: Yes. The total mass loss is changed, through the effect on evolution.

DEMARQUE: My question also has to do with the treatment of the angular momentum - are the effects of internal rotation included on the horizontal branch as well as on the giant branch? Or are you rather considering horizontal branches composed of non-rotating models with core masses consistent with the Mengel-Gross models?

RENZINI: Yes, we used the Sweigart and Gross non-rotating zero age HB models - this is justified considering that rotation begins to affect the stellar structure only when the star is very close to the red giant tip, that is, when the stellar core gets very small. Since the helium flash forces the core to expand (the core radius increases by about a factor of three) we expect that rotation will be much less important in HB stars.

HARDORP: I am still confused about the input angular momentum on the M.S. If I understood correctly, the core of the star is assumed not to be influenced by the slowing down of the rotation of the envelope due to the outer convection zone. So you expect for example the Sun to have a rapidly rotating core now.

RENZINI: What we have shown is that if core angular momentum is conserved then one gets the desired HB spread. On the other hand we are also exploring other alternatives like for instance a composition range within one cluster - concerning the Sun we cannot say anything about one single star.

CONTINUED ON PAGE 406

PHYSICAL PROPERTIES OF CEPHEIDS AND RR LYRAE STARS

J. W. Pel

Max-Planck-Institut für Astronomie, Heidelberg

J. Lub

European Southern Observatory, Genève

1. INTRODUCTION

At the Leiden Southern Station a large program on galactic pulsating variables has been carried out with the Walraven five-channel photometer. For 170 Cepheids and 100 RR Lyrae stars south of $\delta=+15^{\circ}$ there are now available 30000 VBLUW measurements (Walraven et al., 1964; Pel, 1976; Lub, 1977b). These data are used to derive physical properties of the variables. A general discussion of the VBLUW system and its calibration for A-G stars by means of the theoretical fluxes of Kurucz (1975) has been given by Lub and Pel (1977). Following this calibration, reddening and blanketing corrections, as well as effective temperatures, gravities, and their cyclical variations were determined, and luminosity and radius curves were derived (Cepheids: Pel, 1977; RR Lyrae stars: Lub, 1977a). We will present here some interesting results that have been obtained from these data. A more detailed discussion is in preparation.

2. CEPHEIDS

The absence of gravity effects in (B-L) for F-G stars is the basis of the locus method used to derive reddenings for the Cepheids (Pel, 1977). The Cepheid locus in the (V-B)-(B-L) diagram can only be applied to Population I Cepheids, however, as (B-L) is

A. G. Davis Philip and D. S. Hayes (eds.), The HR Diagram, 229-236 and 406.

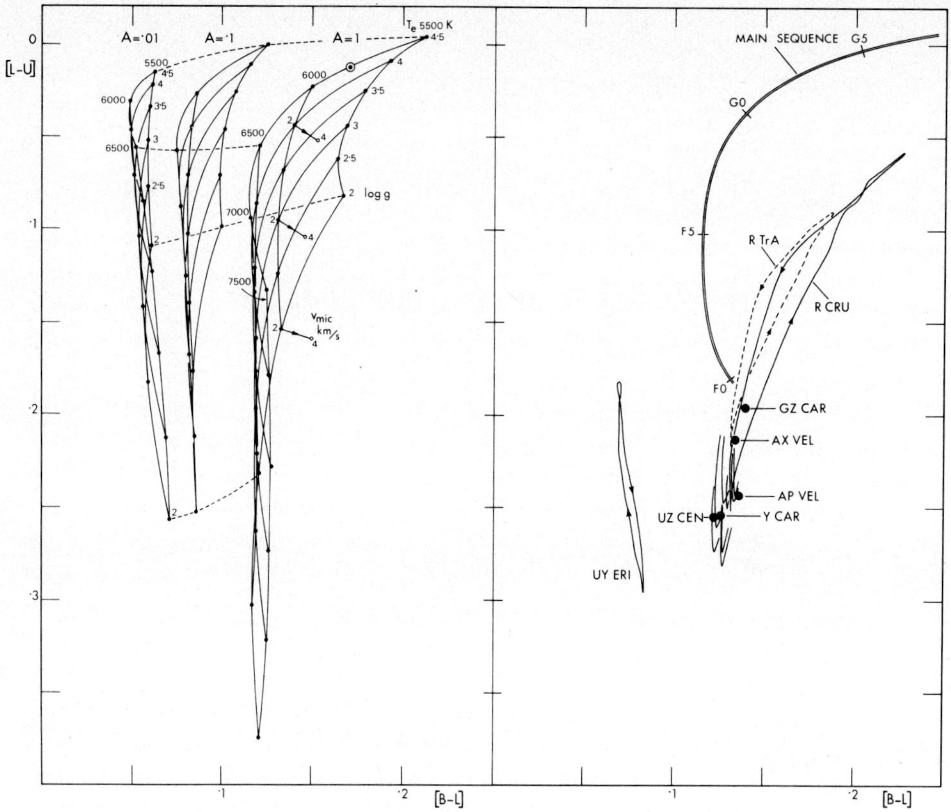

Fig. 1. The [B-L] - [L-U] diagram for F-G stars. Left: theoretical colors for 1x, 0.1x, and 0.01x the solar heavy elements abundance (A=1, 0.1, 0.01), computed from Kurucz (1975). In the A=1 grid the effect of raising the microturbulence from 2 to 4 km/s has been indicated. Right: observed colors for main-sequence stars, the Pop II Cepheid UY Eri, and for short-period Pop I Cepheids with T_{max} 6600 K (two complete loops are shown, for the other 8 Cepheids only the maxima are drawn). Dots indicate the mean maxima of five double-mode Cepheids. The observed main-sequence was used to match the zeropoints of observed and theoretical colors. As is customary in the VBLUW system, all colors are on a $\log_{10}$ (intensity) scale.

sensitive to differences in line blanketing. Extra information about the chemical homogeneity of the Pop I Cepheids is therefore very important. For this purpose the [B-L] - [L-U] diagram (Fig. 1) is useful. [B-L] = (B-L)-0.43 (V-B) and [L-U] = (L-U)-0.21 (V-B) are reddening-independent indices.

The theoretical colors in Fig. 1 indicate that around T_{eff} = 6500-7000 K [B-L] is a reddening-free composition parameter that is virtually gravity and temperature independent. Provided that

a Cepheid reaches the range 6500-7000 K, we can estimate its composition from the bluest [B-L] value along the cycle. Indeed we get a clear separation between UY Eri and the Pop I Cepheids. For UY Eri we estimate A=0.03; a similar value ($\log Z/Z_{\odot}=-1.66$) was used by Böhm-Vitense et al. (1974). The Cepheid loops agree even in detail with the theoretical colors: the Pop I loops are described counter-clockwise and fit well to the cool branch of the A=1 grid, while UY Eri runs in the opposite sense corresponding to the high-temperature branch of the A=0.03 grid. Considering the fact that composition is not the only source of scatter in [B-L], Fig. 1 indicates a remarkably small range in composition for the short-period Pop I Cepheids. This is strong support for the use of one intrinsic locus to derive reddenings for the "classical" Cepheids. Furthermore, Fig. 1 shows that the double-mode Cepheids have Pop I compositions, a result which is important in view of the theoretical problems for these stars (e.g. Petersen, 1973; Cox et al., 1977).

The observed periods and temperatures of the Cepheids can be used to derive luminosities from the pulsation relation, which gives the period as a function of mass M, luminosity L, and temperature. In order to eliminate M, one has to assume a M-L relation. Once T_{eff} and L are known, we can construct the theoretical HR diagram for the Cepheids. This is shown in Fig. 2. Here we used the pulsation relation and the theoretical M-L relation for Cepheids from Iben and Tuggle (1975). For the Cepheids with P=11 days the equilibrium temperatures from Pel (1977) were applied. The theoretical VBLUW colors do not cover the long-period Cepheids over their whole cycle, however. We therefore designed a recipe which gives the colors of the equilibrium configuration directly from the color curves. These equilibrium colors lie within the range of the theoretical colors for all periods. As this algorithm works very accurately for all short-period Cepheids, we applied it also to the long-period Cepheids to obtain the temperatures for those stars.

Both in the HR diagram of Fig. 2 and in the P-T_{eff} diagram (not shown) we get an excellent fit of the data to the fundamental blue edge for Y=0.28 and Z=0.02. A number of stars lie to the left of this blue edge, but most of these are suspected to have blue companions (open symbols) while the others are double-mode or likely first-harmonic pulsators (see below). The Cepheid strip seems to become wider at higher L, an effect also observed by Butler (1976a, 1977) in the Magellanic Clouds (most clearly in the LMC).

The present Cepheid temperatures are significantly cooler than most of the Cepheid temperature scales currently in use. It is therefore interesting to see what the consequences are of our T_{eff} scale for the well-known Cepheid "mass-discrepancy". As an example we take the cluster Cepheid S Nor. Using the cluster data

Fig.2 The Cepheids in the HR diagram. The M-L and pulsation relations of Iben and Tuggle (1975) were used to compute luminosities and lines of constant period. Composite fundamental (F, full curve) and first-harmonic (1H, dashed) blue edges from Iben and Tuggle are given, as well as the fund.blue edge (F, dot-dash) from King et al. (1975); all for Y=0.28, Z=0.02, and the M-L relation of Iben and Tuggle (1975). Open symbols: stars with known or suspected companions or with peculiarities; ● : long-period Cepheids for which the equilibrium-algorithm (see text) was used to derive T_e ; ◉ double-mode Cepheids; ✦ : likely overtone pulsators (see text), plotted at the position corresponding to the fundamental period.

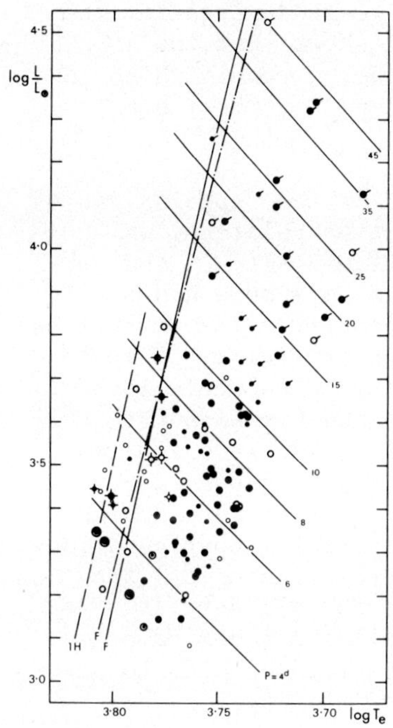

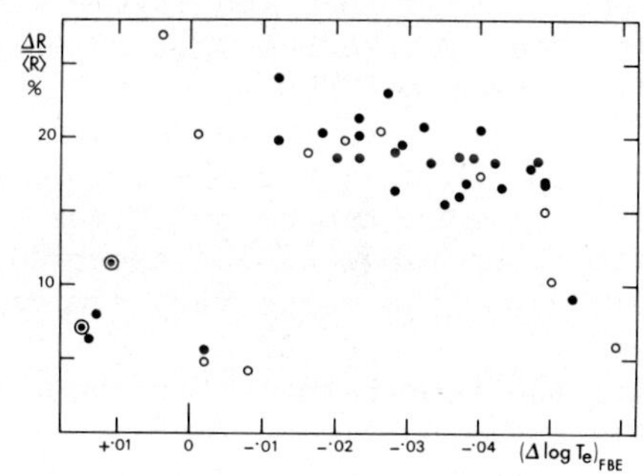

Fig.3 Radius amplitude $\Delta R/\langle R\rangle$ versus $(\Delta \log T_e)_{FBE}$, the distance in T_{eff} measured at constant L from the fundamental blue edge (for Y=0.28, Z=0.02) of Iben and Tuggle (1975). Symbols as in Fig. 2.

---

from Breger (1970), and (m-M) = $3^m\!.20$ as distance modulus for the Hyades, we find an "evolution mass" M=6.48 $M_\odot$ and a "pulsation mass" M=5.73 $M_\odot$ for this Cepheid (again with the M-L and pulsation relations of Iben and Tuggle). This is a mass-discrepancy of only 12%, three times smaller than the discrepancies usually found (e.g. Fricke et al., 1972), and not very disturbing in view of the many uncertainties involved in these mass estimates.

The behavior of the pulsation amplitude in the instability strip is important for several reasons. In particular, many authors have tried to use the pulsation amplitude to locate Cepheids in the strip (e.g. Kraft, 1960; Sandage and Tammann, 1971). Conflicting evidence on this point has been put forward by Madore (1976) and Butler (1976b). In Fig. 3 we have therefore plotted the radius amplitudes from Pel (1977) versus distance in T_{eff} measured at constant L from Iben and Tuggle's fundamental blue edge. Only Cepheids with accurate radius curves (Pel, 1977; Table 5[a]) were used, and only in the interval $3.24 < \log L/L_{\odot} < 3.60$ in order to avoid effects from the bumps around P=9-10 days or from the decrease in amplitude at the shortest periods.

The pronounced amplitude discontinuity in Fig. 3 around the position of the fundamental blue edge reminds us very much of the behaviour of the Type-c and Type-ab RR Lyrae stars, and suggests strongly that the blue small-amplitude Cepheids are first-harmonic pulsators. Contrary to the theoretical computions of Stellingwerf (1975b), Fig. 3 indicates that the double-mode Cepheids are probably related to the fundamental-overtone transition. Finally, we see that the pulsation amplitude is not well suited to measure the position of a Cepheid inside the strip: for the large-amplitude pulsators $\Delta R/\langle R\rangle$ varies too little with temperature, and at the small amplitudes we can not distinguish the red-edge Cepheids from the overtone pulsators by amplitude alone.

3. RR LYRAE STARS

In contrast to the classical Cepheids, the RR Lyrae stars show a wide range in metal abundances. A discussion of their properties is thus preferably done by deblanketed or blanketing-independent colors. Again we use the absence of gravity effects in the (V-B) - (B-L) diagram, but this time to determine the amount of metal line blanketing from [B-L] measured at minimum light (for the RRab stars). A study of the empirical and theorectical blanketing vectors shows that $(V-B)^* = (V-B)-0.370(B-L)$ and (L-U) are the appropriate blanketing-free temperature and gravity indicators in this case (Lub, 1977a). Blanketing is measured with respect to the most metal-poor stars by $\Delta[B-L] = (B-L)-0.51(V-B)^*-0.083$. This $\Delta[B-L]$ correlates very well with conventional blanketing indicators such as ΔS (Preston, 1959) and $\delta(U-B)_S$ (Sturch, 1966); it is however measured to a much higher accuracy (± 0.002 on a total range of 0.1 from A=0.01 to 1).

Using blanketing-independent colors, the reddening is determined from:

1) the correlation between $(V-B)^*$ at minimum and period for RRab stars (cf. Sturch, 1966)

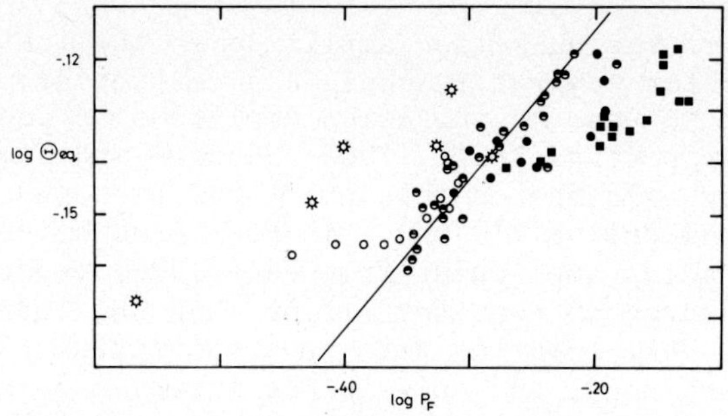

Fig.4 The period-temperature plane for the RRab stars.
■ : $\Delta[B-L]$ < 0.020 , $\Delta S \geq 9$
● : $\Delta[B-L]$=0.020-0.025, $\Delta S \simeq 8$
◓ : $\Delta[B-L]$=0.025-0.045, $\Delta S \simeq 6$
o : $\Delta[B-L]$=0.045-0.075, ΔS =2, 3
☼ : $\Delta[B-L] \geq$ 0.075 , $\Delta S \leq 1$
The line is the theoretically expected relation for $\log M^{0.81}/L = -1.89$.

---

2) the blue envelope in the $\langle V-B\rangle^* - \langle\beta_J\rangle$ plane ($\langle\rangle$: time average) for all variables; $\langle\beta_J\rangle$ has been recalculated from the H_β data of Jones (1973).

Both methods give identical results.

The complete coverage of all light- and color curves (Lub, 1977b) allows a detailed study of the luminosity, temperature, gravity and radius variations. Typical radius amplitudes of 14 and 5% are found for the RRab and RRc stars, respectively, in good agreement with the calculations by Stellingwerf (1975b). The equilibrium position of the variables in the T_{eff}-log g plane is derived by taking the appropriate average over the cycle (van Albada and de Boer, 1975).

The most useful comparison with pulsation theory can be made in the period-temperature plane, which we show in Fig. 4 (here θ_{eff}=5040/T_{eff}. We have grouped the stars in intervals of $\Delta[B-L]$ and a regular trend with metallicity is immediately evident in this diagram. Moreover, from pulsation theory (e.g. van Albada and Baker, 1971) it is predicted that stars of a given mass-luminosity ratio will lie along a linear log θ - log P relation in Fig. 4; the stars with $\Delta S \simeq 6$ are seen to fall almost exactly along such a line with log $(M^{0.81}/L)=-1.89\pm0.01$. It is remarkable that in the inhomogeneous sample of field variables it is possible to isolate groups with such a small internal spread in physical properties. Even though there are transition cases, we are strongly reminded of the Oosterhoff phenomenon (van Albada and Baker, 1973) by the average parameters of the two main groups

present: $\Delta S=6$ and 9, $P_{ab}=0^d.55$ and $0^d.65$, $P_{trans}=0^d.44$ and $0^d.53$, log $(M^{0.81}/L)=-1.89$ and -1.97. A discussion of the best estimates of mass and luminosity, taking into account constraints set by stellar evolution theory, leads to values at $\Delta S=6$ of $M=0.6\pm0.03$, log $L=1.70\pm0.03$, or $M_V=0^m.45\pm0^m.08$ (Lub, 1977a). The agreement with recent results from statistical parallax determinations (Hemenway, 1975; Heck, 1977) seems good.

A second important feature of the period-temperature diagram is the fact that the fundamental and first-harmonic blue edges do not depend upon mass and luminosity (Tuggle and Iben, 1972). This makes this diagram ideal for the determination of the helium abundance in the envelopes of pulsating stars. At $\Delta[B-L] = 0.030$ both from the RRc stars (not shown in Fig. 4) and the RRab stars we find $Y=0.28\pm0.02$. This value agrees with the one derived from a comparison of the mass-luminosity ratio $(M^{0.81}/L)$ with the horizontal-branch evolutionary tracks by Sweigart and Gross (1976). The behavior of the transition period appears to be in agreement with the hypothesis by van Albada and Baker (1973) and with the detailed stability calculations of Stellingwerf (1975a).

REFERENCES

Albada, T.S. van and Baker, N.H. (1971). Astrophys. J. 169, 311.

Albada, T.S. van and Baker, N.H. (1973). Astrophys. J. 185, 477.

Albada, T.S. van and Boer, K.S. de (1975). Astron. Astrophys. 39, 83.

Böhm-Vitense, E., Skody, P., Wallerstein, G. and Iben, I. (1974). Astrophys. J. 194, 125.

Breger, M. (1970). Astron. J. 75, 239.

Butler, C.J. (1976a). Astron. Astrophys. Suppl. 24, 299.

Butler, C.J. (1976b). Roy. Obs. Bull. 182, 159.

Butler, C.J. (1977). preprint.

Cox, A.N., Deupress, R.G., King, D.S. and Hodson, S.W. (1977). Astrophys. J. 214, L127.

Fricke, K., Stobie, R.S. and Strittmatter, P.A. (1972). Astrophys. J. 171, 593.

Heck, A. (1977). preprint.

Hemenway, M.K. (1975). Astron. J. 80, 199.

Iben, I. and Tuggle, R.S. (1975). Astrophys. J. 197, 39.

Jones, D.H.P. (1973). Astrophys. J. Suppl. 25, 487.

King, D.S., Hansen, C.J., Ross, R.R. and Cox, J.P. (1975). Astrophys. J. 195, 467.

Kraft, R.P. (1960). Astrophys. J. 132, 404.

Kurucz, R.L. (1975). In Multicolor Photometry and The Theoretical HR Diagram, A.G.D. Philip and D.S. Hayes, eds., Dudley Obs. Report 9, 271.

Lub, J. (1977a). Thesis, University of Leiden.

Lub, J. (1977b). Astron. Astrophys. Suppl. 29, 345.

Lub, J. and Pel, J.W. (1977). Astron. Astrophys. **54**, 137.
Madore, B.F. (1976). Roy. Obs. Bull. **182**, 151.
Pel, J.W. (1976). Astron. Astrophys. Suppl. **24**, 413.
Pel, J.W. (1977). Astron. Astrophys. (in press).
Petersen, J.O. (1973). Astron. Astrophys. **27**, 89.
Preston, G.W. (1959). Astrophys. J. **130**, 507.
Sandage, A.R. and Tammann, G.A. (1971). Astrophys. J. **167**, 293.
Stellingwerf, R.F. (1975a). Astrophys. J. **195**, 441.
Stellingwerf, R.F. (1975b). Astrophys. J. **199**, 705.
Sturch, C. (1966). Astrophys. J. **143**, 774.
Sweigart, A.V. and Gross, P.G. (1976). Astrophys. J. Suppl. **32**, 367.
Tuggle, R.S. and Iben, I. (1972). Astrophys. J. **178**, 455.
Walraven, J.H., Tinbergen, J. and Walraven, Th. (1964). Bull. Astr. Inst. Neth. **17**, 520.

DISCUSSION

BUSCOMBE: How are masses for classical Cepheids estimated, and what are typical values?

PEL: The so-called evolution masses are taken from the observed luminosities and the theoretical mass-luminosity relation for the Cepheid region. "Pulsation masses" are derived from the observed period, luminosity and temperature and the theoretical pulsation relation.

Both estimates can only be made when direct information about the luminosity is available, which is only for the Cepheids in clusters.

Typical evolution masses are 3 solar masses at the shortest periods to 12 solar masses for the long-period Cepheids.

KRAFT: What do you get for $\langle M_V \rangle$ when you go from $\Delta S = 0$ or $\Delta S = 10$?

LUB: We estimate the absolute magnitude at $\Delta S \sim 10$ to be about $0^m_{\cdot}20$ lower than at $\Delta S = 6$; conversely at $\Delta S \sim 0$, say, to be $0^m_{\cdot}30$ higher than at $\Delta S = 6$.

BUTLER: Do you assume constant mass in deriving these luminosity differences?

LUB: Not quite, stellar evolution theory puts constraints on the possible masses of stars which after reaching the horizontal branch will evolve through the instability strip. This has been taken into account in the values quoted in my answer to Kraft.

DICKENS: Are any of the first harmonic pulsators amongst the Cepheids in clusters and if so are their pulsation masses more in accord with evolutionary masses?

PEL: There is one likely first harmonic Cepheid which is member of an open cluster, and that is EV Sct. I have not determined evolution and pulsation masses for that star yet, but it is clear that the puslation mass-estimate comes out wrong when the period is taken for the wrong mode.

CONTINUED ON PAGE 406

PULSATION MODAL BEHAVIOR OF CLASSICAL CEPHEIDS IN THE HR DIAGRAM

Arthur N. Cox and Stephen W. Hodson

Los Alamos Scientific Laboratory

Mapping of the Cepheid region of the Hertzsprung-Russell diagram has been difficult to do from a theoretical viewpoint because the masses and compositions of the Cepheids have always been very uncertain. It now appears that the various mass anomalies have been solved by distance scale changes and by the realization of very helium rich convection zones. This helium enrichment is caused by a Cepheid wind which blows away more hydrogen than helium, just as in the solar wind. Now with the evolutionary theory masses, radii, luminosities, and our new composition structures, the predicted blue edges of the instability strip and periods throughout the strip agree very well with observations.

Let me first justify the new unconventional composition structure which is the key to my results. Consider the Cepheids with bumps in their light and velocity curves. Their periods range from 5.5 to perhaps 17 days, but the bumps are most visible between 7 and 10 days. Hertzsprung noticed that the bump was very low on decreasing light at 7 days, was at the peak at about 10 days, and crossed over to rising light at longer periods. Simon and Schmidt (1976) have used Stobie (1969) calculations to show that the ratio of the linear theory second overtone to fundamental period ratio Π_2/Π_0 is uniquely correlated with bump phase. Thus the phase is late when Π_2/Π_0 is 0.53 and Π_0 is 5.5 days, and early when Π_2/Π_0 is 0.46 at 17 days.

The first figure shows the period ratio Π_2/Π_0 versus period for models with various masses and compositions. First notice the 7 $M_\odot$ models with the homogeneous King IVa composition. None should give bumps because Π_2/Π_0 is always above 0.53. If the luminosity is retained, but the mass decreased to 5 $M_\odot$ as Christy suggested, there are predicted bumps for at least the longer periods. Use of the new Carson opacities at 7 $M_\odot$ gives bumps at the longer periods as Sastri Vemury (1977) has reported. Actually, we get no bumps at the extension near 9 days using a composite of

A. G. Davis Philip and D. S. Hayes (eds.), The HR Diagram, 237–240. All Rights Reserved.

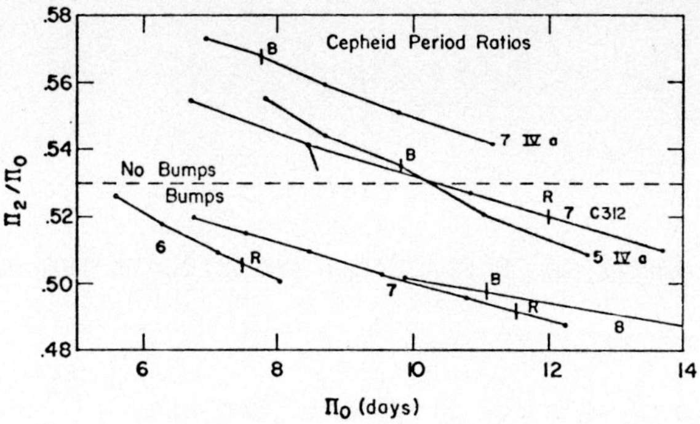

Fig. 1. Π_2/Π_0 versus Π_0 for Cepheids.

opacities, but at a somewhat larger amplitude, Sastri does get a bump. At 6, 7, and 8 $M_\odot$, with Y = 0.75 in the convection zones redward of the blue (B) and blueward of the red (R) edges, the proper Hertzsprung progression is reproduced.

Next we give our theoretical data. The Sandage and Tammann (1969) blue and red edges are the only observational data here. First, notice the 7 $M_\odot$ evolutionary theory track. Next, the masses used are based on an evolutionary theory mass-luminosity law. Linear theory periods and growth rates have been calculated for many models all over this diagram using the helium enrichment for masses up to 8 $M_\odot$. The wind does not blow long enough for enrichment above this mass where evolution is too rapid. The fundamental, first overtone, and second overtone blue edges are indicated. Nonlinear calculations have the second overtone pulsations at too low an amplitude for visibility in the case of the very blue models.

Comparison of periods with observations is in the last figure. The observational and theoretical blue edges appear again. Agreement is very good, but a few interesting discrepancies still exist. The Cepheid wind helium enrichment in the two mixed convection zones seems to be verified by many observations. More nonlinear calculations are underway.

REFERENCES

Christy, R.F. (1968). Qtr. J. Roy. Astron. Soc. 9, 13.
Sandage, A. and Tammann, G.A. (1969). Astrophys. J. 157, 683.
Simon, N.R. and Schmidt, E.G. (1976). Astrophys. J. 205, 162.
Stobie, R.S. (1969). Mon. Not. R. Astron. Soc. 144, 485.
Vemury, S.K. (1977). Bull. Am. Astron. Soc. 9, 360.

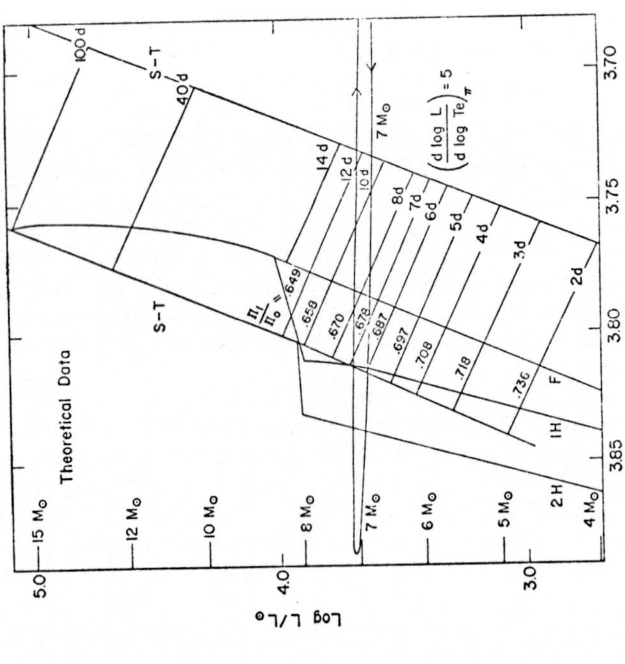

Fig. 2.

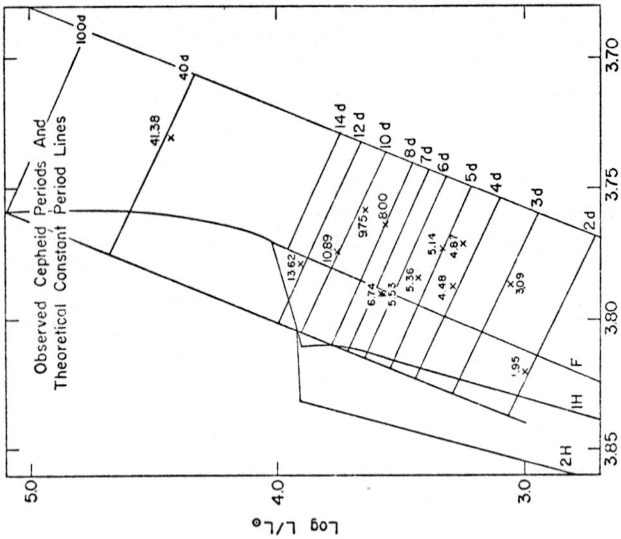

Fig. 3.

Log $L/L_{\odot}$ versus Log T_{eff}.

DISCUSSION

TAYLER: As you know I am worried about whether you could keep a stable high helium content layer on top of a lower helium content interior. There must surely be some instability and mixing. At the least a molecular weight gradient should be set up which is marginally stable. I cannot however make an immediate estimate of the change which this would make in your required Y.

COX: You have brought up one of the main problems to this enrichment. This Rayleigh-Taylor instability might be very fast and prevent the enrichment unless the Cepheid wind is very strong. Short time scales may exist but perhaps a gradient of helium can be set up which mixes slowly enough to give helium enrichment and a reconcilation of the Cepheid mass discrepancies. So far I haven't been able to figure out how to do the calculations for this mixing time scale.

MADORE: Instead of the necessity of invoking mixing to explain the presence of Cepheids without bumps it may be that here we are seeing the first-crossing Cepheids where the wind has not yet had time to act. This of course could be checked by comparing the ratio of Cepheids with and without bumps to the ratio of the second and first crossing times.

COX: I think that there may be good agreement here. Perhaps 20% of the 7-10 day Cepheids do not have bumps, and I believe that the first crossing time is also only about 20% of the total time that a star is a Cepheid.

WALLERSTEIN: If the helium is increased from about Y=0.25 to 0.75, X must have decreased from about 0.72 to 0.22. The effect of this is to increase the metals/hydrogen ratio by a factor of 3 to 4. So far as I know, no such effect has been seen in the comparison of Cepheids with non-variable supergiants.

COX: Of course I don't know if the effect of large changes of Z/X is observable. Remember that all second crossing yellow giants hotter than about 4500 K would have the same effect regardless of whether they pulsate. Construction of a few good model atmospheres could probably answer the question, but, so far, the only ones I know of are by Parsons without properly allowing for blanketing. I belive factors of two or more in Z/X have been easily possible.

APPENZELLER: Assuming your rather high mean molecular weight in the outermost layers of the Cepheids one obviously should expect some change of the stellar radii (as compared to "conventional" models). How big is this effect quantitatively?

COX: In our models we select L and T_{eff} and therefore we hold R fixed for different structures with different helium enrichments. The changes in the periods are due to the changes in the pulsation constant $Q_i=\Pi_i \langle\rho\rangle/\langle\rho_\odot\rangle$ which came about due to different mass concentrations. Periods do change roughly 5-10% making very large relative changes in Π_2/Π_0 and Π_1/Π_0.

THE RADII OF RR LYRAE STARS

L. A. Balona

South African Astronomical Observatory

Balona (1977) has described a modification of the Wesselink technique of radius determination which takes into account the effects of observational errors using the Principle of Maximum Likelihood. This leads to a tighter period - radius relationship for Cepheids than obtained by the classical method and suggests that some of the problems associated with radii estimation of RR Lyraes may be due to the incorrect statistical treatment of the data. With this in mind, we have applied the Maximum Likelihood technique to well-observed RR Lyraes. Except for two stars, the data used were taken from the literature.

The resulting period - radius relationship is shown in Fig. 1. The large scatter must to a great extent reflect the generally poor data, especially the radial velocities. Nevertheless, in contrast to past attempts using the classical Wesselink technique, we obtain meaningful radii for all the stars. We can find no correlation between metallicity and position in the diagram. However, the c-type RR Lyraes (crosses) do seem to have shorter periods than the others (as expected if they are first harmonic pulsators), but the evidence is hardly conclusive.

We obtain the mean quantities $\langle R \rangle = 11.7$ solar radii and $\langle \log P \rangle = -0.28$ after "fundamentalizing" the c-type variables (van Aldaba and Baker 1971). This period corresponds to a mean effective temperature $\langle \log T_{eff} \rangle = 3.85$ (Lub 1977), giving $\langle M_V \rangle = -1.3$. This is considerably brighter than the currently accepted value.

Woolley and Savage (1971) have shown that if the surface

A. G. Davis Philip and D. S. Hayes (eds.), The HR Diagram, 241-243. All Rights Reserved.

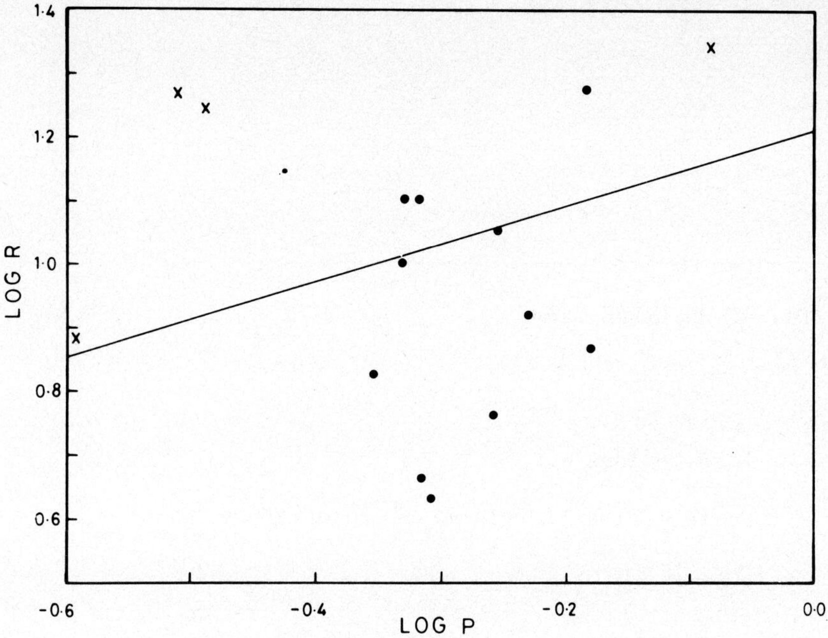

Fig. 1. The period - radius diagram for RR Lyrae stars. The straight line is an extension of the relation for classical Cepheids.

brightness is a function of effective gravity as well as color index, then the Wesselink radius will be an overestimation. This could explain our result, but the theory takes no account of shock wave radiation. One way out of the problem is to exclude the rising branch from the solution for the radius, as it is during this time that rapid accelerations and shock wave formation takes place.

The resulting period - radius relationship when this is done is shown in Fig. 2. Two stars give meaningless radii and have been omitted. We now find $\langle R \rangle = 8.0$, $\langle \log P \rangle = -0.27$ so that $\langle M_V \rangle = -0\overset{m}{.}3$. This is closer to the accepted value, but is still a magnitude too bright.

We conclude that it may be possible to obtain trustworthy radii for RR Lyraes using the Wesselink method provided good velocity curves and accurate photometry can be obtained. To minimize problems associated with shock wave radiation the rising branch must be omitted from the solution. At present the data are too few and too poor to define a reliable period - radius relationship.

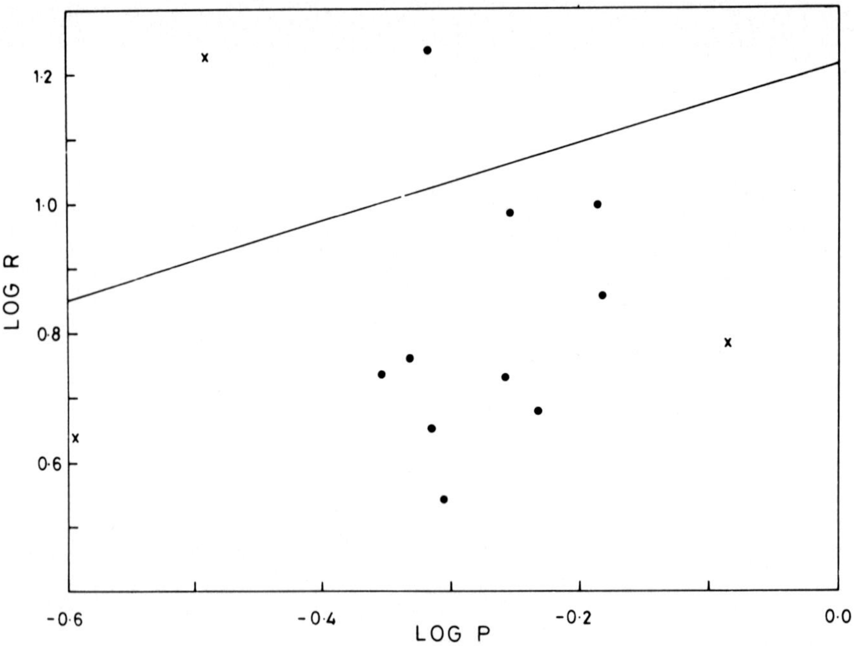

Fig. 2. The period - radius diagram for RR Lyraes obtained by excluding the rising branch from the radius solution. The straight line is an extension of the relation for classical Cepheids.

REFERENCES

Balona, L.A. (1977). Mon. Not. R. Astron. Soc. **178**, 231.
Lub, J. (1977). Unpublished Thesis, Leiden University, Netherlands.
van Aldaba, T.S. and Baker, N.H. (1971). Astrophys. J. **169**, 311.
Woolley, R. and Savage, A. (1971). Roy. Obs. Bulletin **170**, 365.

PART VI

GALAXIES AND MISCELLANEOUS SUBJECTS

Willem J. Luyten making a comment from the floor.

HR DIAGRAMS OF GALAXIES: AGES AND STAGES OF EVOLUTION

Beatrice M. Tinsley

Yale University Observatory

1. HR DIAGRAMS AND STELLAR POPULATIONS

Baade (1944) based his concept of stellar populations in galaxies on the HR diagrams that he inferred from the magnitude at which their brightest stars could be resolved. His type I population had bright blue supergiants like those in the disk of the Milky Way, while the brightest stars in type II were the red giants found in globular clusters. He postulated that the Hubble sequence of galaxy types from irregulars to ellipticals contained increasing proportions of Population II relative to Population I, and that similar differences characterized nuclear bulges of spirals relative to their disks. A very important revision of this picture came with the discovery by Morgan and Mayall (1957; Morgan, 1956, 1959) that the integrated blue light of the nuclear bulges of M31 and the Galaxy is dominated by strong-lined CN giants, not by the weak-lined type found in globular clusters. On the basis of integrated spectra of galaxies, Morgan developed a revised population scheme, in which the extreme types are a young-star rich population, like Baade's extreme Population I, and a young-star deficient population, analogous to Population II but generally metal-rich. Different proportions of these two types are still thought to represent the main differences among stellar populations in different regions of galaxies.

This paper will discuss selected current ideas about stellar populations in galaxies, based on properties of their integrated light, with emphasis on the implications for galactic evolution. Related reviews are those of Roberts (1963), Morgan and Osterbrock (1969), King (1971), Sandage (1975), van den Bergh (1975), and several papers at a recent Yale conference (Tinsley and Larson,

A. G. Davis Philip and D. S. Hayes (eds.), The HR Diagram, 247–257.

1977), as cited below.

Populations may be characterized by three functions: the stellar age distribution, chemical composition, and initial mass function (IMF). Many parameters are needed for an adequate description of these functions. For example, relative abundances of elements heavier than helium probably vary among populations (Faber, 1977), and the IMF varies within galaxies, at least on the scale of star clusters (Freeman, 1977). But, contrary to earlier claims (see King, 1971), there is no longer evidence that the *visible* part of any galaxy has an IMF with an unusually large proportion of red dwarfs, since spectroscopic data now show that late giants dominate the light at infrared wavelengths as well as in the blue (Baldwin et al., 1973; Frogel et al., 1977).

Spectra of galaxies show that, in general, individually bright stars are the main contributors to integrated light. Consequently, very little is known from direct observations about the lower part of the HR diagrams of galaxies - K and M dwarfs. Most of the light comes from luminous main sequence stars or red giants, with $M/L < 1$ (in solar units, and at any wavelength); yet rotation curves and velocity dispersions show that the visible bodies of galaxies have $M/L \sim 10$. At least 90% of their mass is therefore hidden, and the presence of late dwarfs is inferred not from their luminosity but from their weight. (This "hidden mass" problem is much less publicized than the further discrepancy of a factor ~ 10 between the "ordinary" M/L ratios of galaxies and the values ~ 100 derived from virial masses of clusters.) Luckily, the invisible stellar population is the least interesting, because K and M dwarfs hardly evolve in the lifetime of a galaxy. The stars that do show in integrated light give information on many aspects of stellar and galactic evolution, such as the following:

- The rate of evolution of the number of turnoff stars (and their descendent giants) in old populations governs the rate of luminosity evolution, which affects classical "lookback" tests in cosmology. This evolution can be predicted by analysis of the light of nearby galaxies and detected by bluer turnoffs in galaxies at large redshifts (Spinrad, 1977; Tinsley 1978a).

- Low-mass stars radiate a large fraction of their life's energy supply as very late red giants, which therefore contribute bolometrically most of the light from old populations. Their distribution by type can be estimated from galaxy spectra, and used to supplement theories of stellar evolution in stages where the complexities of envelope mixing and mass loss are unpredictable (Faber, 1977; Tinsley, 1978b).

- The color of the low-mass giant branch is sensitive to heavy-element abundances, so integrated colors of old populations

give information about their compositions (see below).

- Upper main sequence stars and supergiants are making the oxygen and carbon responsible for the bulk of chemical enrichment. It would misleading, however, to claim to understand chemical abundances in a galaxy from its present enrichment rate by massive stars; as examples given later will illustrate, the compositions of stars in a system depend on relations between the past star formation rate, gas flows, and gas content.

- The relative numbers of upper main sequence stars and turnoff stars tell how fast stars are forming relative to the past average rate, and so indicate the time scale for star formation in a galaxy. This point will occupy most of the present paper.

We therefore turn to the first function mentioned as characterizing stellar populations: the age distribution. Its effect on HR diagrams can be visualized with Sandage and Eggen's (1969) composite diagram for open clusters. If the stars in a system were coeval, a single cluster diagram would represent its population - e.g. M67 for a metal-rich elliptical galaxy several billion years old (with an important addition of M giants that are missing from the cluster because of its small total population [Baum, 1959; Tinsley and Gunn, 1976]). Similarly, a system with continuous star formation would have the whole spread of clusters. Relative numbers of stars and their contributions to the light can be estimated from simple models: Let the main sequence be populated in accordance with the empirical local IMF (allowing for stellar evolution off the main sequence for stars above $\sim 1\ M_\odot$), and let evolving stars be represented according to the number of their progenitors and their rates of evolution. It is found that turnoff stars contribute at short wavelengths, and, if there is current star formation, B - F (especially B) main sequence stars are important; with increasing wavelength, giants become more dominant, until at 2.2μ (the K band), late M giants contribute most of the light. These results resemble observed spectral types of galaxies (e.g. de Vaucouleurs and de Vaucouleurs, 1959; Morgan and Osterbrock, 1969; Frogel et al., 1977). If the IMF were much steeper than the local function, a young-star rich population would have a later integrated blue spectral type than observed, and late dwarfs rather than the observed giants would dominate in the infrared.

It is useful to concentrate on two main components in the integrated light of galaxies: (1) The young component contributes light from upper main sequence stars and their descendent supergiants; their lifetimes are much shorter than the age of the galaxy, so their contributions are proportional to the present star formation rate (SFR). (2) The old component contributes light mainly from turnoff stars and their descendent red giants;

these stars have accumulated since star formation began, so their contributions depend on the total number formed. The spectrum and colors of a galaxy are sensitive to the ratio of these components, and hence to *the ratio of present SFR to total mass*. The exact age distribution of stars is not critical, because stars of intermediate age (e.g. F dwarfs) contribute less than the two groups mentioned, and because stars with lifetimes > 10^9 yr ($m \lesssim 2\ M_\odot$, A stars and later on the main sequence) have very similar giant branches. As a primary parameter available from integrated light, we therefore have the ratio SFR/mass, or its inverse which gives a *time scale for star formation.* Baade's postulate that the sequence of galaxy types is one of relative amounts of Populations I and II thus becomes a postulate that the time scale for star formation varies along the sequence: The latest types (Im, Sc) form stars over a long period of time, and the earliest (SO, E) made all their stars long ago.

2. STELLAR POPULATIONS FROM UBV COLORS

For stars, the UBV photometric system is closely tied to the two-dimensional classifications of positions in the HR diagram, as well as providing information on metallicity and reddening. UBV colors are also very useful in outlining several broad properties of galaxy populations, although the parameters available are not the same as for individual stars. Factors influencing the integrated UBV colors of galaxies include the ratio SFR/mass, the age of the system, metallicity, the IMF, interstellar reddening (galactic and internal), redshift, and nonstellar sources of light such as gaseous emission lines. Despite the length of this list, we can use evolutionary models, information from different sorts of data, and corrections for the more extraneous factors, to make interesting inferences about stellar populations in galaxies from their UBV colors.

An important fact is that morphologically normal galaxies define a narrow locus, barely wider than observational errors, in the UBV plane, if their colors are corrected for effects of redshift and reddening (e.g. Larson and Tinsley, 1978). Two sequences actually define this apparently one-dimensional distribution, a metallicity sequence and one of time scale for star formation. The near coincidence of these sequences in UBV is simply bad luck; they are separated by spectrophotometric studies, population models, and the absence of HII regions and other signs of star formation in galaxies defining the metallicity sequence.

The lower half of the "normal" UBV line is occupied in part by the central bulges of spirals and by ellipticals and SOs, in a sequence of increasing luminosity towards redder colors (Baum, 1959) that is almost certainly due to a correlation of heavy-

element abundances with luminosity (Faber, 1973). On the assumption that all elements heavier than helium vary together, I have used models with different metallicities to estimate that the color-magnitude relation corresponds to $\Delta\log Z/\Delta M_V \simeq -0.1$ (Tinsley, 1978b). It is possible, however, that only a few elements are responsible for the trend (Faber, 1977), in which case a calibration must await new models for stellar interiors and atmospheres.

With a few exceptions (mainly extreme dwarfs), galaxies do not show orders-of-magnitude variations of mean metallicity among their stellar populations. This uniformity can be understood through very simple models for chemical evolution, on the assumption that the IMF is everywhere the same. Searle and Sargent (1972) pointed out that, if a galaxy evolves as a closed system, the gas acquires a metallicity $Z_g = y \ln \mu^{-1}$, where the yield y is a constant for a given IMF and element and μ is the fractional gas mass; thus only modest variations in Z_g are expected despite huge variations in μ. For a different situation, where gas is accreted by a galaxy at approximately the SFR, Larson (1972) showed that $Z_g \rightarrow y$, independently of the gas content. Moreover, one can show that the mean *stellar* metallicity is about equal to the yield in both types of model. Since information on stellar metallicities is available for very few galaxies, it seems reasonable to neglect the effects of variations on UBV colors when other population parameters appear to be dominant.

Now we consider the UBV locus of normal giant galaxies, which is a narrow sequence running from blue for late spirals to red for giant ellipticals. The color sequence here is surely due mainly to a progression from richness to deficiency in young stars. Theoretical models for evolving stellar systems allow more detailed statements to be made. The sequence can be reproduced by series of models with the same IMF (approximately the local function) and the same age (about 10^{10} yr), and the narrow locus does not allow great variations in IMF or age. The quantity that *does* vary along the sequence is the time scale for star formation. Various authors have reached the above conclusions using different functions to define the time scales: the efficiency of converting gas to stars (Tinsley, 1968), exponential time constants (Searle et al., 1973), and dynamical models (Larson and Tinsley, 1974). Larson and Tinsley (1978) found a more general result, namely that a narrow locus close to the empirical one is occupied by all models with about the local IMF, solar composition, ages between 5 and 20×10^9 yr, and monotonically declining SFRs. Moreover, no matter what the detailed time dependence of its SFR, the position of such a model on the common UBV line depends on only one quantity: the ratio of present SFR to total mass of stars. Thus UBV colors pick out just the ratio of young and old population components discussed above.

3. DYNAMICAL ORIGINS

The color sequence of normal galaxies is, of course, closely related to their sequence of shapes. For example, there is a neat progression of mean colors, from blue to red, along the Revised Hubble Sequence (de Vaucouleurs, 1977). The areas of the UBV diagram occupied by sizeable samples of each type overlap rather widely, which is not surprising since the Hubble Sequence depends on several parameters (Sandage, 1975) that do not correlate perfectly. Van den Bergh (1976) finds that his two-dimensional classification of spirals, in which the strength of spiral arms decreases from "vigorous" spirals through "anemics" to SOs, gives a further correlation, in that the more vigorous types at a given disk/bulge ratio tend to be bluer. (This claim is disputed by de Vaucouleurs [1977], however.) Likewise, the Yerkes classification by central concentration of light also correlates with UBV colors, as expected from its close relation to nuclear spectral types (Morgan and Mayall, 1957), but again there is much overlap among categories. It would obviously be of great interest to understand both the main trends and the scatter in color-morphology relations, in terms of dynamics: Why do time scales for star formation correlate with forms of galaxies? Many studies of this question have led to a coherent working sketch for an answer; its details and even main outlines are often erased, and a definitive picture has by no means emerged! The following sketch is based on many papers, most of which are cited in reviews by King (1971, 1977) and Larson (1976, 1977).

The process of galaxy formation from primordial gas clouds may itself be responsible for the major correlations among morphology, mass, and stellar population; a graphic outline of this part of the picture was given long ago by Baum (1959). The spheroidal parts of galaxies - nuclear bulges of spirals and the entire bodies of ellipticals - must have formed their stars within a few free-fall times, so they contain almost entirely old stars. If the collapse is dissipative, the gas from which stars in the inner regions form will have been enriched in metals by massive stars that died further out; thus the observed metallicity gradients arise naturally. During the collapse and star formation, supernova heating can drive the remaining gas from the system, cutting off further star formation and chemical enrichment; this can occur at an earlier stage in less massive systems, which explains their relatively low metallicities.

Disks, on the other hand, may form very slowly as diffuse, outlying gas gradually dissipates its vertical motions; the less efficiently stars form from this gas, the flatter the system can become. Consequently, disks of galaxies tend to contain gas from which stars are still forming. Some spirals - including our own - may still be accreting gas at a significant rate; M83, with its

warped outer HI that looks dynamically unstable on a short time scale, is suggested as an example (Rogstad et al., 1974). The present SFR in a disk may thus depend both on the efficiency with which gas is compressed to the point where stars form, and on the rate at which new gas is supplied from the outside. Some of the differences in colors among spirals with similar disk/bulge ratios could, in this picture, be due to differences in the ambient intergalactic gas density.

The environment of a galaxy may influence its SFR in other ways. The best-known case is the effect of intracluster gas in sweeping spirals clean of their interstellar matter, thereby turning them into S0s. A galaxy in the process of being stripped would look like the anemic spirals often found in clusters; and within a few billion years of the stripping, the victim would be hardly any bluer than an elliptical that had made all its stars in an initial burst. Conversely, swallowing an intergalactic gas cloud may suddenly enhance the SFR in a galaxy. M82 is the most famous example (Solinger et al., 1977), although its huge burst of star formation is buried in so much dust that the UBV colors are not unusually blue.

Violent interactions with other galaxies can also enhance the SFR, as illustrated by Larson and Tinsley (1978). The interacting systems in Arp's (1966) *Atlas of Peculiar Galaxies* have a very different distribution of UBV colors from the narrow locus of morphologically normal galaxies, and the peculiar colors correspond to population models in which bursts of star formation occurred up to a few times 10^8 years ago. There is even a correlation between time since the burst, indicated by the colors, and time since the interaction began, indicated by the presence or absence of tidal tails: In the dynamical models of Toomre and Toomre (1972), tails develop after the closest passage of the interacting pair, and indeed Arp pairs without tails have colors suggestive of bursts that occurred $\lesssim 10^8$ yr ago. Thus the abnormal color distribution of interacting galaxies can be understood if intense star formation is induced by violent dynamical interactions. Toomre (1977) remarks that ring galaxies are a fine example of this process. Peculiarities of morphology and colors are predicted to become negligible about 10^9 yr after a disturbance, so even so-called normal galaxies could have experienced very ragged histories of star formation in the more distant past.

Unusual distributions of stars in the HR diagram are responsible for the abnormal colors of galaxies with recent bursts (e.g. Searle et al., 1973; Larson and Tinsley, 1978), as follows. If a system has formed stars fairly uniformly for $\gtrsim 10^9$ yrs, the distribution of OB stars varies with both the IMF and their lifetimes (τ_m). But if the majority of massive stars were formed very recently, their distribution is not affected by evolution from the

main sequence, so it is unusually rich in massive stars by a factor $\propto \tau_m^{-1} \sim m^2$. In such a case, the earliest B and O stars are the main contributors to U light, in contrast to the mid-B to F stars seen normally; the B and V bands are less affected, so the upshot is an ultraviolet excess relative to normal galaxies. Later, a second turnoff due to stars made in the burst can give an unusually red (U-B) for a given (B-V).

If the burst of star formation involves more than 10% of the mass in old stars, (B-V) also is so strongly affected that the UBV colors are practically indistinguishable from those of a truly young galaxy, made entirely of young stars. Some chaotic-looking galaxies do indeed have blue enough UBV colors to be young; examples are NGC 4485 and 4490 (Arp 269) and NGC 3395 and 3396 (Arp 270). But in each case, given only UBV colors, we cannot tell if 100% or only 10% of the stars are young, and the appearance and blueness could be due to interaction rather than to youth. Unusually blue (V-K) colors would give a less ambiguous test for the absence of old red giants, although unfortunately the converse conclusion cannot be drawn since there could be 2μ radiation from dust even in a young system (Struck-Marcell and Tinsley, 1978).

Younger galaxies must of course be present at large redshifts, because of the lookback time, which is roughly $\sim 10^{10} z/(1+z)$ yr. Observations and implications of cosmological aspects of galactic evolution have been reviewed recently by Spinrad (1977) and Tinsley (1978a). Among the expected effects of evolution over cosmological time scales are: systematic changes towards bluer colors, especially reflecting the earlier turnoffs of ellipticals at redshifts of a few tenths; much bluer colors among disk galaxies in clusters at times before the spirals have become SOs; and dramatic evidence for rapid star formation in primeval giant elliptical galaxies. The first two effects have almost certainly been found, and prospects for detecting giant primeval galaxies are promising.

4. CONCLUSION

It is obvious from the list given in § 1 of this paper that I have only scratched the surface of the mine of information provided by the integrated light of galaxies. One reason for the lasting value of Baade's original criterion for classifying stellar populations, the nature of individually bright stars in their HR diagrams, is that those same stars provide most of the integrated light from unresolved systems. The colors and spectra of galaxies are therefore sensitive to the presence of OB stars, red giants of the old-disk type, and/or metal-poor giants of the halo type. From

their relative numbers comes information on the age and metallicity distribution of stars in different regions of galaxies, which in turn gives clues about galactic evolution. Despite decades of observational and theoretical work in this field, many interesting details are still out of reach. Several ubiquitous sources of difficulty include the following: the insensitivity of integrated light to details of stellar age distributions; the invisibility of stars comprising most of the mass of galaxies; and the fact that surface temperatures and compositions of the evolved stars depend not only on their original masses and compositions but also on stellar mixing and mass loss. I have concentrated in this brief paper on broad conclusions that can be drawn from fairly coarse indicators of stellar age distributions and metallicities. More detailed studies, referenced above, have suggested that chemical compositions and star formation histories differ in complicated ways, both within and among galaxies. A challenge for the future is to exploit further what *can* be seen in the integrated light, of systems whose HR diagrams cannot be drawn point by point, to obtain a fuller understanding of the evolution of stars and stellar systems.

This work was supported in part by the National Science Foundation (Grant AST76-16329) and the Alfred P. Sloan Foundation.

REFERENCES

Arp. H.C. (1966). Atlas of Peculiar Galaxies, California Institute of technology, Pasadena; Astrophys. J. Suppl. **14**, 1.

Baade, W. (1944). Astrophys. J. 100, 137.

Baldwin, J.R., Danziger, I.J., Frogel, J.A., and Persson, S.E. (1973). Astrophys. Letters **14**, 1.

Baum, W.A. (1959). Publ. Astron. Soc. Pacific **71**, 106.

de Vaucouleurs, G. (1977). In The Evolution of Galaxies and Stellar Populations, B.M. Tinsley and R.B. Larson, eds., Yale Univ. Obs., New York, p. 43.

de Vaucouleurs, G. and de Vaucouleurs, A. (1959). Publ. Astron. Soc. Pacific **71**, 83.

Faber, S.M. (1973). Astrophys. J. **179**, 731.

Faber, S.M. (1977). In The Evolution of Galaxies and Stellar Populations, B.M. Tinsley and R.B. Larson, eds., Yale Univ. Obs., New Haven, p. 157.

Freeman, K.C. (1977). In The Evolution of Galaxies and Stellar Populations, B.M. Tinsley and R.B. Larson, eds., Yale Univ. Obs., New Haven, p. 133.

Frogel, J.A., Persson, S.E., Aaronson, M. and Matthews, K. (1977). preprint.

King, I.R. (1971). Publ. Astron. Soc. Pacific **83**, 377.

King, I.R. (1977). In The Evolution of Galaxies and Stellar Populations, B.M. Tinsley and R.B. Larson, eds., Yale Univ. Obs., New Haven, p. 1.
Larson, R.B. (1972). Nature Phys. Sci. **236**, 7.
Larson, R. B. (1976). in Galaxies, L. Martinet and M. Mayor, eds., Geneva Observatory, Sauverny, p. 69.
Larson, R. B. (1977). American Scientist Vol. **65**, No. 2, p. 188.
Larson, R. B., and Tinsley, B. M. (1974). Astrophys. J. **192**, 293.
Larson, R. B., and Tinsley, B. M. (1978). Astrophys. J. in press.
Morgan, W. W. (1956). Publ. Astron. Soc. Pacific **68**, 509.
Morgan, W. W. (1959). Astron. J. **64**, 432.
Morgan, W. W., and Mayall, N. U. (1957). Publ. Astron. Soc. Pacific **69**, 291.
Morgan, W.W. and Osterbrock, D.E. (1969). Astron. J. 74, 515.
Roberts, M.S. (1963). Ann. Rev. Astron. Astrophys., Vol. **1**. Annual Reviews, Palo Alto, p. 149.
Rogstad, D. H., Lockhart, I. A., and Wright, M. C. H. (1974). Astrophys. J. **193**, 309.
Sandage, A. (1975). in Galaxies and the Universe, A. Sandage, M. Sandage, and K. Kristian, eds., Univsity of Chicago Press, Chicago, p. 1.
Sandage, A., and Eggen, O. J. (1969). Astrophys. J. **158**, 685.
Searle, L., and Sargent, W. L. W. (1972). Astrophys. J. **173**, 25.
Searle, L., Sargent, W. L. W., and Bagnuolo, W. G. (1973). Astrophys. J. **179**, 427.
Solinger, A., Morrison, P., and Markert, T. (1977). Astrophys. J. **211**, 707.
Spinrad, H. (1977). In The Evolution of Galaxies and Stellar Populations, B.M. Tinsley and R.B. Larson, eds., Yale Univ. Obs., New Haven, p. 301.
Struck-Marcell, C., and Tinsley, B. M. (1978). Astrophys. J. in press.
Tinsley, B. M. (1968). Astrophys. J. **151**, 547.
Tinsley, B. M. (1978a). In IAU Symposium No. 79, Large Scale Structure Structure of the Universe, M.S. Longair and J. Einasto, eds., D. Reidel, Dordrecht, p. 343.
Tinsley, B. M. (1978b). preprint.
Tinsley, B. M., and Gunn, J. E. (1976). Astrophys. J. **206**, 525.
Tinsley, B. M. and Larson, R. B. (1977). eds., The Evolution of Galaxies and Stellar Populations, Yale University Observatory, New Haven.
Toomre, A. (1977). In The Evolution of Galaxies and Stellar Populations, B.M. Tinsley and R.B. Larson, eds., Yale Univ. Obs., New Haven, p. 127.
Toomre, A. and Toomre, J. (1972). Astrophys. J. **178**, 623.
van den Bergh, S. (1975). Ann. Rev. Astron. Astrophys. Vol. **13**, Palo Alto, p. 217.
van den Bergh, S. (1976). Astrophys. J. **206**, 883.

DISCUSSION

WEHLAU: Are you able to tell if the extreme colors occur over the entire galaxy or just in some regions? That is, in which parts of the interacting galaxy does the burst of star formation occur?

TINSLEY: I wish this were known. A few have been mapped. In particular, NGC 4038/9, with the long "antennae" tails, has recently been mapped by Schweizer (IAU Symposium No. 77 in press), and he finds H II regions right out to the ends of the tails. It should be done for more, but it isn't easy, because they are mostly faint galaxies.

PARTHASARATHY: Please make a comment on the position of cD galaxies in HR diagrams of galaxies.

TINSLEY: They are red, like giant elliptical galaxies, and of course they are the most luminous of galaxies. There is some suggestion that they may be a little blue for their luminosity, according to the idea that cD's are conglomerates of smaller galaxies.

PARTHASARATHY: With regard to interacting galaxies and cD galaxies, I would like to make a comment. Recently Dr. S.M. Alladin and I studied the problem of the tidal disruption and coalescence of double galaxies. We found that the coalescence rate is higher than the disruption rate. Coalescence of galaxies in clusters can produce supergiant systems like the cD galaxies.

KODAIRA: You have shown us color-color diagrams of galaxies. Could you give us any inferences you can make when you plot galaxies in color-magnitude diagrams?

TINSLEY: For the ellipticals and the S0's the smaller ones are bluer. They would appear in the HR diagram as metal-poor faint blue galaxies. The large ellipticals are red, and metal rich. They would be giants, by spectral type, in the HR diagram. And, of course, in the spiral sequence the smallest galaxies are the little irregulars, which would have a spectral type around B. You can't define a unique spectral type for galaxies, because the spectral type depends upon wavelength. The longer the wavelength you consider, the later the type you pick up. So an elliptical would look like a G8-giant at ultraviolet wavelengths and a very late M-giant in the infrared. A schematic plot of M_V versus (B-V) for galaxies has been given by van den Bergh (1977, The Evolution of Galaxies and Stellar Populations).

SCHOMMER: A comment on NGC 4921. Woody Sullivan, Paul Johnson and I are surveying the HI content of cluster galaxies using the Arecibo radio telescope. Preliminary results show that the spirals in Coma, for example, are hydrogen deficient by at least a factor of 5 or 10 compared to standard field sample and lower concentration clusters. NGC 4921 was one of the few galaxies actually detected initially in Coma, and it appears to be depleted in HI by about a factor of 5 compared to similar galaxies in a less concentrated cluster, (A 1367). Preliminary results will be submitted for publication in the near future.

LUMINOSITIES OF LMC OB STARS

David Crampton*

Dominion Astrophysical Observatory,
Herzberg Institute of Astrophysics,
National Research Council Canada.

1. INTRODUCTION

Our knowledge of the spectra of stars in other galaxies is essentially limited to those which are extremely luminous. In the LMC, slit spectra of ∿ 200 stars have been obtained (Feast, Thackeray and Wesselink 1960, Ardeberg et al. 1972) down to V ∿ $12^m.5$ (M_V ∿ $-6^m.5$). Using an objective prism technique, Brunet et al. (1975) have been able to detect 272 OB stars with $12^m.2 < m_{pg} < 14^m.8$ in the LMC. At the fainter limit this survey should include relatively normal O and early B dwarfs and giants. The principal aim of the present investigation was to obtain spectra of these stars for comparison with galactic stars and to apply the techniques of MK classification and measurement of Hγ equivalent widths to determine their luminosities and hence a distance modulus to the LMC.

The absolute magnitude calibration of the MK classification of O stars has recently been considered by Conti and Alschuler (1971), Walborn (1972, 73) and Crampton and Georgelin (1975); that of the equivalent widths, W, of Hγ by Balona and Crampton (1974). A previous determination of the distance modulus of the LMC from the equivalent widths of luminous B stars by Hutchings (1966) gave $(V_o - M_V) = 17^m.6$, a value considerably smaller than that currently accepted ($18^m.6$, van den Bergh 1976). To explain this anomaly, it has been suggested the Balmer lines are abnormally strong in LMC (and SMC) stars, and perhaps this phenomenon is related, in the case of the SMC, to the weakness of the metal lines.

*Visiting Astronomer, Cerro Tololo Inter-American Observatory
DAO Contribution No. 350 = NRC No. 16109.

A. G. Davis Philip and D. S. Hayes (eds.), The HR Diagram, 259-262. All Rights Reserved.

2. MEASUREMENTS AND RESULTS

Thirty spectra with a dispersion of 47 Å mm^{-1} taken with the Carnegie image tube spectrograph on the CTIO 4m telescope were obtained of 17 of the bluest stars from the list of Brunet et al. Fourteen spectra of 7 galactic stars to act as MK and W standards were obtained with the same equipment. Although these standards served reasonably well for the spectral classification, the transformation between the equivalent widths measured from the image tube plates and Petrie's system could not be determined confidently. Because of this, and the small number of spectra per star, the results are of a preliminary nature.

The average distance moduli and their dispersion for the 17 stars resulting from the use of various calibrations are:

Method	(V_o-M_V)
MK, Blaauw (1963)	$18^m.91 \pm 0^m.59$ (11 stars only)
MK, Walborn (1972, 73)	18.82 ± 0.53
MK, Conti & Alschuler (1971)	18.67 ± 0.56
MK, Crampton & Georgelin (1975)	18.52 ± 0.60
W, Balona & Crampton (1974)	18.57 ± 0.83
van den Bergh (1976)	18.58

Unfortunately, the majority of the stars are classified as 07.5-09.5 III stars, so that the distance moduli rest on the calibrations of only a few MK types. More observations are required to explore the calibrations in detail but it appears that both the calibration of Blaauw and that of Walborn assign a somewhat high luminosity to these types of stars if the true distance modulus is $\sim 18^m.6$.

The $H\gamma$ distance modulus agrees well with the other determinations but its error is large and since the W's are not yet reliably tied to galactic standards, this may be fortuitous. Measurements of the He I and Si IV lines in the spectra indicate that these are of normal strength so perhaps then it is not surprising that the H lines appear normal too.

It is worthwhile exploring what the various calibrations are based on. All of the MK calibrations are based on main sequence fitting of galactic clusters and are ultimately related to the Hyades, (with a distance modulus of $3^m.0$). The $H\gamma$ calibration is tied to the distance modulus of the Pleiades which was taken to be $5^m.55$. This is the distance derived by Blaauw (1963), also from a main sequence fit to the Hyades. However, Crawford (1977) finds (V_o-M_V) (Pleiades) = $5^m.51$ based

largely on trigonometric parallaxes of A and F stars, and Petrie (1964) found $5^m.37$ again independently of the Hyades. Thus, it is not clear that the Pleiades distance modulus is seriously in error, and since all of the calibrations are related to the Pleiades, perhaps the Hyades problem does not affect them at all.

The mean value of the distance modulus to the LMC from the three recent calibrations of the MK types of O stars is $18^m.67$. Since this value agrees well with the $H\gamma$ equivalent width modulus, and with the modulus determined by a variety of other methods, it appears that the O stars in the LMC are quite similar to those of our Galaxy.

REFERENCES

Ardeberg, A., Brunet, J.-P, Maurice, E. and Prevot, L. (1972). Astron. Astrophys. Supp. 6, 249.

Balona, L. and Crampton, D. (1974). Mon. Not. R. Astron. Soc. 166, 203.

Blaauw, A. (1963). In Basic Astronomical Data, K. Aa. Strand, ed., U. of Chicago Press, Chicago, p. 383.

Brunet, J.P., Imbert, M., Martin, N., Mianes, P., Prevot, L., Rebeirot, E. and Rousseau, J. (1975). Astron. Astrophys. Suppl. 21, 109.

Conti, P.S., and Alschuler, W.R. (1971). Astrophys. J. 170, 315.

Crampton, D. and Georgelin, Y.M. (1975). Astron. Astrophys. 40, 317.

Crawford, D.L. (1971). In Dark Nebulae, Globules and Protostars, B.T. Lynds, ed., U. of Arizona Press, Tucson, p. 27.

Crawford, D.L. (1977). preprint.

Feast, M.W., Thackeray, A.D. and Wesselink, A.J. (1960). Mon. Not. R. Astron. Soc. 121, 25.

FitzGerald, M.P. (1970). Astron. Astrophys. 4, 234.

Hutchings, J.B. (1966). Mon. Not. R. Astron. Soc. 132, 433.

van den Bergh, S. (1976). IAU Colloquium No. 37, Redshift and Expansion of the Universe, B.E. Westerlund and C. Balkowski, eds., to be published.

Walborn, N.R. (1972). Astron. J. 77, 312.

Walborn, N.R. (1973). Astron. J. 78, 1071.

DISCUSSION

PAYNE-GAPOSCHKIN: What would be the result of applying the same method to the Small Cloud? I believe the luminosities of Small Cloud supergiants show some anomalies.

CRAMPTON: It will be interesting to find out. I have no data yet, but I believe Azzopardi and his collaborators in France are presently working on this problem.

WALBORN (to PAYNE-GAPOSCHKIN): The Of star Sanduleak 80, the only one known in the SMC, has a reasonably normal M_V of $-7^{m}_{\cdot}3$ on the assumption of a true modulus of $19^{m}_{\cdot}3$.

(to CRAMPTON): In your compilation of LMC modulus determinations, one of the most discrepant was that from novae. It is interesting that the well-observed nova 1977b studied by Canterna and Schwartz gives $18^{m}_{\cdot}6 \pm 0^{m}_{\cdot}2$.

Can you suggest any explanation of Hutchings' earlier result? Did you measure Hγ for any of the same stars Hutchings studied?

CRAMPTON: The discrepancy with the result by Hutchings probably arises mainly from a) problems with the calibration of super-supergiants for which there are few galactic ounterparts, and b) problems with the measurement of Hγ equivalent widths at low dispersion, or on image tube spectra.

PARTHASARATHY: Absolute magnitudes of some of your program stars may be in error. Some of them must be binaries.

CRAMPTON: Duplicity is certainly a problem, particularly since corrections for duplicity were made in the Balona and Crampton calibration.

LESH: Isn't it true that calibration of MK spectral types for B stars - especially the Blaauw calibration - is based mainly on the distances of the stars in Scorpio-Centaurus, and so is essentially independent of the controversy surrounding the distance modulus of the Hyades?

CRAMPTON: The Blaauw calibration for B stars is in good agreement with the results obtained from the use of Sco-Cen - but the distance of Sco-Cen itself is the subject of controversey.

THE HR DIAGRAM FOR LUMINOUS STARS IN NEARBY GALAXIES

Roberta M. Humphreys

University of Minnesota

HR diagrams for the stellar populations in other galaxies play a fundamental role in our understanding of the progress of stellar evolution and the effects of possible variations in chemical composition. It is important to compare what little information we have about stars in other galaxies with the same types of stars in our own Milky Way. Basically we are asking - are they the same, and how universal are the processes we observe in our Galaxy?

Due to the extreme faintness of stars in other galaxies it is only possible to sample the brightest stars in the nearest galaxies. The observations must then be compared with comparable data for the brightest stars, the supergiants and O-type stars, in the Milky Way.

The data for the luminous stars are most complete for the Milky Way and the Large Magellanic Cloud. The luminosities for the stars in our Galaxy are based on their membership in associations and clusters, and consequently are representative of Population I within ≈3kpc of the Sun. The data for the stars in the LMC with spectral types O to G8 come from published observations, and the M supergiants are from my recent observations of red stars in the LMC. This is the first time that the M supergiants have been included in an HR diagram of the Large Cloud. The presence of the red stars is important for any discussion of the evolution of the massive stars.

Comparison of the HR diagrams shows that the supergiants in these two galaxies are remarkably similar. There are several significant features common to both galaxies - 1) a group of very

A. G. Davis Philip and D. S. Hayes (eds.), The HR Diagram, 263-267.

luminous O-type stars with M_{bol} between -10^m and -12^m, 2) a lack of supergiants of later spectral type at these very high luminosities, and 3) an obvious upper envelope to the luminosities at $M_{bol} \simeq -9^m.5$ beginning at B5 to A0 which extends across the diagram to the M supergiants. The only obvious difference between the two galaxies is a greater number of the most luminous B and early A-type supergiants in the LMC.

Since the most luminous stars have great potential as extragalactic distance indicators, another useful comparison is the visual luminosities of the supergiants of different spectral types which is summarized in Table I.

TABLE I MAXIMUM VISUAL LUMINOSITIES FOR DIFFERENT SPECTRAL TYPES

Spectral Type Interval	Milky Way M_V	LMC M_V
O	1. $-8^m.5$ 2. -7.8 3. -7.6	1. $-8^m.2$ 2. -8.1 3. -7.5
B,A	1. -9.9: 2. -8.8 3. -8.7	1. -9.8 2. -9.2 3. -9.0
F,G,K	1. -9.5 2. -9.2 (3)	1. -9.2 2. -9.1 (3)
M	1. -7.6 (-8(3)) 2. -7.5 (-7.8(3)) 3. -7.4	1. -7.5 (-8.3(2)) 2. -7.3 (-7.7(3)) 3. -7.1

The magnitudes in parenthesis are their luminosities at maximum light and the number of stars that reach that brightness. The most luminous B to A-type supergiant in the Milky Way may be #12 in Cyg OB2 although there is some doubt about its membership. If it is not a member of Cyg OB2, there will be a significant difference in maximum luminosity between the B and A supergiants in the LMC and the Galaxy. Otherwise, the luminosities of the brightest supergiants confirm the close resemblance of the Population I stars in the galaxies.

The blue/red supergiant ratio (including O stars) is an important indicator for checking evolutionary models and may also be sensitive to abundance variations. The B/R ratio for different luminosity intervals is given in Table II for the Galaxy and the LMC.

TABLE II BLUE TO RED STAR RATIOS

M_{bol} range	Milky Way			LMC		
	B	R	B/R	B	R	B/R
$-5^m.5$ to $-6^m.5$	19	10	1.9	26	-	-
- 6.5 to - 7.5	96	27	3.4	81	3	incomplete data
- 7.5 to - 8.5	155	18	8.6	133	14	9.5
- 8.5 to - 9.5	119	7	17.0	76	5	15.2
- 9.5 to -10.5	30	1	30	21	-	>21
-10.5 to -11.5	15	-	-	8	-	-
≤ -11.5	3	-	-	1	-	-

The data for the M supergiants in the LMC are definitely incomplete fainter than $M_{bol} \approx -7^m.5$, and the other numbers may also change somewhat when additional observations are available. With this in mind, it is clear that the B/R ratios are comparable, confirming that the evolution of the most massive stars in these two galaxies have proceeded in similar ways.

Since the LMC is a companion to the Milky Way it is perhaps not too surprising that the data for the supergiants are so much alike. Therefore it is important to study the luminous stars in more distant members of the Local Group.

In this paper I will just briefly present some preliminary results for the luminous stars in Baade's Field IV in M31 and in NGC 6822. Field IV, one of four along the major axis of M31 is the only one for which a color-magnitude diagram exists from which candidate supergiants can be selected. The outermost spiral arm of M31 passes through Field IV and Baade and Swope have identified four associations in the arm. Spectra for classification have been obtained for fourteen stars in this field and five supergiants have been identified including three B-type supergiants, an F supergiant and an M supergiant. It should be emphasized that Field IV is only one part of one spiral arm and these supergiants are only a small sample of the luminous stars in M31. The HR diagram discussed here is a composite of the four associations and a small cluster in Field IV from which all confirmed foreground stars have been eliminated.

Field IV is ≈ 18.5 kpc from the center of M31 which scaled to the Milky Way corresponds to ≈ 14 kpc from the galactic center, so the most appropriate comparison is with our outer spiral arm - the Perseus arm (R ≈ 12 - 13 kpc). For example, the B/R ratio for Field IV is about 3.0 and the mean ratio for the Perseus arm is 3.2, which is good agreement between the outer parts of two spiral galaxies.

Preliminary results for the supergiants in the irregular galaxy NGC 6822 suggest that there may be some differences for the hot stars. Sixteen stars have been confirmed as members on the basis of spectra and photoelectric photometry of stars selected from Susan Kayser's color-magnitude diagram. In Table I the visual luminosities for the stars in NGC 6822 are compared with results for the Galaxy and the LMC. It is obvious that there may be some difference in maximum luminosity for the O and B to A-type supergiants, however the observations are still incomplete. It is significant that the M supergiants have the same maximum luminosity as in the Milky Way and LMC. Initially stars bluer than (B-V) = $0^{m}.2$ and redder than (B-V) = $1^{m}.7$ were selected for observation and the lack of data for the F, G and K-type stars is due to these selection criteria. It is also possible that the visually very luminous late B and early A-type supergiants have not been observed yet. There is some evidence that the reddening determined from some of the early-type members is higher than the foreground reddening, so it is probably necessary to observe somewhat redder stars to identify the A-type supergiants.

The data for the Milky Way and the LMC show that the luminous star populations are remarkably similar. It is not possible to make any definite statement about M31 because only a limited sample of supergiants is available, while for NGC 6822 there is evidence that there are some differences for the early-type supergiants. Further observations are scheduled for the red stars in the LMC and observations are also planned of additional stars in NGC 6822. Other Local Group galaxies included in this study which were not discussed here are M33, IC 1613 and the SMC.

DISCUSSION

McCARTHY: I would like to ask Dr. Philip or Dr. Sanduleak if their recent survey of OB stars in the LMC confirms the lacuna observed in the realm of the supergiants by Dr. Humphreys.

SANDULEAK: The gap in the region of very high luminosity F and G-type stars shown by Dr. Humphreys conforms with our objective-prism survey results for the Magellanic Clouds.

APPENZELLER: At least most (if not all) of the most luminous stars of types A and B in the LMC seem to be variable in light as well as in spectrum. Are such stars included in your HR diagrams or have they been disregarded?

HUMPHREYS: All known variable and peculiar stars have been eliminated in both the LMC and the Milky Way.

CHIOSI: From what was said here one may perhaps get the impression that neutrino energy losses in the advanced phases (core carbon burning, etc.) are the leading factor accounting for the observed lack of very luminous red and perhaps also intermediate spectral type supergiants. This could be the case if the core He-burning phase should entirely or almost entirely occur at high effective temperature. On the contrary the bulk of present evolutionary tracks in the range of massive stars gives quite a different picture, as will be summarized later in my talk. In fact too many red luminous supergiants are forseen by adopting both the Ledoux and Schwarzschild criteria. Moreover the classical results do not account for the observed distribution of the blue supergiants - on the other hand, recent evolutionary computations of stars in the mass range from 20 to 100 $M_{\odot}$, where mass loss in the blue and red phases is taken into account, give a better interpretation of the observations (Chiosi, Nasi and Sreenivasn 1977, Astron. and Astrophys., in press). The results easily account for both the lack of red supergiants and the bulk of the blue ones. Finally I wish to remark that the very close similarity of the HR diagrams for our own galaxy and LMC is a further indication of close simil arity of the chemical composition (Y and Z) of supergiant stars in the two galaxies.

LAMB: I would like to mention that recently Iben, Howard and I (Astrophys. J. 1976, 207, 209) have evolved 15 $M_{\odot}$ and 25 $M_{\odot}$ stars, using the Schwarzschild criterion for convection, and including no mass loss. We find good agreement between the evolutionary tracks and time scales and Humphreys' data on the distribution of red and blue supergiants.

A PRELIMINARY COLOR-MAGNITUDE DIAGRAM FOR THE URSA MINOR DWARF GALAXY

Robert A. Schommer and Edward W. Olszewski

University of Washington

William E. Kunkel

Observatorio Nacional de Brasil

The Ursa Minor dwarf galaxy (RA = $15^h08^m.2$, Dec = +67°18'(1950)) was one of four dwarf ellipticals discovered by Wilson (1955) in a search of the Palomar Sky survey plates. The general properties of these galaxies resemble those of galactic globular clusters, linear size and density excepted (see Hodge, 1971, for a review). The characteristics of Ursa Minor are very similar to those of the Draco dwarf, for which a color magnitude diagram was published by Baade and Swope (1961). Van Agt (1967) published a study of the variable stars in Ursa Minor based on Baade's 200-inch plates.

Recent studies of distant globular clusters and dwarf galaxies have used these systems as probes of the galactic halo. Studies of the metallicity of individual stars in these systems based on spectra (Cowley, Hartwick and Sargent, 1977) and on photometry (Canterna and Schommer, 1977) have shown that they are extremely metal poor (similar to M92) with some evidence for [CNO] enrichment. The low stellar densities and lack of continual star formation imply an absence of extensive chemical processing in these systems, and thus their current conditions and abundances may reflect those of the proto-Galactic cloud (Eggen, Lynden-Bell, and Sandage, 1962). Most of these systems, including Draco, show evidence for anomalous red horizontal branches (Sandage and Wildey, 1967) and this has led to investigations of possible [He] or age differences (Hartwick, 1968; Rood, 1973). The unique spatial distribution of these systems has inspired Magellanic Plane models (Kunkel and Demers, 1977; Lynden-Bell, 1976).

A. G. Davis Philip and D. S. Hayes (eds.), The HR Diagram, 269-272. All Rights Reserved.

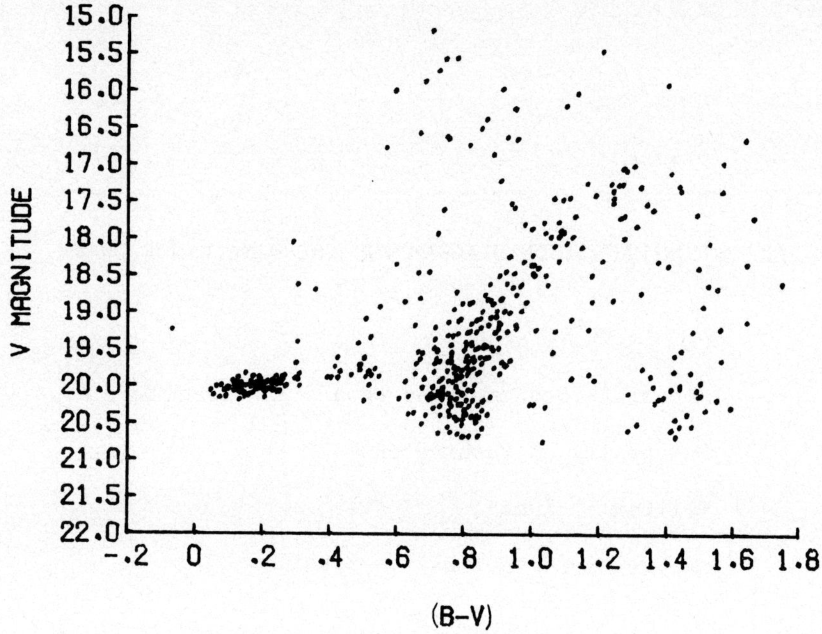

Fig. 1. CM diagram for Ursa Minor, 457 stars measured, excluding variables identified by Van Agt (1967)

Data for the present investigation consist of two prime focus plate pairs taken on the 4-meter Mayall telescope at KPNO, and photometry of standard stars done on the 2.1 meter telescope at KPNO and on the University of Washington 76 cm telescope. The standard sequence (down to $V=18\overset{m}{.}7$) was extended to $V=22\overset{m}{.}0$ on the plates through the use of a Racine wedge (Racine, 1969) which forms secondary images 6.91 (±.05) magnitudes fainter than the primary. This calibration was checked on the standard sequence of M13.

Parameters derived from this color magnitude diagram (Fig. 1) are $\Delta V=2\overset{m}{.}9 \pm 0\overset{m}{.}2$, $S=5.7 \pm 0.4$, and $(B-V)_{o,g} = 0\overset{m}{.}69 \pm 0\overset{m}{.}04$. These results and the blue horizontal-branch morphology indicate a metal poor system, and agree well with the value of [Fe/H] = -2.1 recently determined by Canterna and Schommer (1977). The reddening, determined by UBVr field star photometry and the blue edge of the RR Lyrae gap, is found to be $E(B-V) = 0\overset{m}{.}06 \pm \overset{m}{.}03$. By assuming $M_V = 0\overset{m}{.}6$ for the RR Lyrae stars, we derive a true distance modulus of 19.2 (69 kpc). The color magnitude diagram looks remarkably like that of M92 and the luminosity function resembles that of M3 (Sandage 1954).

Ursa Minor exhibits a horizontal-branch morphology different than the other halo objects and dwarf galaxies. Essentially all the systems with $R_{gc} > 20$ kpc show the red horizontal-branch anomaly and several of the galaxies have unusual variable stars (Swope 1968). The postulated Magellanic Plane tied these systems together and possibly pointed to a different epoch of formation. The seemingly "normal" properties of the Ursa Minor system, however, complicate this interesting explanation and the recent radial velocity determinations (Hartwick and Sargent 1977) seriously damage the dynamical model of Lynden-Bell (1976).

The similarity between the extreme Population II globular clusters and Ursa Minor has implications for pre-Galactic element enrichment. Hodge and Mitchie (1969) have argued that Ursa Minor formed much farther from the Galaxy than its present position, so the apparent uniformity of chemical abundances implies the same or at least very similar initial enrichments over a spatial scale of several hundred kpcs.

REFERENCES

Baade, W. and Swope, H. (1961). Astron. J. **66**, 300.
Cowley, A., Hartwick, F.D.A. and Sargent, W.L.W. (1977). preprint.
Canterna, R. and Schommer, R. (1977). in press.
Eggen, O., Lynden-Bell, D. and Sandage, A. (1962). Astrophys. J. **136**, 748.
Hartwick, F.D.A. (1968). Astrophys. J. **154**, 475.
Hartwick, F.D.A. and Sargent, W.L.W. (1977). preprint.
Hodge, P. (1971). Ann. Rev. of Astron. and Astrophys. **9**, 35.
Hodge, P. and Mitchie, R. (1969). Astron. J. **74**, 587.
Kunkel, W. and Demers, S. (1976). Greenwich Obs. Bulletin #182.
Lynden-Bell, D. (1976). Monthly Notices of the R.A.S. **174**, 695.
Racine, R. (1969). Astron. J. **74**, 1073.
Rood, R. (1973). Astrophys. J. **184**, 815.
Sandage, A. (1954). Astron. J. **59**, 192.
Sandage, A. and Wildey, R. (1967). Astrophys. J. **150**, 469.
Swope, H. (1968). Astron. J. **73**, 2504.
Van Agt, S.L. Th. (1967). Bull. Astron. Neth. **19**, 275.
Wilson, A. (1955). Publ. Astron. Soc. Pacific **67**, 27.

DISCUSSION

TINSLEY: You remarked at the end that this system gives information on the scale of chemical enrichment. Can you say more about the homogeneity of abundances in the UMi system?

SCHOMMER: Not really at this preliminary stage. The diagram here samples ∿500 stars in the central field surveyed for variables by van Agt. We are in the process of reducing a separate field in the outer regions of this galaxy. I think the width of the giant branch in our diagram here can be attributed to observational error and field star contamination.

THE HERTZSPRUNG-RUSSELL DIAGRAM OF METAL-POOR DISK STARS

Sidney van den Bergh

David Dunlap Observatory

A quarter of a century ago Keenan and Keller (1953) showed that the majority of high-velocity stars near the Sun outline a Hertzsprung-Russell diagram similar to that of old Population I. This result, which did not appear to fit into Baade's (1944) two-population model of the Galaxy was ignored (except by Roman 1965) for the next two decades. Striking confirmation of the results of Keenan and Keller was, however, obtained by Hartwick and Hesser (1972). Their work appears to show that high-velocity field stars with an ultraviolet excess (which measures Fe/H) of $\delta(U-B) \simeq +0\overset{m}{.}11$ lie on a red giant branch that is more than a magnitude fainter than the giant branch of the strong-lined globular cluster 47 Tuc for which $\delta(U-B) \simeq +0\overset{m}{.}10$. Furthermore Demarque and McClure (1977) show that the red giants in the old metal poor $[\delta(U-B) \simeq +0\overset{m}{.}11]$ open cluster NGC 2420 are significantly fainter than are those in 47 Tuc. Calculations by these authors show that the observed differences between the giants in 47 Tuc and in NGC 2420 can be explained if either (1) 47 Tuc is richer in helium than NGC 2420 by $\Delta Y \simeq 0.1$ or (2) if 47 Tuc has a ten times lower value of Z(CNO) than does NGC 2420.

Presently available chemical abundance determinations for galactic disk and halo stars do not allow one to make a clear-cut choice between these two alternatives. Nevertheless the remarkable uniformity of Y values observed in different galaxies tends to favour the suspicion that the peculiarity of NGC 2420 should probably be interpreted in terms of an anomalously low Fe/CNO ratio. At this point it should perhaps be emphasized again that Z is largely determined by the abundance of the most plentiful elements i.e. C, N, O and Ne, whereas the ultraviolet excess $\delta(U-B)$ is

A. G. Davis Philip and D. S. Hayes (eds.), The HR Diagram, 273-276.

mainly a function of the Fe abundance.

The bright red giants that Baade (1944) discovered in the disk of M31 are embedded in a much richer population of faint giants which contribute most of the light. Baade referred to the bright Population II giants in the Andromeda Nebula as "the frosting on the cake".

Recently Sandage (1977) has emphasized the fact that the red giants in the disk of M33, in NGC 6822 and in IC 1613 (Sandage 1971) are all bright ($V \sim 21^{m}.5$) and easily resolvable. Sandage points out that the surface density of bright red stars in M33 is ~100 times greater than it is in the galactic disk near the Sun. It follows that these bright red giants in M33 constitute the dominant old population in the Triangulum Nebula; not a "frosting on the cake" as they do in M31 and the Galaxy.

Probably the most straightforward interpretation of this result is that the old red giants in M33, NGC 147, NGC 185, NGC 205 and NGC 6822 lie on tall red giant branches because they have low Z values. The alternative hypothesis that these stars lie on tall branches because they have high Y values appears to be ruled out by the observation that the present helium abundance in M33 and in NGC 6822 (Peimbert and Spinrad 1970a, b) is similar to or slightly below that prevailing in galactic HII regions.

The results obtained above may be summarized as follows: The galaxies of the Local Group may be divided into two groups according to the properties of their dominant red giant populations. In M33, the LMC, NGC 6822, NGC 147, NGC 185, NGC 205 and probably in the SMC and IC 1613 as well the red giants are bright indicating a relatively low heavy element abundance. In the Galaxy and M31 the dominant red giant population consists of old faint giants that are composed of material with a relatively high Z value. The red giants in M32 seem to have characteristics that are intermediate between those of the high-luminosity and low-luminosity members of the Local Group. This result is consistent with the work by Faber (1973a, b) who finds that M32 has lost a considerable amount of mass by tidal stripping resulting from encounters with M31. Its CN + Mg index suggests that M32 may now be ~5 times less luminous than it was originally.

REFERENCES

Baade, W. (1944). Astrophys. J. **100**, 137.
Demarque, P. and McClure, R.D. (1977). Astrophys. J. **213**, 716.
Faber, S.M. (1973a). Astrophys. J. **179**, 423.
Faber, S.M. (1973b). Astrophys. J. **179**, 731.
Hartwick, R.D.A. and Hesser, J.E. (1972). Publ. Astron. Soc. Pacific **84**, 813.

Keenan, P.C. and Keller, G. (1953). Astrophys. J. 117, 241.
Peimbert, M. and Spinrad, H. (1970a). Astrophys. J. 159, 809.
Peimbert, M. and Spinrad, H. (1970b). Astron. Astrophys. 7, 311.
Roman, N. G. (1965). In Stars and Stellar Systems, Vol. V, Galactic Structure, A. Blaauw and M. Schmidt, eds., Univ. of Chicago Press, Chicago, p. 345.
Sandage, A.R. (1971). Astrophys. J. 166, 13.
Sandage, A.R. (1977). Paper presented at the Pomona meeting of the Astron. Soc. of the Pacific, June 21, 1977.

DISCUSSION

BUSCOMBE: Augensen and Buscombe (this symposium, page 413) obtained an HR diagram for 800 southern high-velocity stars (with UBV photometry, individual parallaxes and space motion data) showing a close fit to the diagram for the old galactic cluster M67.

THE USE OF PRAESEPE FOR THE DEFINITION OF THE LOWER PART OF THE ZERO-AGE MAIN SEQUENCE

M. Golay

Observatoire de Geneve

In an attempt to determine the Hyades distance (Golay, 1973), it was assumed that stars of the same "photometric $0^m.01$ box" (see Golay, 1977a) have the same visual absolute magnitude. The large amount of photometric data in the UBV $B_1B_2V_1G$ photometric system allows a discussion on this hypothesis (Golay, 1977b). We have 60 "photometric $0^m.01$ boxes", each containing a central star of known trigonometric parallax and at least one Praesepe star. We select the 16 boxes (Table I) containing single stars or binaries with an estimated mass ratio, a relative probable error < 30% for the parallaxes and a standard deviation for colors $<0^m.007$. The UBV $B_1B_2V_1G$ colors, the indices (B-V), (B_2-V_1) and the magnitude m_V are taken from the Second Catalogue (Rufener, 1976) and the internal catalogue of the Geneva Observatory. The color index (B-V) is taken from Johnson (1952, 1957), Johnson and Knuckles (1955), the trigonometric parallax from Jenkins (1952, 1963) and Gliese (1969) and the spectral type for Hyades stars from Morgan and Hiltner (1965). The listings of all $0^m.01$ photometric star boxes in the UBV $B_1B_2V_1G$ system are given by Golay (1977c). The parallax obtained for Praesepe is $\pi(0''.001) = 6.175 \pm$ p.e. 0.1, i.e. a distance modulus $(m-M) = 6^m.05$ and a distance of 162 parsecs. Golay (1977c) published the differences of the distance moduli for pairs of clusters having stars in the same box. The distances of these clusters is given in Table III, assuming a distance of 162 pc for Praesepe. The accuracy of this method is independent of both the distance magnitude and the chemical composition of the stars of a cluster since the stars have to be in the same box as a star with a known trigonometric parallax. The main sequence of Praesepe and a sample of Hyades stars, in the same photometric box with a Praesepe star is given in Table II. The depth effect in

A. G. Davis Philip and D. S. Hayes (eds.), The HR Diagram, 277–280.

Praesepe being very small, the main sequence is very thin and the main sequence fitting procedure is better starting from Praesepe than from the Hyades.

TABLE I

SELECTION OF STARS WITH TRIGONOMETRIC PARALLAX IN THE SAME SEVEN COLOR PHOTOMETRIC "STAR BOX" AS PRAESEPE'S STARS

HD	π 0".001	Sp	Praesepe Star
10476	133 ± 6	K1V	KW246
12230	26 ± 7	F0	KW124
17051	71 ± 7	G3IV	KW418
25680	69 ± 5	G5V	KW399
25893	45(± 7)	K2	KW297
30501	46 ± 7	K0V	KW533
34101	62 ± 12	G6	KW334
42250	41 ± 11	G5	KW 9*
43834	115 ± 8	G5V	KW543
48682	68 ± 6	G0V	KW258
61994	45 ± 9	G5	KW 30
65371	35 ± 11	K0	KW 9*
86661	32 ± 12	G8IV-V	KW368
101501	110 ± 8	G8V	KW539
151044	35 ± 6	F8V	KW341
160269	67 ± 4	G1V	KW325

KW = Klein-Wassink number (1927). * = Same star in two boxes.

TABLE II

ABSOLUTE MAGNITUDE FOR THE REGION OF THE FITS BETWEEN PRAESEPE AND HYADES IN THE COLOR-LUMINOSITY DIAGRAM

Praesepe KW	Sp	(B-V)	(B2-V1)	m_V	(B-V)	M_V (162 pc) (m-M) = 6^m.05
KW124	F2V	-^{m}.567	+^{m}.112	8^m.987	+^{m}.321	2^m.94
KW439	F5V	-.485	+.183	9.415		3.375
KW478	F4V	-.423	+.218	9.673	+.43	3.62
KW421	F9	-.327	+.287	10.160	+.516	4.11
KW341	F8	-.300	+.289	10.269	+.520	4.22
KW418	G4	-.268	+.324	10.481	+.565	4.43
KW288	G0	-.239	+.350	10.695	+.583	4.65
KW392	G5	-.229	+.349	10.702	+.595	4.65
KW466	G2	-.167	+.382	10.979	+.649	4.92
KW432	G	-.139	+.391	11.057	+.646	5.00
KW326	G	-.084	+.423	11.353	+.716	5.30
KW 32	G	+.011	+.479	11.634	+.775	5.58
KW263	K	+.061	+.517	11.972	+.814	5.92

Hyades VB	Sp	(B-V)	(B2-V1)	m_V	(B-V)	M_V (162 pc) (m-M) = $6^{m}.05$
VB 44	F5	$-^{m}.396$	$+^{m}.228$	$7^{m}.159$	$+^{m}.450$	$3^{m}.73$
VB 65	F8V	-.298	+.300	7.416	+.535	4.23 R1
VB113	F5	-.276	+.311	7.246	+.549	4.20
VB 52*	G1V	-.225	+.343	7.804	+.597	4.62 R2
VB 50	G1V	-.217	+.346	7.612	+.601	4.62
VB 52*	G1V	-.225	+.343	7.804	+.597	4.74 R2
VB 18	G0	-.171	+.381	8.033	+.638	4.93
VB 15	G3V	-.147	+.387	8.072	+.658	5.00
VB 69	G8V	-.046	+.446	8.593	+.746	5.39

VB = van Bueren number (1952). R1, M_V for main component
*R2, same star in two boxes

TABLE III

DISTANCES OF GALACTIC CLUSTERS (FIRST APPROACH)

Cluster	Modulus	Distance (parsec)
Praesepe	$6^{m}.05$	162
Coma	4.50	79
Pleiades	5.93	153
NGC 6475	a) 7.27	284
NGC 6475	b) 7.21	277
NGC 752	a) 8.16	428
NGC 752	b) 7.90	380
NGC 7092	7.38	299

REFERENCES

Bueren, H.G. van (1952). Bull. Astron. Inst. Neth. 11, 385.
Gliese, W. (1969). Veröffentl. Astron. Rechen-Inst. Heidelberg No.22.
Golay, M. (1973). IAU Symposium No. 54, Problems of Calibration of Absolute Magnitudes and Temperature of Stars, B. Hauck and B. E. Westerlund, eds., D. Reidel, Dordrecht, p. 27.
Golay, M. (1977a). Astron. Astrophys. 60, 181.
Golay, M. (1977b). Astron. Astrophys. in press.
Golay, M. (1977c). Publ. Obs. Genève Serie B, No. 3 and 4.
Jenkins, L.F. (1952). General Catalogue of Trigonometric Parallaxes, Yale Univ. Obs.
Jenkins, L.F. (1963). Suppl. to the General Catalogue, Yale Univ. Obs.
Johnson, H.L. (1952). Astrophys. J. 116, 640.
Johnson, H.L., Knuckles, C.F. (1955). Astrophys. J. 122, 209.
Johnson, H.L. (1957). Astrophys. J. 126, 121.

Klein-Wassink, W.J. (1927). Publ. Kapteyn Astron. Lab. No 41.
Morgan, W.W., Hiltner, W.A. (1965). Astrophys. J. **141**, 177.
Rufener, F. (1976). Astron. Astrophys. Suppl. **26**, 275.

DISCUSSION

GLIESE: How many stars with reliable trigonometric parallaxes are available for calibrating the absolute magnitudes of stars in the same box?

GOLAY: I very often have one star, and sometimes two.

GLIESE: What seems to be the dispersion in M_V among the stars in one box?

GOLAY: The dispersion in M_V is $0^m.1$.

CAYREL de STROBEL: How can you put stars of the Hyades cluster and stars of the Praesepe cluster in the same box? They probably do not have exactly the same chemical composition, and not the same age. I understand that the stars in a $0^m.01$ Geneva photometric box must have the same age.

GOLAY: With respect to age, it will be true that the stars in a box have nearly the same age only for young clusters. I don't, myself, put Hyades and Praesepe stars in the same photometric box. The photometric data show that we can have a photometric box containing both Hyades and Praesepe stars. The stars in a box have the same energy distribution, and the spectroscopist has to confirm that these stars have the same chemical composition. With the published values of [Fe/H] we can confirm that stars in the same box have the same [Fe/H] (within the errors of determination of [Fe/H]).

A NEW KIND OF HR DIAGRAM?

David S. Evans

McDonald Observatory and University
of Texas at Austin

When the system of correlations between stellar parameters that we know as the HR diagram was discovered, the physical basis for them was unknown and remained essentially unknown during four decades of eager application.

Two years ago Thomas Barnes and I found a relationship which we dare to hope may have a similar history. A parameter related to the visual surface brightness of stars is expressible as

$$F_V = \log T_{eff} + 0.1C = 4.2207 - 0.1V_o - 0.5 \log \phi'$$

where T_{eff} = effective temperature, $C = M_{bol} - V_o$ = bolometric correction, V_o = apparent visual magnitude corrected for extinction, and ϕ' = angular diameter of the star in arc milliseconds. The formula is equivalent to one given in Russell, Dugan and Stewart (Vol. II, p. 738, 1927 edn). Barnes and I found that F_V was extremely closely related to the Johnson $(V-R)_o$, (and not to any other color index), that it is almost completely immune to the effects of interstellar extinction, and independent of luminosity class (Barnes and Evans, 1976; Barnes et al, 1976). The relation started from a consideration of occultation angular diameters but went on to include all determinations. In the latest version to be published by Barnes, Evans and T.J. Moffett (1978) there are 92 measures of angular diameter by all methods and much new photometry. The result shown in Fig. 1 is closely similar to a theoretical black body color index computed for 550 nm and 700 nm with non-trivial deviations between types A and early M, suggested to be due to the effects of negative hydrogen ions. It seems unlikely that it will be much changed by the

A. G. Davis Philip and D. S. Hayes (eds.), The HR Diagram, 281-284.

addition of new data.

It invites applications, though the first that struck me was one of the least obvious. Two of the points refer to (speckle measures) of Mira at different phases and it was conjectured that variable stars stayed on the relation during pulsation. George Wallerstein (1977) disbelieved the result on the ground that the radial velocity variations of Miras in the visual were too small but is apparently convinced by the much greater velocity range found by Hinkle (1977) in the near infrared.

The relationship allows us to calculate angular diameters from photometric observations and, if the conjecture is correct, for a Cepheid at any phase. The angular diameter variation from photometry ought to match the linear diameter variation calculated from radial velocity measures at each phase, with the scaling factor giving the distance of the Cepheid. Tom Barnes and I put our variable star class to work on this and, with Sidney Parsons as consultant, produced distances for the nine Cepheids for which suitable data were available (Barnes et al, 1977). The results are practically unaffected by the assumed values of interstellar extinction and reproduce the Fernie-Hube distance scale within a few per cent. This has occasioned criticism from theoretical quarters as being inconsistent with the new Hyades distance. The distance results are critically dependent on the adopted slope of the relation and when we apply the revised version it seems likely that the scale, at least for all except the longest period Cepheids will be expanded by a few per cent. We should know by early in the New Year. So far this seems a reliable technique for the determination of true geometrical distances up to possibly 2000 parsecs, but once established it should go vastly further.

Energetic application to the RR Lyrae stars and dwarf Cepheids by Barnes, Tom Moffett and Wayne van Citters indicates that we are on sure ground there. Practical difficulties have got in the way of meeting the ideal of simultaneous photometric and spectroscopic observations of RR Lyraes and we proceed with caution before committing ourselves to definitive statments in the critical area of RR Lyrae absolute magnitudes.

Although white dwarfs have degenerate interiors they must radiate through an outermost layer which is normal. We have tried an application here with apparent success (Moffett et al, 1977) judged by consistency with results by other methods. New photometry mainly by Barnes and Moffett covered a limited number of white dwarfs but the realization that old photometry could be transformed to the Johnson (VR) system extends the coverage to 71 stars. The mean radius of DA stars with known parallaxes is

0.0132 ± 0.0005 solar radii with only a small scatter, and adoption of this figure identifies 14 new cases which should be within 25 parsecs. Other topics include the computation of the mean density of 40 Eridani B as near 2×10^5 g cm^{-3}, in reasonable accord with the observed gravitational redshift.

Preoccupied with variables, we failed to notice the fact that any star for which a linear radius (in kilometers) is known can yield a distance, while for any star with a good parallax the linear radius can be found. Claud Lacy had the wit to realize this and has produced a preliminary paper (Lacy, 1977a) dealing with the distances of nine eclipsing binary systems--a number he hopes to increase to 500-- and a paper in press (Lacy, 1977b) dealing with the radii of 150 nearby stars. He finds an intriguing radius deficiency of 25% for M dwarfs as compared with theoretical models, and presents independent evidence in support of this conclusion.

The number of possible applications seems very large, ranging from estimation of the distance of a nova by Barnes (1976), new estimates of effective temperature, direct determination of the sizes of spots on M-dwarfs and even the estimation of the dimensions of stellar flares (Evans and Lacy, 1977). I am sure that we have nowhere nearly exhausted the possibilities and are happy to see others making use of the relationship.

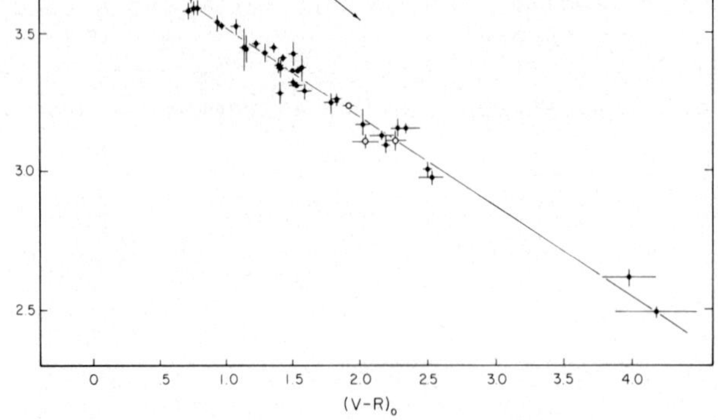

Fig. 1. The Visual Surface Brightness Relation. The arrow showing the effect of one magnitude of visual extinction is almost parallel to the curve.

REFERENCES

Barnes, T.G. (1976). Mon. Not. Roy. Astron. Soc. 177, 53P.
Barnes, T.G. and Evans, D.S. (1976). Mon. Not. Roy. Astron. Soc. 174, 489.
Barnes, T.G., Evans, D.S. and Parsons, S.B. (1976). Mon. Not. Roy. Astron. Soc. 174, 503.
Barnes, T.G., Dominy, J.F., Evans, D.S., Kelton, P.W., Parsons, S.B. and Stover, R.J. (1977). Mon. Not. Roy. Astron. Soc. 178, 661.
Barnes, T.G., Evans, D.S. and Moffett, T.J. (1978). Mon. Not. Roy. Astron. Soc. (submitted).
Evans, D.S. and Lacy, C.H. (1977). Nature (submitted).
Hinkle, K.H. (1977). Astrophys. J. (in press).
Lacy, C.H. (1977a). Astrophys. J. 213, 458.
Lacy, C.H. (1977b). Astrophys. J. Suppl. 34, 479.
Moffett, T.J., Barnes, T.G. and Evans, D.S. (1977). Astron. J. (submitted).
Wallerstein, G. (1977). Jl. Roy. Astron. Soc. of Canada (in press).

DISCUSSION

WEIDEMANN: Does it not all boil down to the fact that (V-R) is taken as a (single valued) measure of effective temperature? It works in the white dwarf case since the gravity scatter is small. But in principle (V-R) will depend not only on T_{eff} but also on surface gravity.

D. EVANS: It is somewhat more complicated than that since for much of the range bolometric corrections are large, but this certainly plays some part.

BELL: Have you included limb darkening corrections in your angular diameter data e.g., in the first slide that you showed?

D. EVANS: Yes. These have been included for all the angular diameter data using results from Hanbury Brown and studies for later type stars, part theoretical and part from double reductions assuming uniform disk and full darkening. The range in values comes out at about 12% between the two extremes.

DETECTION OF Am AND Ap STARS BY PHOTOMETRIC MEANS

B. Hauck

Institute of Astronomy of the University of
Lausanne and Geneva Observatory

The detection of Am and Ap stars is possible in various photometric systems. Those most suitable for this purpose are certainly the uvbyβ (Strömgren 1967, Cameron 1967), the Barbier Morguleff system completed by Gerbaldi (1972, 1977) and the Geneva system. It seems important to me that the Am and Ap stars should be detected with a system which also allows one to classify normal stars, and to obtain quantitative information about metallicity or other peculiarity. The Geneva system satisfies both these criteria and numerous Am (about 400) and Ap (about 250) stars are in a catalogue prepared on magnetic tape by Rufener, which follows the one published in 1976 (Rufener 1976).

The method by which one can obtain a tri-dimensional classification (temperature, absolute magnitude and chemical composition) for the A and F stars is given in two previous papers (Hauck 1968a, 1973).

The parameter m_2 = (B1-B2) - 0.457(B2-V1) is used as a blanketing parameter. The distance Δm_2 to the Hyades sequence in a diagram m_2 vs. (B2-V1) is an indicator of chemical composition. For the Am stars, we have been able to check the Δm_2 is correlated satisfactorily with the g = sp(ml) - sp(K) parameter defined by Hack (1959). Thus Δm_2 will be the parameter by which we can detect the Am stars. Two interesting results were obtained in previous studies: (a) Δm_2, for a given (B2-V1) is more important for an Am star with a low V sin i value than for an Am star with a higher V sin i value (Hauck 1968b); (b) in a Δm_2 vs. (B2-V1) diagram we see that the redder Am stars have a higher degree of metallicism (Hauck 1970).

A. G. Davis Philip and D. S. Hayes (eds.), The HR Diagram, 285-287. All Rights Reserved.

The criteria of detection of the Ap stars in the Geneva system (Hauck 1974) are the following:

(a) in a d vs. (B2-V1) diagram (Gerbaldi et al. 1974) the Ap stars are generally below the sequence of reference, and a negative value of Δd can be considered as an Ap characteristic.

(b) in a (V1-G) vs. (B2-G) diagram, the Ap stars are below the relation $(V1-G) = 0.289(B2-G) - 0^m.302$ for the normal stars (Steiger 1974).

Δ(V1-G), the difference for a given (B2-G) between the observed value and the value of the sequence of reference, is used as a photometric parameter of peculiarity. This results from the fact that V1 is affected by the 5300 Å feature. At present, 234 Ap stars are measured in the Geneva system and a list updating the first one (Hauck 1976) is in preparation.

Lately Jamar et al. (1977) have shown that a feature was present at 1400 Å in the continuum of the Ap stars (with the exception of Mn stars) and that it is possible to use this feature as a criterion to select Ap stars. There is a good correlation between the δ_{1400} parameter of Jamar et al. and our Δ(V1-G) parameter. Jamar et al. have also examined the origin of the observed feature in the continuum of the Ap stars. They reach the conclusion that SiII autoionization lines and FeII lines could be the main agents, the first ones for the hotter Ap stars and the second ones for the cool Ap stars, both effects being added in Si stars where Fe and Si are overabundant.

We have looked in the Geneva catalogue for stars which are photometric Ap star candidates and find that HD 41367, HD 72524, HD 214433, HD 215304 and BD $1^{o}2668$ could be candidates. We also use the criteria for Am and Ap stars to examine the stars mentioned as Am or Ap stars in the Bertaud and Floquet (1974) catalogue. Thirteen stars of this kind have measurements in the Geneva catalogue and in many cases we are in agreement with the most recent spectral type determination of Bertaud and Floquet (1967). The results of this study will be published in a forthcoming paper.

REFERENCES

Bertaud, C. and Floquet, M. (1967). J. Obs. 50, 425.
Bertaud, C. and Floquet, M. (1974). Astron. Astrophys. Suppl., 16, 71.
Cameron, R.C. (1967). in The Magnetic and Related Stars, R.C. Cameron, ed., Mono Book Corp. Baltimore, p. 471.

Gerbaldi, M. (1972). Compt. Rend. Acad. Sci. Paris 275, ser. B, 295.
Gerbaldi, M. (1977). Thesis, Institut d'Astrophysique de Paris.
Gerbaldi, M., Hauck, B. and Morguleff, N. (1974). Astron. Astrophys. 30, 105.
Hack, M. (1959). Mem. Soc. Astron. Ital. 30, No. 1-2.
Hauck, B. (1968a). Publ. Obs. Geneva Ser. A. 75.
Hauck, B. (1968b) Publ. Obs. Geneva Ser. A. 74.
Hauck, B. (1970). Publ. Obs. Geneva Ser. A. 77.
Hauck, B. (1974). Astron. Astrophys. 32, 447.
Hauck, B. (1976). Astron. Astrophys. Suppl. 26, 49.
Jamar, C., Macau-Hercot, D. and Praderie, F. (1977). Submitted for publication.
Rufener, F. (1976). Astron. Astrophys. Suppl. 36, 275.
Strömgren, B. (1967). in The Magnetic and Related Stars, R.C. Cameron (ed.), Mono Book Corp. Baltimore, p. 461.
Steiger, P. (1974). Publ. Obs. Geneva Ser. A. 81.

THE HR DIAGRAM OF ABNORMAL A-TYPE STARS

Eugenio E. Mendoza V.

Instituto de Astronomia, Universidad Nacional
Autonoma de Mexico

We have observed over a hundred A-type stars in the $\alpha(16)\Lambda(9)$-photometric system defined by Mendoza (1976a, b; 1977). These stars also have UBVRI measurements. Sixty five stars have been observed in the thirteen color-photometric system (Mendoza 1971, and Johnson and Mitchell 1975).

The analysis of these narrow, intermediate and broad-band photometric measures could include many comparisons of the several color indices. Thus, for instance, Am stars are separated in the (V-I, U-V)-plot (Mendoza et al. 1977). A-type supergiants are separated in the (40-52, 37-52)-plane (Mendoza 1971). A-type supergiants, Ae (Herbig 1960), Am and Ap stars are separated very neatly in the $\alpha(16)\Lambda(9)$-diagram as illustrated in Fig. 1.

The Am and Ap stars are located on the main sequence in Hα- photometry ($\alpha(16)$-index). An interpretation of the neutral oxygen triplet (7774 Å)-photometry ($\Lambda(9)$-index) of Am and Ap stars is that they are oxygen underabundant at the surface layers, as compared with normal main sequence A-stars. Ap stars, under this interpretation will have less oxygen than Am stars in the observed stellar atmospheric layers.

The photometric measures under discussion could be affected by emissions and, or by interstellar extinction. In many cases interstellar extinction can be corrected.

A correction to the $\alpha(16)$-index, for high luminosity class-stars can be derived when it is contaminated by emission,

A. G. Davis Philip and D. S. Hayes (eds.), The HR Diagram, 289-291.

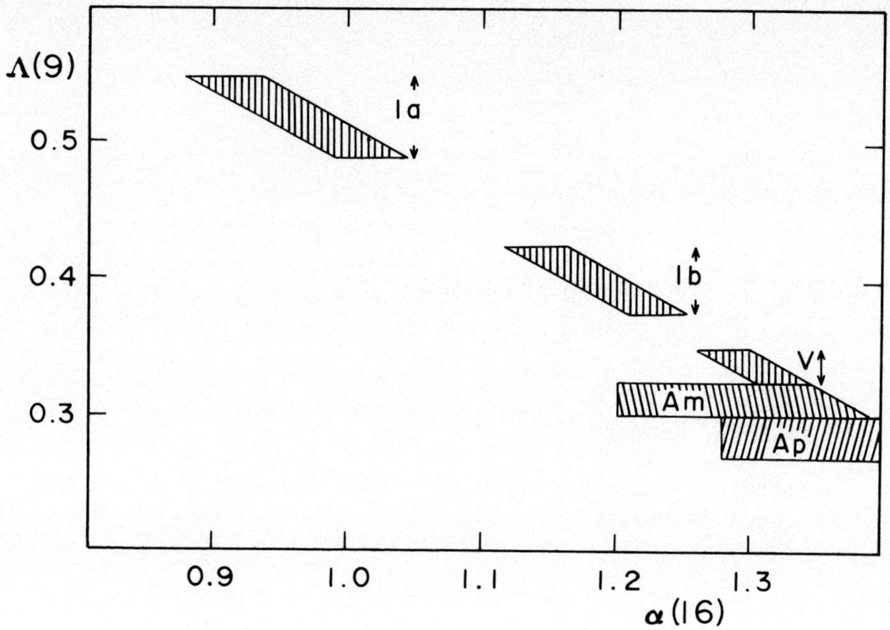

Fig. 1. A schematic diagram for A-type stars in the α(16)Λ(9)-plane. The Ae stars lie far to the left and below the Ap sequence.

provided that there is no emission contamination in the Λ(9)-index. For this purpose, we can use the relationship between spectral type, luminosity class and the (16)-index given schematically in Fig. 2. Thus, to locate an A-type supergiant star on the HR diagram we proceed as follows:

1) Correct the photometric data for interestellar extinction, if any. Standard procedures may be used (see Johnson 1968).

2) The OI-photometry gives the correct luminosity, provided that the Λ(9)-index is not contaminated by emission.

3) Correct the Hα-line for emission, if any, through the Λ(9)-index.

The Ae (Herbig 1960) stars such as V380 Orionis are more difficult to locate in the HR diagram with precision because both indices α(16) and Λ(9) are contaminated by emission.

REFERENCES

Herbig, G.H. (1960). Astrophys. J. Suppl. 4, 337.

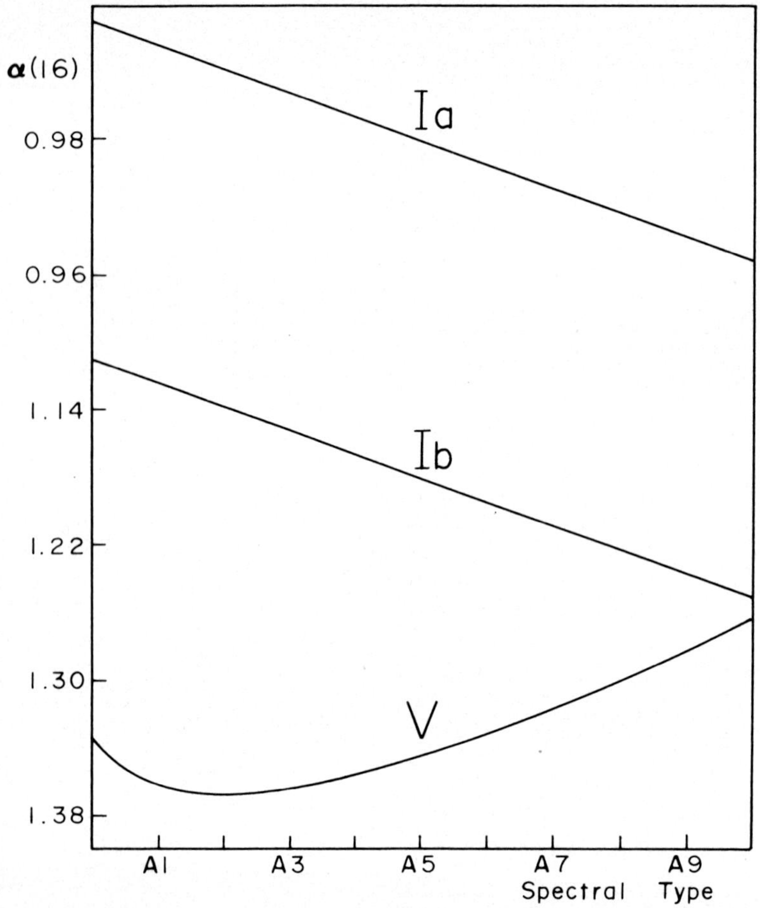

Fig. 2. Mean (16)-index values for 3 luminosity classes.

Johnson, H.L. (1968). In Stars and Stellar Systems, Vol. VII, Nebulae and Interstellar Matter, B.M. Middlehurst and L.H. Aller, eds., University of Chicago Press, Chicago, p. 167.
Johnson, H.L. and Mitchell, R.I. (1975). Rev. Mex. Astron. Astrof. 1, 299.
Mendoza, E.E. (1971). In The Magellanic Clouds, A.B. Muller, ed., D. Reidel, Dordrecht, p. 69.
Mendoza, E.E. (1976a). Rev. Mex. Astron. Astrof. 2, 29.
Mendoza, E.E. (1976b). Rev. Mex. Astron. Astrof. 2, 33.
Mendoza, E.E. (1977). Rev. Mex. Astron. Astrof. 2, 287.
Mendoza, E.E., Gomez, T. and Gonzalez, S. (1977). In preparation.

FROM MULTICOLOR PHOTOMETRIC INDICES TO THE HR DIAGRAM

V. Straižys

Vilnius Astronomical Observatory

ABSTRACT

The Vilnius photometric system permits one to determine photometrically temperatures and log g's or spectral types and M_V's of stars in spite of the presence of interstellar reddening. The position of any normal star in the HR diagram can be determined by its direct calibration in five reddening-free quantities Q. The position of the star in the HR diagram corresponds to the point at which five Q-isolines intersect. This is the most direct way to find the position of the star in the HR diagram from photometry.

The Vilnius photometric system, which consists of seven bands at 3450 (U), 3750 (P), 4050 (X), 4660 (Y), 5160 (Z), 5440 (V) and 6560 Å (S) has been developed for two-dimensional classification of both unreddened and reddened stars of spectral types from B to M. Later on the system was shown to be applicable in detecting photometrically some types of stars with certain peculiarities, e.g. subdwarfs, late-type metal-deficient giants, Am- Ap- and Be-type stars, white dwarfs, double stars, etc.

The general classification scheme proposed by Straižys (1970) can be applied to stars of various spectral types, luminosities, chemical compositions, peculiarities and various values of interstellar reddening. It is based on Q, Q diagrams calibrated in spectral types and absolute magnitudes (Sviderskienė and Straižys, 1971; Straižys and Sviderskienė, 1972). Later on Straižys

A. G. Davis Philip and D. S. Hayes (eds.), The HR Diagram, 293–300.

(1974a, b, 1975) proposed another method of classification based on reddening-free energy curves Q_{mVS}, λ, where

$$Q_{mVS} = (m-V) - \frac{E_{m-V}}{E_{V-S}} (V-S).$$

In this paper the calibration of quantities, Q in temperature and log g is given and the new method of quantitative classification or quantification as suggested by Barry et al. (1977), is described. The calibration of quantities Q in spectral types and M_V can be made and used in a similar way. However, one should not forget that photometric quantities actually depend on gravity, not on absolute magnitude.

The color indices of the Vilnius photometric system (U-V), (P-V), (X-V), (Y-V), (Z-V) and (V-S) were calibrated in T_{eff} and log g in two different ways: (1) by using model stellar atmospheres and (2) by transferring the calibration from the MK classification system. In the first case the model stellar atmospheres by Kurucz (1976) and Peytremann (1974) and the response curves of the system by Straižys and Zdanavicius (1970) were used. In the second case the color indices of stars with different T_{eff} and log g were obtained from intrinsic color indices of different MK types determined from observations by Kurilienė (1977). The calibration of B-type stars against T_{eff} is based on Code et al. (1976). The T_{eff} for A-M-type stars were averaged from different sources. The log g values are based on the data of 280 main sequence stars, 140 giants and 80 supergiants compiled from literature.

The Kurucz models in general show satisfactory agreement with real B-type stars of different luminosities including supergiants. However, the Kurucz energy curves for A-F-G V models in the blue and ultraviolet parts of the spectrum are too bright and this discrepancy between models and stars increases with later spectral types. A similar effect has been found by Philip and Relyea (see page 443 of this volume). This may mean that Kurucz models need some additional line blocking corrections in short wavelengths. Indices (Y-V), (Z-V) and (V-S) of the main sequence stars representing longer wavelengths also show some systematic differences. The same is true for all the indices of A-F supergiants.

Peytremann models of spectral types A-F-G V in the ultraviolet show much better agreement with real stars than Kurucz models do. However, for supergiants we observe the opposite situation: The indices (U-V) and (P-V) of A-G-F I models are reddened by $0^m.1$-$0^m.2$ in comparison with real stars.

It can be concluded that the energy distribution curves of Kurucz and Peytremann model stellar atmospheres may be used for the calibration of color indices and other photometric quantities

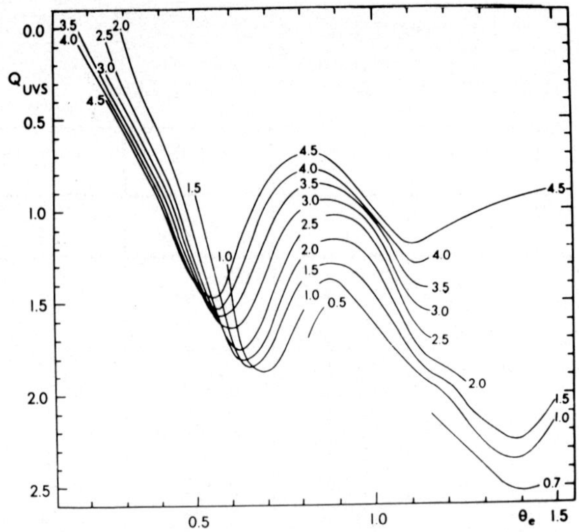

Fig. 1. Q_{UVS} versus Θ_{eff} for different values of log g.

for B-type stars and A-F stars of luminosities V-III but certain precautions must be taken. Until better models appear for A-F supergiants and stars of later types it is more reasonable to transfer the calibration in T_{eff} and log g from the MK system. This method has the advantage of being free not only from the errors introduced by inaccuracy of model atmospheres, but from the errors of response curves and integration errors as well. It can be used to calibrate all the spectral types including late ones for which model energy curves are absent.

Color indices calibrated in T_{eff} and log g can be used for two-dimensional quantification of unreddened stars only. If interstellar reddening is present one should use the five independent reddening-free quantities Q_{UVS}, Q_{PVS}, Q_{XVS}, Q_{YVS} and Q_{ZVS}. These quantities can be calibrated in T_{eff} and log g in the same way as the color indices. An example of the calibrated diagram of the Q_{mVS}, θ_{eff} type with isolines of log g is shown in Fig. 1 (see the monograph Straižys, 1977 for more details). It is obtained by combining both methods of calibration described above. The calibration of B- and early A-type stars is based on Kurucz models with only small corrections added. The calibration of G-K-M stars is transfered from the MK system as shown in Table I.

The parameters θ_{eff} and log g of a star can be derived in the following way:

1) Enter the diagram Q_{UVS}, θ_{eff} with the observed value Q_{UVS} and read out a series of pairs of log g and θ_{eff} at the points where the line Q_{UVS}= const intersects the lines log g = const.

TABLE I

CALIBRATION OF MK TYPES IN TEMPERATURE AND LOG g

Sp Type	T_{eff}	log g	log g	T_{eff}	log g	log g
	V	V	III	I	Ib	Ia
O7	35400	4.00	3.5			
O8	34100	4.00	3.5			
O9	33600	4.02	3.5	30000		
B0	31400	4.03	3.5	27000	3.60	3.20
B1	26000	4.04	3.5	23500	3.15	2.90
B2	22700	4.06	3.5	19500	2.90	2.65
B3	18100	4.08	3.5	16500	2.70	2.40
B5	14800	4.10	3.5	14000	2.50	2.00
B6	13800	4.13	3.5	13000	2.30	1.80
B7	12900	4.16	3.5	12000	2.20	1.70
B8	12100	4.18	3.5	11500	2.12	1.55
B9	10400	4.20	3.5	10400	2.04	1.45
A0	9600	4.20	3.5	9800	1.98	1.33
A1	9400	4.20	3.5	9500	1.92	1.24
A2	9150	4.20	3.5	9200	1.88	1.15
A3	8900	4.20	3.5	8900	1.83	1.08
A5	8400	4.20	3.5	8300	1.74	0.97
A7	8000	4.20	3.4	7900	1.67	0.88
F0	7300	4.20	3.1	7400	1.57	0.78
F2	7000	4.20		7100	1.50	0.72
F3	6850	4.20		6900	1.47	0.70
F5	6500	4.20		6500	1.40	0.65
F8	6150	4.20		6000	1.30	0.56
G0	5950	4.30		5700	1.23	0.50
G2	5800	4.40		5400	1.17	0.45
G5	5600	4.45	2.7	5000	1.07	0.37
G8	5440	4.50	2.7	4600	0.97	0.29
K0	5240	4.55	2.7	4400	0.90	0.24
K1	5120	4.57	2.6	4300	0.86	0.21
K2	4960	4.60	2.5	4200	0.83	0.19
K3	4780	4.62	2.2	4100	0.80	0.16
K4	4600	4.64	1.9	3950	0.77	0.14
K5	4400	4.65	1.7	3850	0.74	0.11
K7	4000	4.65	-	-	-	-
M0	3750	4.65	1.6	3700	0.70	0.08
M1	3600		1.5	3600		
M2	3400		1.3	3500		
M3	3300		1.1	3300		
M4	3200		0.9	3100		
M5	3100		0.7	2950		
M6	2950		0.4			
M7	(2850)					
M8	(2750)					

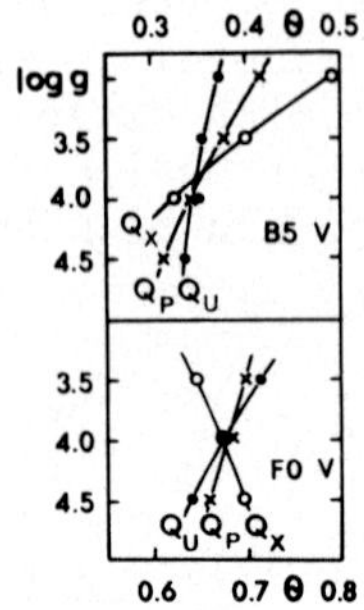

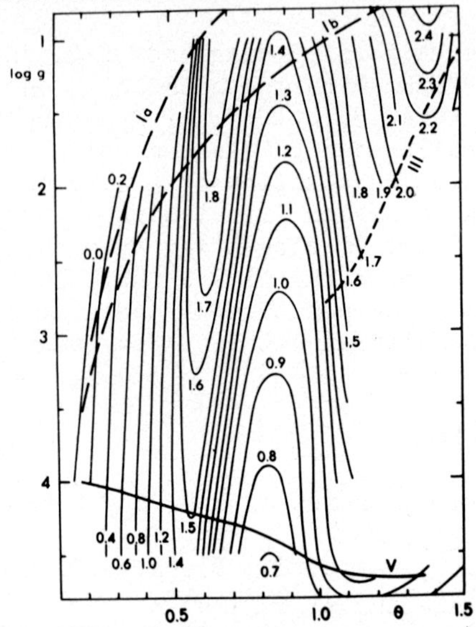

Fig. 2. Log g versus θ_{eff} for a B5 and F0 main sequence stars. The log g, θ_{eff} value for each star is indicated by the intersection of the three lines.

Fig. 3. Log g versus θ_{eff}. Luminosity classes Ia, Ib, III and V are indicated as well as lines of constant Q

---

2) Plot this series of points on the diagram log g, θ_{eff} and join them with a line.
3) Repeat the same procedure for Q_{PVS}, Q_{XVS}, Q_{YVS} and Q_{ZVS}.

If the calibration of diagrams, Q, θ_{eff} is correct and if the star is a normal one and observed with sufficient accuracy, all five lines in the diagram log g, θ_{eff} must intersect at one point. The coordinates of this point give the temperature and gravity for the star. Two examples of such a diagram are shown in Fig. 2. Only three Q isolines are plotted. The remaining two give us information for early-type stars. Due to errors of calibration and observation the lines in the diagram log g, θ_{eff} will not intersect at one point. Optimum values of both the parameters log g and Θ_{eff} can be evaluated taking into account error bars of different Q quantities.

Also the lines log g, θ_{eff} will not intersect at one point in the case of stars having certain peculiarities in their spectra. The character of deviations of different lines in the diagram log g, θ_{eff} can be used to evaluate the peculiarity type of the star: Looking only at these five lines a trained investigator will be able to decide - is the star a subdwarf, metal-deficient giant, metallic-line star, Ap-star, white dwarf or double star? A computer makes it possible to determine the degree of peculiarity or metallicity.

The calibrated diagrams of Q, θ_{eff} can be transformed into another form, with log g and Θ_{eff} on the axes and with isolines Q=const (Fig. 3). If M_V or M_{bol} are substituted for log g we have a calibrated HR diagram which can be used for quantification of stars. The position of a star on this diagram corresponds to the point at which five isolines Q_{UVS}, Q_{PVS}, Q_{XVS}, etc. intersect. This is the most direct way from photometry to the HR diagram. If one is observing with a photoelectric photometer on-line with a computer the position of the star just measured can be seen on the HR diagram within a few seconds.

REFERENCES

Barry, D.C., Cromwell, R.H. and Schoolman, S.A. (1977). Astrophys. J. 212, 462.
Code, A.D., Davis, J., Bless, R.C. and Hanbury Brown, R. (1976). Astrophys. J. 203, 417.
Kurilienė, G. (1977). Bull. Vilnius Obs. No. 44, 15.
Kurucz, R.L. (1976). Personal communication.
Peytremann, E. (1974). Astron. Astrophys. Supp. 18, 81.
Straižys, V. (1970). Dissertation Thesis, Vilnius.
Straižys, V. (1974a). Astron. Circular, Moscow, No. 831, 3.
Straižys, V. (1974b). Astron. Astrophys. 36, 435.
Straižys, V. (1975). In Multicolor Photometry and the Theoretical HR Diagram, A.G.D. Philip and D.S. Hayes, eds., Dudley Obs. Report No 9, p. 65.
Straižys, V. (1977). Multicolor Stellar Photometry, Mokslas Publishers, Vilnius.
Straižys, V. and Sviderskienė, Z. (1972). Astron. Astrophys. 17, 312.
Straižys, V. and Zdanavičius, K. (1970). Bull. Vilnius Obs. No. 29, 15.
Sviderskienė, Z. and Straižys, V. (1971). Bull. Vilnius Obs. No. 31, 3.

DISCUSSION

KEENAN: Did I understand you to say that you can detect barium stars by your photometry? How do you do it?

STRAIŽYS: The barium stars show a weakening in the blue-violet part of the spectrum in comparison with ordinary K-giants of the same temperature. As a result the Vilnius color index (X-V) is larger for barium stars. This problem is discussed in greater detail in my monograph Multicolor Stellar Photometry, which was just published by Mokslas Publishers in Vilnius. This book is available from the Vilnius Astronomical Observatory.

HAUCK: Yesterday, Dr. Greenstein made a remark concerning a possible gap in the distribution of late B stars in the diagrams presented by Dr. Crawford. We have observed such a gap at B7 in the Geneva system, and a paper by J.C. Mermilliod was published in Astronomy and Astrophysics (**53**, 289, 1976). Do you observe the same gap in your diagram?

STRAIŽYS: I can't answer this question because we have not looked for the gap in our diagrams.

BELL: Have you made similar calculations for metal deficient stars and do you have observations of such stars?

STRAIŽYS: The Vilnius Observatory astronomers, A. Bartkevicius and J. Sperauskas have published three lists of Pop II stars observed in the Vilnius system (Bull. Vilnius Obs. Nos. 26, 36, and 40). These lists contain the F-G subdwarfs, late-type metal deficient giants, horizontal-branch stars, hot subdwarfs, etc. We have developed the method of detection and determination of [Fe/H] for subdwarfs and MDG's when interstellar reddening is present. We are continuing our observations of metal-deficient stars and a new catalogue of photometric data containing about 250 stars will soon be published.

GLIESE: How well is it possible to determine absolute magnitudes M_V from data in the Vilnius system? Especially, is it possible also for late type main-sequence stars and has it been done already?

STRAIŽYS: For the determination of M_V we usually use the Q, Q diagrams calibrated in spectral classes and M_V. The absolute magnitudes determined in this way for about 700 stars have been published in Bull. Vilnius Obs. Nos. 34 and 36.

Considering late-type dwarfs the problem is complicated because we have not investigated the effects of chemical composition differences on our colors. However, we estimate that the method permits us to determine M_V for red dwarfs with a mean square error of $+0^m.5$ (Bull. Vilnius Obs. No. 31).

PARTHASARATHY: There are now many photometric systems. With the improved detectors, the broad band systems are not so necessary. Is it not better to unify some of the photometric systems which have some bands in common?

STRAIŽYS: Yes, I agree with you. Some systems could be unified. We are making the first steps in this direction by trying to join the Vilnius and Geneva Observatory systems. Discussions with Dr. Golay and Dr. Hauck are in progress and we have agreed to introduce a joint seven-color system which we shall call the VILGEN system. This new system will have all the best properties of both of our systems.

D. EVANS: Why not call it the GENIUS system?

PART VII

THEORETICAL PAPERS AND SUMMARY

Martin Schwarzschild in the Lecture Hall, National Academy of Sciences.

THE VOGT-RUSSELL THEOREM, AND NEW RESULTS ON AN OLD PROBLEM

Helmuth Kähler

Hamburger Sternwarte

1. THE VOGT-RUSSELL THEOREM

Half a centry ago Henry Norris Russell and Heinrich Vogt independently made a conjecture concerning the structure of spherical stars which are in hydrostatic and thermal equilibrium (Russell, 1927; Vogt, 1926). This conjecture has later come to be known as the Vogt-Russell theorem and is usually formulated as follows: The structure of a star is uniquely determined by the mass and the composition. In other words, the statement claims the existence and uniqueness of a stellar equilibrium configuration for given parameters mass and composition, and you may find what is called a mathematical proof in many textbooks on stellar structure.

In the last decade, however, there have been found many counter-examples which disprove the theorem. In fact, both the existence and the uniqueness part of the theorem are violated. Let me give you some examples. Consider first the existence part: For a given mass and composition there should exist an equilibrium model. But you certainly can't find a model composed of iron with a mass of two solar masses: Since iron doesn't burn there could be at best a cold degenerate model, but for such models there is an upper mass limit at about 1.2 solar masses, the so-called Chandrasekhar limiting mass.

Consider next the uniqueness part: For given parameters there should be only one model. In numerical calculations however, up to 9 different models have been encountered. Let me give you an example of such multiple solutions which is closely connected with stellar evolution. The HR diagram in Fig. 1 shows the post-main-sequence

A. G. Davis Philip and D. S. Hayes (eds.), The HR Diagram, 303-311.

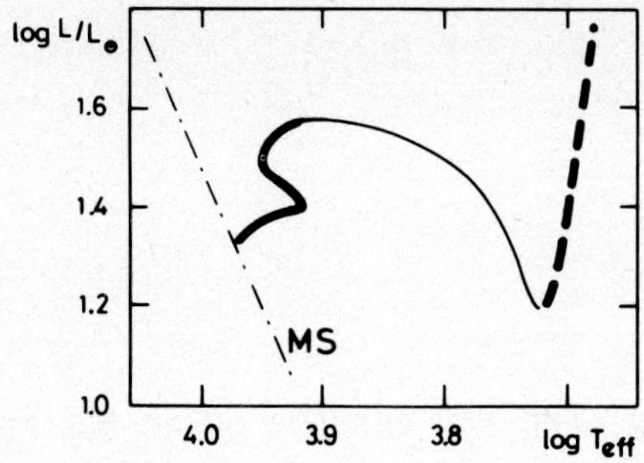

Fig. 1. Evolution of a star of two solar masses (see text).

evolution of a star of two solar masses (Roth, 1973). The heavy solid part of the evolutionary path corresponds to a slow phase in which the star evolves on a nuclear time scale. Thereafter the evolution speeds up and the star moves on a thermal time scale through the Hertzsprung gap, until it goes again through a slow phase near the Hayashi line which is indicated by a dashed line. During these phases the star has a burnt-out helium core and a hydrogen-burning shell source. Consider now the corresponding equilibrium models. Fig. 2a shows a sequence of models with different values of the core mass (Roth, 1973). I have plotted the core radius against the core mass. You can see that for a certain range of the core mass there are three different solutions, in contradiction to the uniqueness theorem. In the HR diagram (Fig. 2b) this sequence of equilibrium models appears as a curve which is similar to the evolutionary path, and indeed there is a simple connection. The equilibrium models on the solid part of the curve have an isothermal He-core and simulate stars in the slow phase just after leaving the main sequence. During evolution the core mass is growing. Correspondingly, in Fig. 2a the stars move to the right until, at the point SC, the Schönberg-Chandrasekhar limit for isothermal cores is reached. Now the star can no longer stay in thermal equilibrium but has to move on a thermal time scale, until

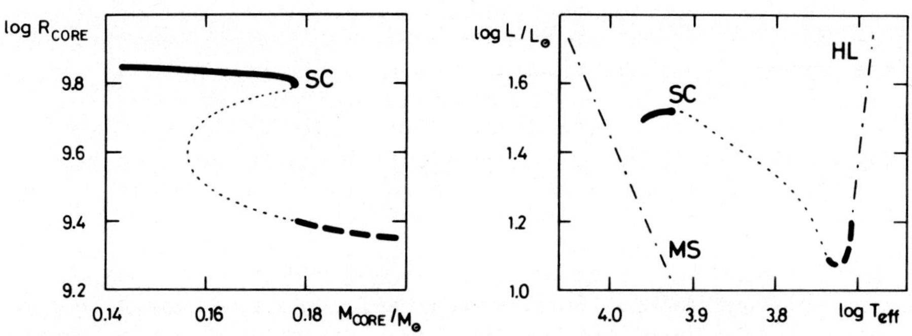

Fig. 2a and b. Equilibrium models of two solar masses (see text).

it finds a new equilibrium state on the dashed part of the sequence. This dashed part indeed simulates evolutionary models in the slow phase near the Hayashi line. The rapid crossing of the Hertzsprung gap can therefore be interpreted as a transition between two different equilibrium configurations for the same mass and composition. This example shows therefore not only the existence of multiple solutions, but also their relevance for the interpretation of evolutionary calculations.

We have therefore seen in explicit examples that both the existence and uniqueness parts of the so-called Vogt-Russell theorem are violated. On the other hand proofs of the theorem have been published. What can we say about these proofs? In fact there has been no proof but in essence only the following plausibility argument: We have a system of four differential equations for stellar structure together with four boundary conditions (two at the stellar center and two at the surface). Since the number of differential equations is equal to the number of boundary conditions, the solution should be completely determined. This classical arqument however applies only to linear equations, and even then it may fail as we shall see.

In order to make this point quite clear, let us comprise the parameters (i.e. mass and composition) in the symbol p, and a particular stellar model (i.e. a particular solution of the structure equations) in the symbol S. So far we have asked for the global existence and uniqueness of a solution: Is there always, for given p, one and only one stellar model S?

<u>global</u> existence and uniqueness: $p \to S$?

The answer is no, the classical argument cannot say anything because the corresponding equations are highly nonlinear. If however we confine ourselves to a small neighborhood of a given solution, we may work with linearized equations, and then the classical argument will usually be applicable. So we are led to the problem of the local existence and uniqueness of a given solution: Do small changes of the parameters lead uniquely to small changes of the model?

<u>local</u> existence and uniqueness: $\Delta p \to \Delta S$?

The answer will usually be yes because this is a linear problem.

At this point let us stop for a moment and recall what Russell actually claimed about the manifold of stellar models. He wrote in his book "that a star of given mass and composition will usually be in equilibrium for only one value of the radius..." (as an example he considered main sequence stars), but he admitted that "in more complicated cases there might be two or more different configurations". Accordingly Russell did not claim the global uniqueness

as a strict theorem. He admitted exceptions in complicated cases; today we would add that most cases are complicated indeed. Summarizing, Russell's original conjecture is cautious, but the modified version which was later formulated in lectures and textbooks as a so-called theorem is definitely wrong and did Russell no service.

Next let me report about recent work which has given an exact local formulation of the Vogt-Russell theorem (Kähler, 1972; Kähler and Weigert, 1973). The problem is to give precise conditions for the local existence and uniqueness of stellar models, in the vicinty of a given model. First let me illustrate the problem with an example. Consider again a star with a burnt-out He-core (Fig. 2a). We have already observed that within a certain range of the core mass there are three different solutions. With growing core mass two of them approach each other, and finally they merge. At this point (where the Schönberg-Chandrasekhar limit has been reached) the local existence and uniqueness is violated. For a somewhat larger core mass there is no neighboring solution at all, and for a somewhat smaller core mass there are two different neighboring solutions. It will become clear later that the solution at the turning point itself should be considered as a double solution.

Let us now obtain a criterion for the local existence and uniqueness. For given parameters p, a model is asserted to be completely determined by the values of luminosity and effective temperature. Starting with trial values for L and T_{eff}, we may use the surface conditions for stars which are provided by the theory of stellar atmospheres, and may then perform an inward integration of the differential equations for the stellar interior. In this way we obtain a solution of the differential equations which satisfies the outer boundary conditions, and which depends on p, L and T_{eff}. This solution gives a stellar model if and only if the two boundary conditions at the stellar center are also satisfied. It can be shown that these two conditions can be written in the form $g_1 = 0$ and $g_2 = 0$, where the g_i are functions of p, L and T_{eff}. Summarizing, a stellar equilibrium model corresponds to a common zero of two functions g_1 and g_2 which are defined by the differential equations and boundary conditions for stellar structure, and which depend on the arguments mass, composition, luminosity and effective temperature.

Let a model now be given, and consider the existence and uniqueness of neighboring models. An arbitrary but small variation of the mass or the composition should yield one and only one neighboring model. This means that an infinitesimal change δp should lead to a unique solution δL, δT_{eff} of the equations $\delta g_i = 0$. By linearization we find the following vector equation

$$\begin{vmatrix} \partial g_1/\partial p \\ \partial g_2/\partial p \end{vmatrix} \delta p + \begin{vmatrix} \partial g_1/\partial L & \partial g_1/\partial T_{eff} \\ \partial g_2/\partial L & \partial g_2/\partial T_{eff} \end{vmatrix} \cdot \begin{vmatrix} \delta L \\ \delta T_{eff} \end{vmatrix} = 0$$

in which the first vector stands for the change in the stellar parameters, and the last vector shows the reaction of the star in the HR diagram. Let the 2x2 matrix be called G. If the determinant of G is non-zero, the reaction of the star is uniquely determined:

$$|G| = 0: \quad \begin{vmatrix} \delta L \\ \delta T_{eff} \end{vmatrix} = -G^{-1} \cdot \begin{vmatrix} \partial g_1/\partial p \\ \partial g_2/\partial p \end{vmatrix} \cdot \delta p$$

If on the other hand the determinant vanishes, there may be no solution at all, or there may be several solutions; this means that the local existence and uniqueness is violated. Thus we have arrived at a rigorous criterion: The local existence and uniqueness of stellar models, in the vicinity of a given model, is equivalent to the non-vanishing of the determinant of G. This statement may be considered to be the Vogt-Russell theorem in a local sense.

What is the physical meaning of this criterion? It can be shown that the vanishing of the determinant of G is equivalent to the occurrence of a zero eigenvalue in the stellar stability problem. From the viewpoint of stellar evolution, a model can be stable only if all eigenvalues have a negative real part, which excludes zero eigenvalues. We thus deduce that the existence and uniqueness of neighboring models holds automatically for each model which is stable. Finally it can be shown that in numerical calculations the sign of the determinant of G is closely connected with the sign of the Henyey determinant. This provides a simple means for testing computed models.

2. GLOBAL STATEMENTS ON THE MANIFOLD OF STELLAR MODELS

We have already observed that classical arguments cannot say anything about the total number of equilibrium models for the given parameters, and that indeed global uniqueness statements are out of the question. Accordingly it seemed to be hopeless to look for general statements on the number and the stability of the different models. Surprisingly, recent work has shown that such statements are nevertheless possible when appropriate algebraical and topological methods are applied (Kähler, 1975). For this purpose two new properties of equilibrium models have been defined, the multiplicity m and the charge c.

Consider first the multiplicity. If we ask for the number of models, we have to think about cases of mathematical degeneracy in

which several solutions coincide. For example, we have seen that at the Schönberg-Chandrasekhar limit two solutions have merged, so we might speak of a double solution. In order to give a corresponding definition we recall that, for given parameters p, a stellar model is completely determined by its location in the HR diagram, and that the values of L and T_{eff} satisfy the two equations $g_i = 0$. Each of these equations defines a curve in the HR diagram, and the model corresponds to an intersection of the two curves. Let us define the multiplicity m of the model as the corresponding intersection number in the sense of algebraic geometry. The multiplicity so-defined has a simple physical meaning. A model which is locally unique has m = 1. This is the usual case; such a model reacts uniquely to parameter changes. If on the other hand the local uniqueness is violated, we have m > 1. In this case m solutions coincide. This mathematical degeneracy can be resolved by a small change of a parameter; up to m different models can thereby be obtained. In the HR diagram, the usual case of m = 1 corresponds to a simple intersection of the two curves, as is shown in Fig. 3a. A higher multiplicity may for example correspond to a contact of the two curves (Fig. 3b), or to the occurrence of singular points (Fig. 3c).

Next, let us consider the charge of a stellar model. We have observed that the vanishing of the determinant of G is accompanied by the occurrence of a zero eigenvalue in the stellar stability problem. Hence we may expect that the sign of this determinant (or equivalently of the Henyey determinant) is important for stellar stability. This has indeed been shown by Paczynski (1972). The charge of a model is now defined as a property which allows for this sign. The definition involves topological properties of

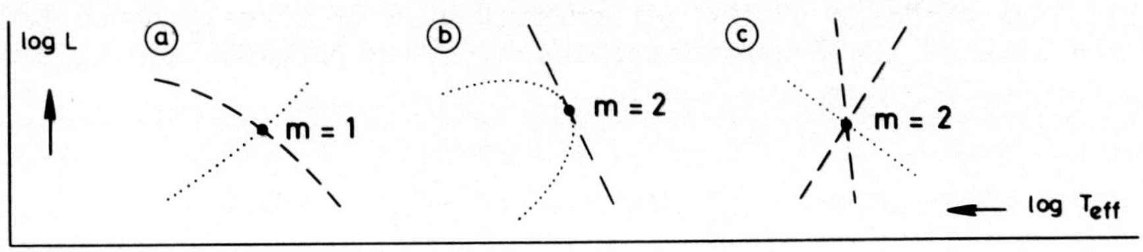

Fig. 3. Definition of the multiplicity (see text). The dashed and dotted lines represent the curves $g_1 = 0$ and $g_2 = 0$, respectively.

the functions g_i; let me omit details. The charge c is equal to the sign of det G if this sign is defined, that means if the determinant is non-zero. Otherwise the charge is usually equal to either 0 or ± 1, depending on whether m is even or odd. You may ask why this property of a model has been called, in particular, charge. The reason is that the total charge of all models in a given region in the HR diagram depends only on the boundary of the region and can be expressed by a certain line integral. This indicates an

analogy to electrostatics where the total electric charge in a given volume can be determined from the electric field on the surface.

So far we have defined two properties of a model, multiplicity and charge, which are small integers and which might be considered to be quantum numbers for the model. Next let us ask for applications. Consider first the stability of a model with given quantum numbers. If $m > 1$, the model has a zero eigenvalue and is therefore unstable from the viewpoint of stellar evolution. Stellar stability, therefore, requires $m = 1$. Furthermore, it can be shown that stability requires a definite value of the charge. This value turns out to be unity. Summarizing, a model can be stable only if $m = c = 1$. This condition is necessary but not sufficient.

As a second application of this formalism let us ask for the models which can be obtained from a given model by continuous changes of the parameters. (In other words, let us consider a linear series of stellar models.) Suppose we have, for the parameter p, a model of multiplicity m' and charge c'. What are the total number and the total charge of the different neighboring models obtained by a small change δp? The general result is that the number of models (defined as the sum over the multiplicities) is either conserved or it decreases by an even number, and that the total charge is always conserved:

	p	$p+\delta p$
Σm	m'	$m' - 2r$ (r integer ≥ 0)
Σc	c'	c'

This result may be considered to be a selection rule for the quantum numbers m and c. As an example consider again stars with a burnt-out helium core and a total mass of two solar masses (Fig. 4a). The number of models with given core mass is either 1 or 3; it may change by an even number. But the total charge of the different models is thereby conserved. A more complicated example is the bifurcation point in Fig. 4b which has been reported by Paczynski (1972).

We are now ready to ask for a global statement on the manifold of equilibrium models with given mass and composition. These basic parameters are again comprised in the symbol p. Let C_1 denote the total charge of all models which do exist for $p = p_1$. Similarly, let C_2 be the total charge of all models which do exist for $p = p_2$. Take a continuous sequence of parameters which connects p_1 and p_2. Along this sequence, the total charge of all models is conserved provided that no model escapes to 'infinity', which means to extreme values of pressure or temperature for which no local thermodynamic equilibrium is possible for the given composition. This assumption is strong but physically plausible if we confine ourselves to

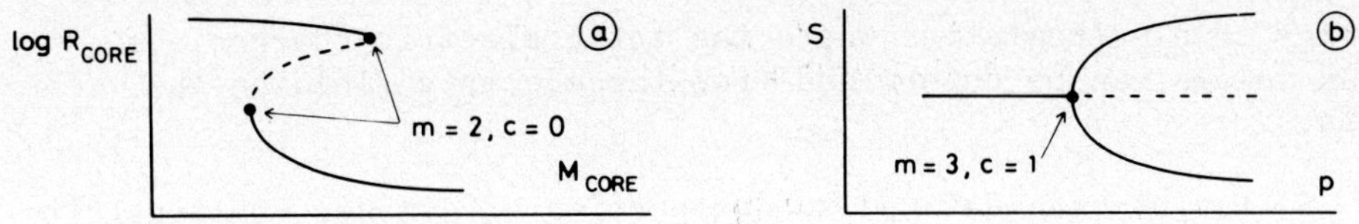

Fig. 4. Examples of linear series of stellar models (see text). Solid and dashed lines denote models with m = c = 1 and m = -c = 1, respectively.

compositions consisting mainly of lighter elements (say, up to carbon). Let us make the above assumption. The conservation of charge implies then $C_1 = C_2$. In other words, the total charge C of all models does not depend on the mass and the composition. The value of C can therefore be obtained from a simple special case, the limit of very small stellar mass. In this limit there is only one degenerate model (white dwarf or planet) which has c = 1. Accordingly the total charge C of all models has the universal value +1.

Consider now the total number N of models for given mass and composition. We may write N as the sum of the contributions N_+ and N_- which refer to models with positive and negative charge, respectively. The requirement that the total charge be unity implies the following equations:

$$N_+ = (N + 1)/2, \text{ and } N_- = (N - 1)/2.$$

These equations show that N is odd, Accordingly N ≥ 1, and we arrive at an existence theorem: For given mass and composition there is always one equilibrium model, but additional models may appear in pairs of opposite charge. Moreover we obtain some information about the stability of the models. Since stable models are of positive charge, up to (N+1)/2 models can be stable but the remaining ones are necessarily unstable.

Summarizing, we have made a plausible assumption which excludes stars consisting of heavy elements. Under this assumption, a general statement has been derived which concerns the old problem Russell dealt with. The statement gives some information about the number and the stability of the different equilibrium configurations, and in particular it guarantees the existence of at least one model.

REFERENCES

Kähler, H. (1972). Astron. Astrophys. **20**, 105.
Kähler, H. and Weigert, A. (1973). Astron. Astrophys. **30**, 431.
Kähler, H. (1975). Astron. Astrophys. **43**, 443.
Paczyński, B. (1972). Acta Astron. **22**, 163.
Roth, M.L. (1973). Thesis, Hamburg.

Russell, H.N. (1927). In Astronomy, Vol. 2, by H.N. Russell, R.S. Dugan and J.Q. Stewart, Ginn and Co., Boston, p. 910.
Vogt, H. (1926). Astron. Nachr. 226, 301.

COMPARISON BETWEEN OBSERVATIONAL AND THEORETICAL (log T_{eff}, M_{bol}) DIAGRAMS

G. Cayrel de Strobel and M.N. Perrin

Paris-Meudon Observatory

In a paper by Perrin et al. (1977), we have constructed an empirical HR diagram for 138 nearby F, G and K stars, for which we had: i) an effective temperature and a metal content derived from a detailed analysis; ii) a reliable bolometric magnitude obtained from an absolute magnitude M_V, based on a large parallax and a rather small bolometric correction.

The main purpose of this paper was to see if one obtains a "cleaner" HR diagram when grouping stars of same metal/hydrogen ratio, and if this "cleaner" HR diagram was evolving with the value of the metal/hydrogen ratio as expected from theoretical grids of evolutionary tracks or isochrones. This was indeed the case, when comparing strongly metal poor stars (halo stars) with disk stars having a solar chemical composition.

But a puzzling result was that in our observational ZAMS, stars of

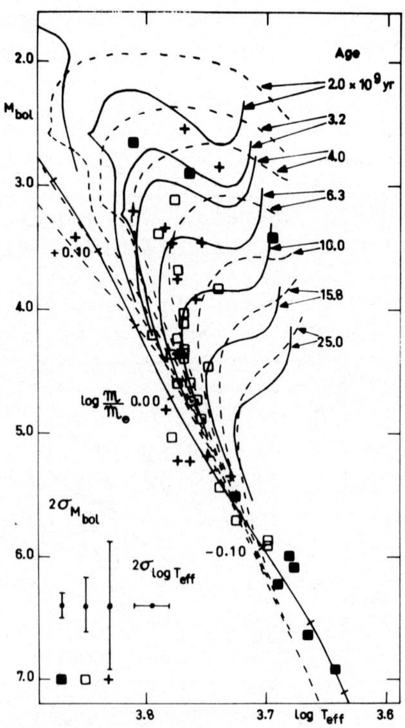

Fig. 1. Observational and theoretical (log T_{eff}, M_{bol}) diagrams for normal metal content stars.

A. G. Davis Philip and D. S. Hayes (eds.), The HR Diagram, 313–318. All Rights Reserved.

Table I : Comparison of age determinations from two grids of isochrones (Hejlesen 1975, Demarque 1977) for stars in the lane between 0.5 and 2.0 mag above the main sequence and in the T_{eff} interval : 5500 - 6500 K.

		Hejlesen			Demarque					Hejlesen			Demarque		
HD	$[Fe/H]^*_\odot$	Y	Z	Age(10^9yr)	Y	Z	Age(10^9yr)	HD	$[Fe/H]^*_\odot$	Y	Z	Age(10^9yr)	Y	Z	Age(10^9yr)
121370	+0.50	0.26	0.04	2.5	0.30	0.04	2.5	22484	-0.16	0.29	0.01	5.6	0.30	0.01	5.5
102870	+0.27	"	"	4.5	"	0.04	2.0	16895	-0.26	"	"	7.9	"	"	6.0
"	"	"	"	4.5	"	0.01	6.0	2151	-0.31	"	"	10.0	"	0.01	8.5
34411	+0.22	"	"	6.3	"	0.04	2.5	"	"	"	"	10.0	"	0.004	12.0
"	"	"	"	6.3	"	0.01	7.5	142373	-0.35	"	"	6.3	"	0.01	6.5
161797	+0.15	0.28	0.02	10.0	"	0.01	11.0	"	"	"	"	6.3	"	0.004	8.5
19373	+0.05	"	"	7.9	"	"	7.0	165908	-0.40	"	"	11.2	"	0.004	12.0
158614	+0.02	"	"	15.9	"	"	12.5	157214	-0.43	"	"	22.4	"	"	20.0
136064	-0.03	"	"	6.3	"	"	6.0	136352	-0.50	"	"	22.4	"	"	22.0
84737	-0.04	"	"	10.0	"	"	7.0	222368	-0.51	"	"	7.1	"	"	8.0
61421	-0.06	"	"	2.8	"	"	3.5	69897	-0.52	"	"	10.0	"	"	8.5
38393	-0.07	"	"	10.0	"	"	7.5	203608	-0.68	0.30	0.004	25	"	0.004	16
142860	-0.09	"	"	5.6	"	"	5.0	"	"	"	"	25	"	0.001	25+
210027	-0.10	"	"	4.5	"	"	5.5	63077	-0.80	"	"	25	"	0.004	18
9826	-0.16	0.29	0.01	5.0	"	"	5.0	"	"	"	"	25	"	0.001	25+

solar composition, metal-rich stars and moderately metal-poor disk stars are well mixed and do not segregate on different ZAMS as extreme metal-poor stars do.

We have interpreted this result by suggesting that in these stars there is a concomitant variation of Y and Z, tending to cancel the displacement of the ZAMS that one should observe if Z was varying alone. Based on Hejlesen's evolutionary models (Hejlesen, 1975), that would be the case with a related variation of Y with Z: $\Delta Y \simeq 5\Delta Z$.

Age determinations have been obtained for subgiants lying 0.5 to 2.0 magnitudes above the ZAMS. Needless to say, the knowledge of the proper ZAMS to be used in age determinations is quite important, especially for the subgiants close to the lower limit ($0^m\!.5$). Therefore we have decided to see how our results were affected by using a different theoretical grid.

We seized the opportunity to check our results with the publication of the tables of isochrones by Ciardullo and Demarque (1977).

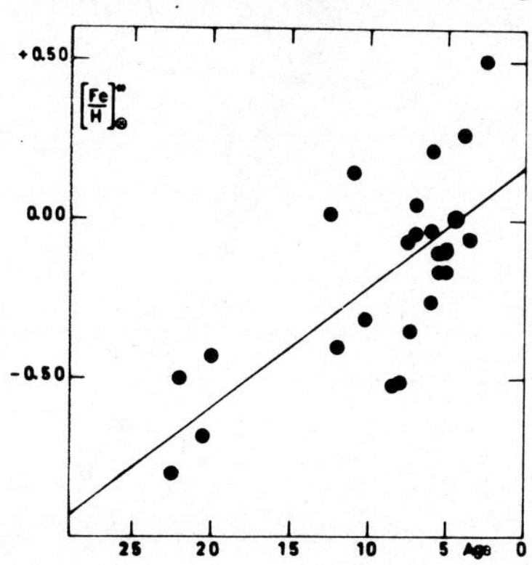

Fig. 2a. $[Fe/H]^*_\odot$ versus Age diagram (Demarque).

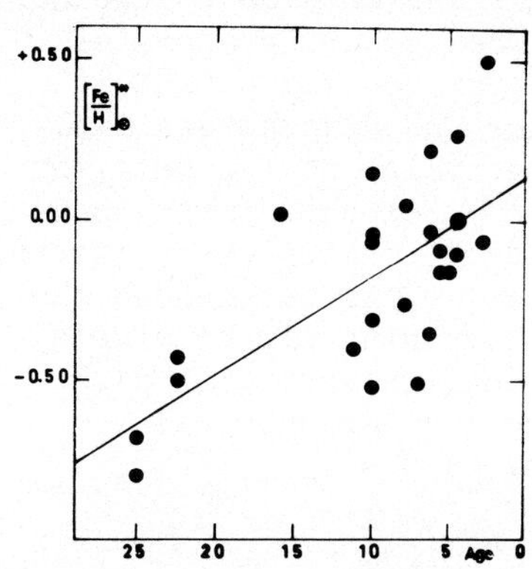

Fig. 2b. $[Fe/H]^*_\odot$ versus Age diagram (Hejlesen).

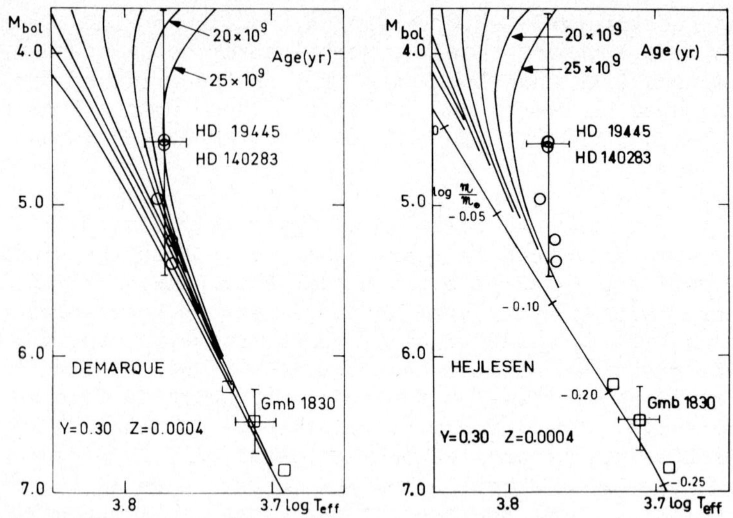

Fig. 3a and 3 b. Observational and theoretical ($\log T_{eff}$, M_{bol}) diagrams for halo stars.

In Fig. 1, a comparison is made between the grid of isochrones computed by Hejlesen with that by Demarque. The full lines represent the isochrones of Hejlesen, the broken lines those of Demarque. The ages for which these grids of models have been computed are given in units of 10^9 years. In Demarque's grid we took the 200 million year isochrone to be equivalent to a zero-age main sequence. The two grids of isochrones have not been computed with exactly the same values of Y and Z in Hejlesen and Demarque, (Y = 0.28, Z = 0.02), (Y = 0.30, Z = 0.01), respectively, because for the models of Demarque we had at our disposal only the grid (Y = 0.30, Z = 0.01) which sets the Sun at an acceptable place. The two sets of isochrones have been drawn showing the beginning of the giant branches. The differences between the two grids should be studied by specialists of internal structure.

We have studied them in an empirical way with the help of an observational HR diagram, containing stars of normal metal abundance falling in the interval: $-0.15 < [Fe/H] < +0.15$ and in the effective temperature interval: 5500 - 6500 K. The representative points of the stars have different symbols corresponding to the errors of their absolute bolometric magnitude. At the lower left side of the diagram the $2\sigma_{M_{bol}}$ bars and the average $2\sigma_{\log T_{eff}}$ bar are drawn.

We can see on the diagram that the slope of the observational ZAMS agrees somewhat better with the slope of the theoretical ZAMS of Hejlesen.

We have constructed more such composite diagrams for different chemical compositions. We have stimated the ages for stars falling in the strip $0^m.5-2^m$ above the main sequence, and in a

metal abundance interval: $-0.80 < [Fe/H] < +0.50$, 6 times less and 3 times more abundant than the normal solar metal abundance. These age estimations have been made once on the grids of Hejlesen and once on those of Demarque. In Table I these estimations are given in columns 5 and 8; column 2 contains the [Fe/H] values for the stars.

The [Fe/H] versus age relation of Demarque (left) and of Hejlesen (right) are shown in Figs. 2a and 2b. Both the diagrams are rather well populated by stars having ages between 5 and 10 billion years. For these stars the dispersion in heavy-element content is as high as a factor of 5. Even though the spread in [Fe/H] is great, an attempt has been made to derive the average enrichment of stars in the solar neighborhood from a least squares fit. The enrichment relation we obtained is very similar for Demarque and Hejlesen: The metal content grows roughly by a factor of 2 per 10 billion years. Here is is interesting to point out that the ages of the oldest disk stars with Demarque's estimation are not very much smaller than the ages with Hejlesen's.

The same can be said for the ages of the two halo subgiants HD 19445 and HD 140283: Demarque's models attribute to them a very large age, both the stars fall on the 25 billion year isochrone, as we can see in Fig. 3a (left). These two stars fall outside of this isochrone on Hejlesen's models (Fig. 3b, right), but their parallaxes are not very good.

Using the Demarque models it seems also necessary to assume the existence in the disk stars of a simultaneous variation of the helium content with the metal content, roughly: $\Delta Y \simeq 5\Delta Z$.

In conclusion, we can assert that all the results given in Perrin et al. (1977) based on the theoretical grid of evolutionary models of Hejlesen remain valid if we study the observational ($\log T_{eff}$, M_{bol}) diagram with Demarque's evolutionary models.

REFERENCES

Ciardullo, R.B. and Demarque, P. (1977). Transactions of the Astronomical Observatory of Yale University, Vol. **33**.
Hejlesen, P.M. (1975). private communication.
Perrin, M.N., Hejlesen, P.M., Cayrel de Strobel, G. and Cayrel, R. (1977). Astron. Astrophys. **54**, 779.

DISCUSSION

DEMARQUE: I think the discrepancy you find between the main sequences can be understood in terms of the choice of mixing lengths.

CAYREL de STROBEL: Yes.

DEMARQUE: I still believe that the oldest disk clusters, such as NGC 188, appear to be much younger than the globular clusters. If you correct our evolutionary tracks to a mixing length of about 1.5 scale heights, you get excellent agreement and you find an age which is only about 5 billion years.

CAYREL de STROBEL: I don't believe that. I can get the age down by 20 to 30%, but to not less than 12 billion years. Reeves also finds an age of greater than 12 billion years for the old disk stars.

SEARS: With regard to the difference in the oldest age of clusters (Demarque), the clusters are fitted to the Hyades and the field stars have directly determined parallaxes. Perhaps that is part of the discrepancy.

McNAMARA: How do you reconcile your results, which show a dependence of metal content on age, with those of Dr. McClure, whose study of galactic clusters does not show such a correlation?

CAYREL de STROBEL: I reconciled my results (1) when I showed the great dispersion of metal content in stars having ages between ~ 5 to 10 billion years; and (2) when I showed immediately after, the age <u>versus</u> [Fe/H] relation between these stars and the very much older disk stars. Between these two samples there is a significant metal content enriched relationship.

TAYLER: If we really take your ages literally, even Sandage and Tammann would have to reduce their Hubble constant. Are you willing to admit that your ages may be sufficiently over-estimated to be consistent with present estimates of the Hubble constant and an open universe?

With reference to main sequences of stars with different chemical composition, if we use a mixing length theory of convection (which we know is only an approximation) can we be sure that the mixing length does not depend on chemical composition?

CAYREL de STROBEL: In answer to your first question, surely my error bars for the oldest stars are so great that I wouldn't dare touch the actual value of the Hubble constant.

With respect to your second question, I would be happy to try again to estimate the ages of the oldest disk stars and the halo stars using isochrones computed with several values of the mixing length. But they are very difficult to acquire.

BELL: The question of the mixing length for metal deficient main sequence stars could be checked using Groombridge 1830, which is known to be metal poor.

KRAFT: Is it possible that any galactic clusters which might have ages > 6×10^9 yr. would have been dissolved? After all, very old galactic clusters such as NGC 188 are rare. Perhaps there are field disk stars much older than the oldest known galactic cluster.

CAYREL de STROBEL: I agree with you that certainly there are very old disk stars with ages about the same as those of halo stars. But Dr. Buscombe has pointed out that we are probably wrong about the scarcity of old galactic clusters, because we can't see them very far away.

FOY: I'd like to mention that I have analyzed two stars in M67, and I have found that they are marginally metal deficient relative to the Sun. I remember that in Lausanne Mayor said that, from the point of view of dynamics, old open clusters could be peculiar, because normal ones must already have evaporated. Consequently, it would be better not to use them to determine the metal-enrichment law of the galactic disk.

IS IT POSSIBLE TO DERIVE THE POSITION OF A STAR IN THE HR DIAGRAM FROM MULTICOLOR PHOTOMETRY?

M. N. Perrin and G. Cayrel de Strobel

Paris-Meudon Observatory

In the preceeding paper of this Symposium by Cayrel and Perrin (1978), HR diagrams have been discussed, in which the [Fe/H] ratio and effective temperatures are derived from model atmosphere analyses and absolute bolometric magnitude from apparent visual magnitude, parallax and bolometric correction. The use of parallaxes imposes a very severe restriction on the number of objects, as parallaxes of at least 0''050 are needed to have an absolute magnitude good to $\pm 0^m.4$.

We have therefore given great attention to two recent papers by Golay et al. (1977) and Golay (1978) suggesting that the concept of photometric boxes could make a similar study possible without appealing either to detailed analyses nor to luminosities derived from parallaxes, but getting all this information from the seven color photometry of the Geneva system. In particular, one of our aims, i.e. to obtain the luminosity function and a non-biased age-distribution in the galactic disk, would be possible with purely photometric data available and in a much greater volume of space.

Indeed, Golay, as he has explained in part VI of this Symposium (Golay, 1978), claims that stars in a same photometric box (having Geneva colors, not more different than $0^m.01$ from the colors of the central star of the box) have also the same physical parameters as those of the central star of the box, such as: spectral type, visual absolute magnitude, chemical composition, equatorial rotation, multiplicity, interstellar extinction, etc., hence perhaps, mass, radius and age.

To check the physical reality of this concept, we have tried to find stars belonging to photometric boxes, having also available

A. G. Davis Philip and D. S. Hayes (eds.), The HR Diagram, 319–322.

the other type of data as [Fe/H] ratio, effective temperature and absolute bolometric magnitude. We have used the Catalogue of Lausanne (Morel *et al.*, 1977), but unfortunately we have found stars belonging to only two boxes (three belonging to the box of 16 CygB and two to the box of HD 10307). These five stars have also been studied by Hardorp (1976, 1977). Hardorp agrees with Golay that the continua of the stars in the same box are almost identical. The various data concerning the stars are given in Table I and II.

TABLE I: THE GENEVA PHOTOMETRIC DATA (RUFENER, 1976)

HD	Name	Sp.T	U	V	B_1	B_2	V_1	G
186427	16 CygB	G5V	1.420	0.129	1.107	1.291	0.889	1.174
76151	-	dG3	1.419	0.138	1.102	1.290	0.889	1.177
20630	ϰ Cet	G5V	1.418	0.126	1.103	1.285	0.880	1.162
10307	-	G2V	1.366	0.198	1.081	1.313	0.951	1.250
34411	λ Aur	G0V	1.378	0.199	1.076	1.308	0.947	1.246

TABLE II: VARIOUS PARAMETERS CONCERNING THE STARS (MOREL *et al.* 1977, PERRIN *et al*. 1977).

HD	B_2-V_1	d	m_2	π"	M_{bol}	$\log T_{eff}$	log g	[Fe/H]
186427	0.402	0.576	-0.368	0.036	3.92	3.763	-	+0.10
76151	0.401	0.586	-0.371	0.085	5.58	3.748	4.5	0
20630	0.405	0.575	-0.367	0.107	4.88	3.753	4.45	+0.08
10307	0.362	0.617	-0.397	0.087	4.59	3.763	4.38	+0.12
34411	0.361	0.634	-0.400	0.067	3.77	3.768	4.11	+0.22

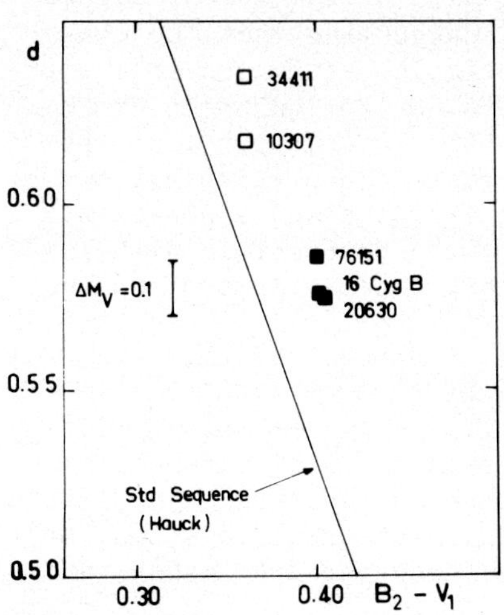

Fig. 1. Geneva luminosity indicator d *versus* (B_2-V_1).

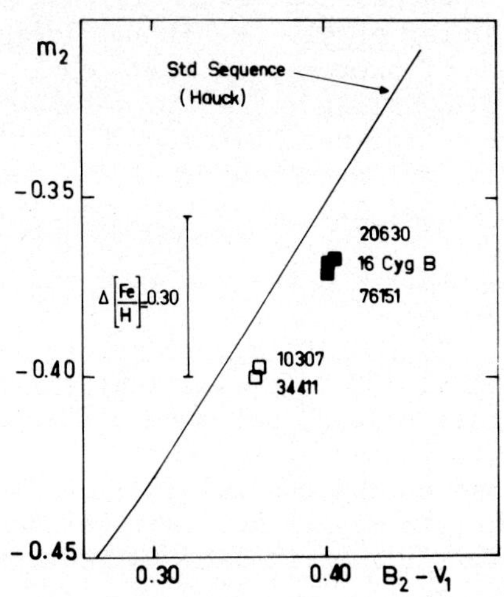

Fig. 2. Geneva metallicity indicator m_2 *versus* (B_2-V_1).

The stars have been plotted in the (d, B_2-V_1) and (m_2, B_2-V_1) diagrams of the Geneva photometry (Fig. 1 and 2). The differences $\Delta M_V = 0^{m}.1$ and $\Delta[Fe/H] = 0.30$, indicated respectively in Fig. 1 and 2 by a bar, come from the calibration of Hauck (1973). The stars belonging to the same box are tightly clustered. They may be supposed to have the same luminosity and metal content.

The stars have been plotted in an observational ($\log T_{eff}$, M_{bol}) diagram (Fig. 3). They are represented with their proper error of M_{bol}. The average error of T_{eff} is drawn at the left side of the diagram. The position of the Sun is also shown. These stars have been compared with five ZAMS computed by Hejlesen (1975) for several chemical compositions (Z or Y) as shown in the diagram.

The position of the stars in this diagram is very different from that of Fig. 1. Here the difference in magnitude between the less evolved and the more evolved star of the same box reaches $1^{m}.7$ whereas it does not exceed $0^{m}.1$ in Fig. 1. The main fact which seems arise from this comparison is the insensitivity of the Geneva photometry in the range of spectral type considered here (middle G) to rather large evolutionary effects. This may come from a concomitance of several physical effects acting in opposite way, and/or, may it come from the fact that the luminosity criterion of the Geneva photometric system (the Balmer jump) loses its sensitivity for solar and later type stars?

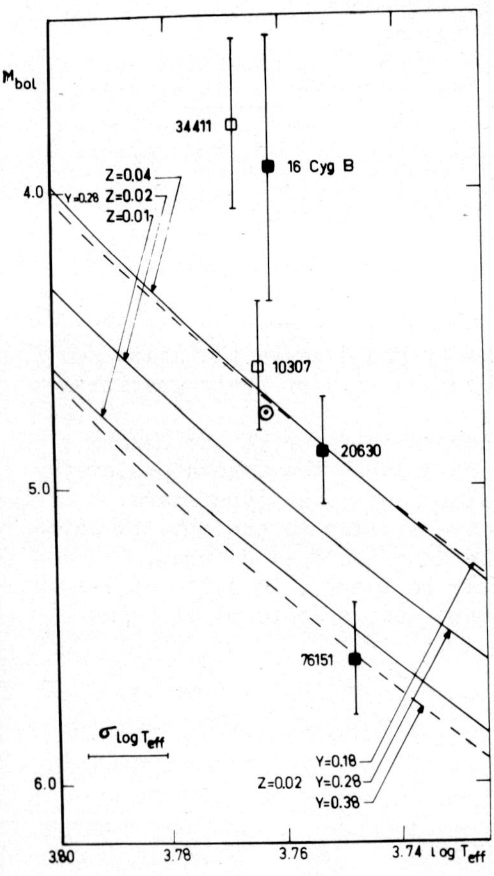

Fig. 3. Observational ($\log T_{eff}$, M_{bol}) diagram.

A complete discussion of the position of the stars in the observational HR diagram would imply a discussion of the helium content of the stars (a parameter which is not taken into account in the Geneva photometry). Such a discussion has been attempted for cooler, non evolved stars in Perrin _et al_. (1977). Let us just indicate here the case of HD 76151 and HD 20630, which are of solar metal content. One of these stars is surely not evolved and the other, according also to its spectroscopic gravity, is probably not evolved. From their position in the HR diagram these

two stars could differ in He-content by an amount as high as 0.33 dex.

REFERENCES

Cayrel de Strobel, G. and Perrin, M.N. (1977). In IAU Symposium No. 80, The HR Diagram, A.G.D. Philip and D.S. Hayes, eds., Reidel, Dordrecht, p. 313.
Golay, M. (1978). Astron. Astrophys. (in press).
Golay, M. (1978). In IAU Sumposium No. 80, The HR Diagram, A.G.D. Philip and D.S. Hayes, eds., Reidel, Dordrecht, p. 277.
Golay, M., Mandwewala, N. and Bartholdi, P. (1977). Astron. Astrophys. **60**, 181.
Hardorp, J. (1976). Mitt. Astron. Ges. **38**, 143.
Hardorp, J. (1977). Astron. Astrophys. (in press).
Hauck, B. (1973). In IAU Symposium No. 54, Problems of Calibration of Absolute Magnitudes and Temperature of Stars, B. Hauck and B.E. Westerlund, eds., Reidel, Dordrecht, p. 117.
Hejlesen, P.M. (1975). private communication.
Morel, M., Bentolila, C., Cayrel de Strobel, G. and Hauck, B. (1977). In IAU Symposium No. 72, Abundance Effects in Classification, B. Hauck and P.C. Keenan, eds., Reidel, Dordrecht, p. 223.
Perrin, M.N., Hejlesen, P.M., Cayrel de Strobel, G. and Cayrel, R. (1977). Astron. Astrophys. **54**, 779.
Rufener, F.G. (1976). Astron. Astrophys. Suppl. **26**, 275.

DISCUSSION

BELL: Probably only the Geneva U depends on gravity for these stars. A small error in this quantity could cause the stars you mention to fall in the same box, whereas in part their gravities could be different.

HARDORP: I should like to say a word in favor of the Geneva system. There is a good correlation between the Geneva photometry and the ultraviolet absorption features of solar type stars: Stars in the same box as 16 Cyg B are similar to the Sun, in spite of their different locations in the M_{bol}/Log T_{eff} diagram, whereas HD 10307 differs appreciably in the UV, in spite of its closeness to the Sun in this diagram, and in spite of it being an MK standard for G2V.

ON MAGNETIC MIXING AND ROTATION IN STELLAR EVOLUTION

Peter G. Gross

Case Western Reserve University

Warner and Swasey Observatory

1. INTRODUCTION

In this paper some thoughts and problems are presented from the viewpoint that the evolution of stars may play a key role in generating magnetic fields which, in turn, may affect the mixing of nuclearly processed elements from the stellar interior to the surface. The relevant parameter is stellar rotation which, upon interaction with convective turbulence driven by thermal instabilities, leads to the generation of magnetic fields. A possible connection to Bidelman's hypothesis on the evolutionary status of Ap stars is also discussed in the context of a post-core-helium-flash hypothesis.

2. PROCESSING OF MAGNETIC FIELDS IN STELLAR EVOLUTION?

Historically, the assumption of magnetic flux conservation during stellar evolution has been the simplest so perhaps the most adequate approach to the problem. Despite theoretical complications, this view may eventually be abandoned because of increasing observational evidence for non-conservation of flux and steady advances in the theory of stellar dynamos (cf. Weiss, 1971). From the latter one infers that the fates of B-fields and rotation must be intertwined. Of fundamental importance in this respect is the existence of thermal instabilities associated with shell helium burning as shown by Schwarzschild and Härm (1965), and with nuclear ignitions in electron degenerate cores. The former has been recently emphasized by Levy and Rose (1974) who considered the B-fields for the problem of dissipating angular momentum from the main sequence to white dwarfs and neutron stars. In turn,

A. G. Davis Philip and D. S. Hayes (eds.), The HR Diagram, 323-331.

the latter is inexorably connected with mass loss precesses which may sometimes proceed during the late stages of stellar evolution at rather high rates as shown by Weidemann (1977).

To estimate the strengths of the B-fields, one may adopt the approximate formula of Malkus (1959) which gives $B^2 \sim 4\pi\rho \langle v \rangle \Omega r$, for a density ρ, turbulent velocity $\langle v \rangle$, angular velocity Ω at distance r from the center. Assuming conservation of angular momentum in the radiative zones, one has $\Omega(r) = \Omega_o r_o^2/r^2$; for a double shell burning red giant Ω_o and r_o would be referred to the inner edge of the envelope convective zone and r to the intershell convective zone (ICZ). Taking typical physical values from the shell-flash computations of Schwarzschild and Harm (1967, hereafter SH), one obtains $B^2 \sim \pi \times 10^{22} \Omega_o$ or $B \sim 2 \times 10^8$ G for $\Omega_o \sim 7 \times 10^{-7}\,s^{-1}$ (~3 revolutions/yr) at $r_o \sim 3 \times 10^{11}$ cm and $\rho \sim 10^3\,gcm^{-3}$, $\langle v \rangle \sim 5 \times 10^4\,cm\,s^{-1}$ and $r \sim 2 \times 10^9$ cm for the ICZ of a double shell burning red giant. For thermal instabilities in electron degenerate cores one may expect even higher values, especially for cases of rapid core rotation.

An indirect argument can be based on a hypothesis of flux ejection or destruction during dynamical collapse. It is well known that during the collapse of a protostellar cloud the conservation of magnetic flux is violated by many orders of magnitude. The usual explanation is that significant flux loss must occur (cf. Mestel and Weiss 1974). Similarly, one may expect the dynamic collapse to the neutron star phase to lead to B-fields significantly greater than 10^5-10^8G in the presupernova giant for pulsar fields of 10^{11}-10^{14}G (cf. Glasser and Kaplan 1975, Sarazin and Bahcall 1977). In any case, one might expect an upper limit to the field from the requirement that the pressure be not very different in the presence and absence of B-fields, i.e., $B^2/8\pi \lesssim 10^{17}$-$10^{18}$ erg cm^{-3}, or $B \lesssim 1.5 \times 10^9$-$5 \times 10^9$G in the red giant preceeding the dynamical transition to the end phase of evolution. If one thinks of stellar B-fields being returned to the interstellar medium (cf. Michel and Yahil 1973), the above numbers indicate flux violation during dynamical ejection as well. In this case flux generation would have to occur if one ascribed a stellar origin to fields in the interstellar medium, particularly for fields in the Crab Nebula (cf Woltjer 1972).

Independently of any theories of dynamo generated B-fields, the observations of white dwarf (WD) fields in excess of 10^8G (cf. Angel 1977, Tapia 1977) indicate that such fields must have been generated in the pre-WD red giant, not too long before the onset of the dynamical planetary nebula ejection, since otherwise such fields would have floated upward from the surface of the red giant core (cf. Parker 1974). Thus, the epoch of the onset of the final dynamical instability relative to the previous shell thermal instability may be responsible for the strength of the

B-field on the subsequent WD or neutron star. The small number of WD's with fields over 10^8G might be consistent with this, if the B-fields in the red giant progenitor can move upward fast enough. Direct observations of fields of ~400 G on the Cepheid W Sgr by Wood and Weiss (1977) are consistent with an origin in strong interior fields that floated to the surface. The only way such high WD or redgiant fields could come from the main sequence via flux conservation is from Ap stars. While this possibility cannot yet be excluded, the converse inference that the fields of the magnetic main sequence stars (MSS) may have originated in interior stellar dynamos, much like the fields of the magnetic WD's and red giants, should be considered within the context of Bidelman's hypothesis discussed in Section 4.

Finally, the complicated effects of anomalous abundances in low-mass stars (cf. Dickens and Bell 1976, Bessell and Norris 1976, Auer and Demarque 1977, and references therein) suggest that a partial mixing mechanism is at work. In light of the difficulty of bringing nuclearly processed material to the surfaces of low-mass stars via ordinary convective theories (Sweigart 1974), the idea of transport by buoyant magnetic bubbles is an attractive one, especially if future studies will succeed in combining it with the complications of s-process nucleosynthesis recently investigated by Truran and Iben (1977).

3. THE MAGNETIC STIRRER APPROXIMATION

The author's interest in this subject was originally stimulated by the magnetostrictive force on the interface of media with different magnetic susceptibilities and two remarkable theorems by Woltjer (1958) and Cowling (cf. Ferraro and Plumpton 1966) on force-free fields. However, the magnetic susceptibility of any Boltzmann plasma is of the same magnitude as either the spin paramagnetic or the Landau diamagnetic susceptibility. This makes the magnetostrictive forces negligible compared to the magnetic pressure forces. A number of investigations (Parker 1955, Gurm and Wentzel 1967, Finzi and Wolf 1968, Kippenhahn 1974) have shown that the effect of the latter force is to make a magnetic tube of flux buoyant, so that it will rise.

Consider a prominence-like magnetic perturbation at the upper edge of a convective (dynamo) zone, so that the magnetic tube of force is lifting material of mean molecular weight μ' into a radiative zone of mean molecular weight μ. In most stellar cases of interest $\mu'>\mu$ due to nuclear reactions. Then the conditions of pressure and thermal equilibrium of the prominence with its surroundings imply that

$$(\rho-\rho')/\rho = (P_B/P)(\mu'/\mu)-(\mu'-\mu)/\mu, \qquad (1)$$

where primes denote "inside" quantities, P is the outside gas pressure and $P_B = B^2/8\pi$ is the magnetic pressure. According to Archimedes' principle, if $\rho > \rho'$ the magnetic strand will be buoyant.

There are a number of possibilities for detaching such a buoyant prominence: a) cutting of the lines of force by convective elements from below, b) dynamical instabilities (cf. Landquist 1951, Parker 1955), c) aerodynamic drag due to differential rotation in the radiative zone above and d) the buoyancy overcoming the tension anchoring the prominence when its length L increases to the extent that $Lg(\rho-\rho') > B^2/4\pi$ (Parker 1955), where g is the local acceleration due to gravity. Combining this with condition (1) above shows that the magnetic buoyancy will rip out the prominence when $L \gtrsim 2\, H_p$, where H_p is the pressure scale height.

It appears that c) and d) together can easily lead to the formation of detached toroidal magnetic bubbles with size $\sim H_p$ and some thickness D. The bubble will rise initially on a rapid, dynamic time scale. However, the adiabatic cooling of the bubble will decrease the inside temperature T' relative to T outside. Thereafter, the bubble rises quasistatically, with inside and outside densities nearly equal ($\rho \approx \rho'$) at a rate controlled by the time scale on which the bubble can be heated. Following Parker (1974), one obtains in this case an approximate rising velocity

$$v \sim 5\gamma^2(\gamma-1)^{-2}\, F/(g^2\rho^2D^2\varepsilon)\, P\,[(P_B/P)(\mu'/\mu) - (\mu'-\mu)/\mu], \qquad (2)$$

where γ is the adiabatic exponent, F the energy flux and $\varepsilon = |\nabla T - \nabla T_{ad}|/|\nabla T_{ad}|$, with ∇T the actual, and ∇T_{ad} the adiabatic temperature gradients. One notes that the bracket in Eqn. (2) is a small quantity by virtue of Eqn. (1) and the condition $\rho \approx \rho'$; what is important is the algebraic sign of this bracket which shows that the condition for the bubble to keep rising is that $P_B/P \approx (\mu'-\mu)/\mu'$, which at the outer edge of the ICZ of a double shell burning star becomes $P_B/P \approx 0.1$. From the flash computations of SH one then has $B \sim 8.7\times10^8$, 3.8×10^8 and 1.4×10^8G for the 1st, 7th and 13th He -shell flash cycles of a 1 $M_\odot$ star, respectively. These are in the appropriate range for the observations discussed in the last section. If the bubble rises, it rises on the thermal adjustment or Kelvin-Helmholtz time scale $\tau_{KH} \sim 3c_p\kappa\rho^2D^2/(8acT^3)$, where κ is the opacity, c_p the specific heat, a the radiation constant and c the speed of light (Kippenhahn 1974). Whether the rising bubble slows down or speeds up depends on what happens to P_B/P during the ascent; Gurm and Wentzel (1967) found this ratio to be approximately constant (as can also be deduced from a dimensional argument and flux conservation), while in a slightly different analysis Kippenhahn (1974) deduced that the bubbles must slow down. In any case, in the absence of detailed numerical

calculations, it appears that the magnetic toroids will rise 1-2 pressure scale heights, which is all one needs to produce mixing of hydrogen-rich material into the ICZ.

The transport of ICZ material to the surface can be estimated by considering that during any flash cycle there will be $N \sim f4\pi r^2/LD$ prominences being formed, where f is a "magnetic activity" number of order unity. These contain an amount of mass $N\rho LD^2$, so $M \sim f4\pi r^2 \rho D N_f$ will be transported upward after N_f flashes. With $r \sim 3x10^9$cm, $\rho \sim 100$ gcm^{-3}, $D \sim 10^7$cm ($\sim 0.1 H_p$), $N_f \sim 100$ one has $M \sim 10^{31}$g$\simeq 0.005 M_\odot$, quite a sufficient amount.

In the limit of this approximation, one might consider the rising toroids in the differentially rotating radiative zone as rigid stirrers. The toroids must then rotate with the curl of the velocity field. If there is negligible spin down so that $\Omega = \Omega_o r_o^2/r^2$, the toroids will stir with a period of a few days, so that one may expect the hydrogen-helium interface above the ICZ to be well mixed. The differential rotation is thus converted into turbulence; however, the magnetic mixing (and dynamo) can be effective only if significant differential rotation can be maintained. This can be achieved if two conditions are met: 1) the H and He burning shells should "burn" through an amount of mass comparable to that between the ICZ and the envelope convection zone during the span of a few flashes and 2) the spin-down in the "fresh" radiative zone should be negligible during this time. For the 1$M_\odot$ star of SH, condition 1) is met after 13 cycles of relaxation oscillations. However, the spin-down criterion is less well understood. If the Ekman layer below the envelope convection zone is not turbulent (<u>cf.</u> Goldreich and Schubert 1967, Howard, Moore and Spiegel 1967); the spin-down time for the radiative region below it would be $\tau_s \sim r_o/(\nu\Omega)^{\frac{1}{2}} \sim 10^6$-$10^7$yrs, and is approximately constant for double shell burning stars of practically all masses (0.6 - 7 $M_\odot$); this is because the morphology as a function of radial distance from the center is insensitive to the total mass. On the other hand, the amount of mass ΔM in the radiative zone above the ICZ decreases with increasing stellar mass (<u>cf</u>. Paczynski 1974), as does the time needed to burn through a given amount of mass. The time scale in which the H-shell burns through the mass ΔM for the 1 $M_\odot$ model of SH is τ_e $\sim 5x10^6$ yrs. Since τ_e increases with decreasing total mass, comparison with τ_s indicates that magnetic mixing might be effective for stars down to below 1 $M_\odot$ and may be applicable to globular cluster stars.

More appropriately, the problem of magnetic fields and Eddington-Sweet currents should be treated together. The principal virtue of the magnetic-stirrer approximation is that it can be adapted for use in simple, one-dimensional, Henyey-type evolutionary calculations.

4. ON BIDELMAN'S HYPOTHESIS FOR Ap STARS

Bidelman (1960a,b) suggested that the peculiar A stars, metallic-line stars and possibly other sharp-line stars of classes O,B,A, may be in a phase of evolution analogous to the horizontal-branch stars found among Population II stars. After many conferences on the magnetic and peculiar stars, the complexity of the current state of affairs is well indicated by the fact that none of the handful of theories proposed have been rejected and there is a strong possibility that several processes may be responsible for these phenomena (cf. Cameron 1971). Fowler Burbidge, Burbidge and Hoyle (1965) have shown that the internal nucleosynthesis may be consistent with Bidelman's hypothesis for sufficiently complicated evolutionary histories. The theoretical HR diagrams of Fig. 1 illustrate the complications and ambiguities in the evolutionary status of stars falling near the intersections of main sequences and horizontal branches.

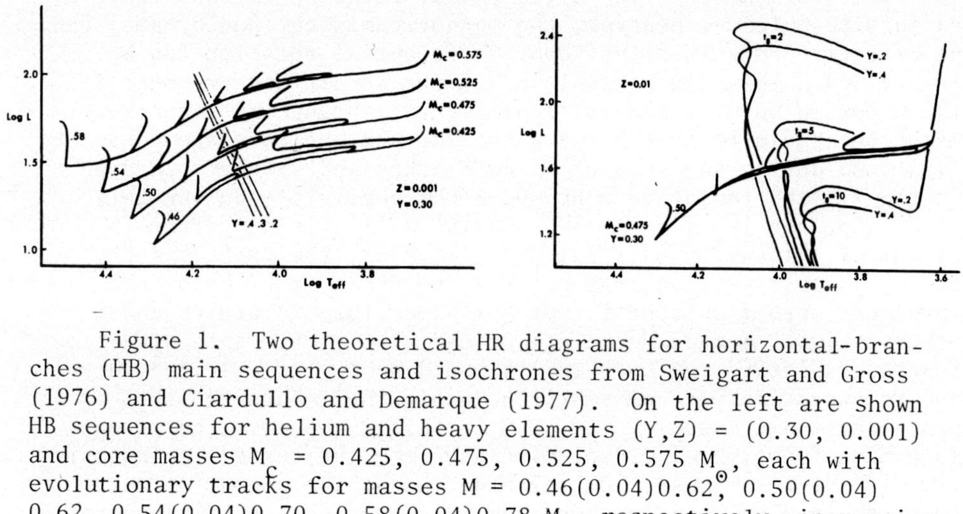

Figure 1. Two theoretical HR diagrams for horizontal-branches (HB) main sequences and isochrones from Sweigart and Gross (1976) and Ciardullo and Demarque (1977). On the left are shown HB sequences for helium and heavy elements (Y,Z) = (0.30, 0.001) and core masses M_c = 0.425, 0.475, 0.525, 0.575 $M_\odot$, each with evolutionary tracks for masses M = 0.46(0.04)0.62, 0.50(0.04) 0.62, 0.54(0.04)0.70, 0.58(0.04)0.78 $M_\odot$, respectively, increasing from the left. Three zero-age main sequences are shown for Y=0.4, 0.3, 0.2 and Z = 0.001. On the right, the HB sequence is for (M_c,Y,Z) = (0.475, 0.30, 0.01) and M = 0.50(0.04)0.62 $M_\odot$, and three pairs (Y = 0.2, 0.4) of isochrones with Z = 0.01 are shown for ages t_8 = 2,5,10, in units of 10^8 yr. The luminosity L is in solar units.

What do the zero-age HB models falling on the main sequence have in common? In Fig. 1 they all have the same inhomogeneity parameter $q = M_c/M \approx 0.85$. The importance of q for the theoretical HR diagram has been particularly emphasized by Giannone (1967); his more approximate computations gave $q \approx 0.87$ for HB stars along the main sequence. The HB models computed by Rood (1970)

and Gross (1973) are in general agreement with this constant-q hypothesis.

In the context of the mixing by magnetic bubbles discussed above, the Bidelman hypothesis must be connected with the details of the thermal instabilities during the preceeding core helium flash. Rapid internal rotation of the degenerate helium core can then provide magnetic dynamo mechanisms to create buoyant magnetic strands that will bring both magnetic fields and nuclearly processed matter to the surface of the star. The implied spread in M_c for the constant-q, HB models can also be produced by the internal rotation as shown by Mengel and Gross (1976) who found M_c as large as 0.6 for a rotation rate of $6 \times 10^{-4}\ s^{-1}$ for solid-body rotation on the main sequence. In more realistic calculations, the approximation of solid-body rotation would have to be replaced by rapid internal rotation on the main sequence. Even higher core masses will result if the energy of rotation is included into the equations, since this becomes comparable to the thermal energy near the core flash (cf. Tuominen and Musylev 1974). Within this context of higher core masses, one might be able to represent stars like HR 7129 (Wolff and Wolff, 1976) which is a helium variable with a large magnetic field ($\sim 6 \times 10^3$G), as an extension of the group of magnetic Ap stars to higher temperatures (log $T_{eff} \sim 4.3$). The dissipation of rotation in the core-flash dynamo could also explain the origin of the magnetic fields and slow rotation of Ap stars.

Mass loss on the red giant branch (cf. Fusi-Pecci and Renzini, 1975) will also affect the value of $q = M_c/M$. This is further complicated by the effect of an influx of magnetic energy into the outer envelope as a result of the rising magnetic strands. It is not evident how the influx of magnetic energy might change the adiabatic exponents (cf. Kippenhahn 1970), thereby altering the mean rate of mass loss. Is the mass-loss rate coupled to the internal rotation, as is the mixing to the surface? The correlation between total mass and M_c implied by a constant-q hypothesis is the weakest part in the Bidelman hypothesis, but this possibility cannot be dismissed and needs theoretical and observational investigation. The possible importance of fast rotating cores for explaining the gaps on the giant branch of M15 has been emphasized by Demarque, Mengel and Sweigart (1972). Magnetic mixing in the core flash may also be important in the Balmer-jump explanation of gap 1 of Newell (1973) proposed by Auer and Demarque (1977). It is not yet possible to say whether Newell's BC stars are related to the $q \approx 0.85$ phenomenon.

The general importance of rotation in theoretical stellar evolution in the context of magnetic fields and differential rotation has been discussed by Kippenhahn (1973), with particular emphasis on the variation of rotation along the zero-age main

sequence. In this context, it is important to note that the constant-q hypothesis with magnetic mixing will work only if the main sequence progenitors of the peculiar stars were among the fast main sequence rotators, but their main sequence masses were sufficiently small for degenerate electron cores to form. The latter limit is about 2.5 $M_\odot$ (Iben, 1967) and might be higher for MSS with rapid internal rotation. This would imply that Bidelman's hypothesis can be correct, only if Ap stars were found predominantly in clusters which also have normal (fast rotating) A stars. This necessary (not sufficient) condition for Bidelman's hypothesis seems to be supported by observations.

5. CONCLUSION

This investigation has emphasized the interconnections between the magnetic enrichment of the ISM through stellar dynamos, which must exist if the high magnetic fields of some white dwarfs and pulsars are to be explained within the context of "conventional" stellar evolution. The approximation of mixing by magnetic toroids is evidently a crude one and, although the numbers fall into place, much work in magnetohydrodynamics remains to be done. Bidelman's hypothesis needs to be thoroughly investigated, especially in regard to the interior nucleosynthesis and the constant-q hypothesis. The upper mass limit to electron-degenerate helium ignition for MSS with rapid internal rotation has to be determined from theoretical computations, and the masses of Ap stars which are members of binaries, the rotation of main sequence stars in globular clusters, time variations of red giant peculiarities, etc. need to be observed.

The author thanks Dr. R.L. Garwin, I. Iben, G. Lasher and V. Weidemann for critical comments. Partial support by the Cleveland Foundation and NSF grant AST76-14759 is gratefully acknowledged.

REFERENCES

Angel, J.R.P. (1977). Astrophys. J. 216, 1.
Auer, L.H., and Demarque, P. (1977). Astrophys. J. 216, 791.
Bessell, M.S., and Norris, J. (1976). Astrophys. J. 208, 369.
Bidelman, W.P. (1960a). Trans. I.A.U. 10, 677.
Bidelman, W.P. (1960b). Publ. Astron. Soc. Pacific. 72, 24.
Cameron, A.G.W. (1971). Publ. Astron. Soc. Pacific. 83, 585.
Ciardullo, R.B., and Demarque, P. (1977). Trans. Yale Obs. 33.
Demarque, P., Mengel, J.G., and Sweigart, A.V. (1972). Astrophys.J. 173, L27.
Dickens, R.J., and Bell, R.A. (1976). Astrophys. J. 207, 506.

Ferraro, V.C.A. and Plumpton, C. (1966). Introduction to Magneto-Fluid Mechanics, Clarendon Press, Oxford, p. 39.
Finzi, A. and Wolf, R.A. (1968). Astrophys. J. 153, 865.
Fowler, W.A., Burbidge, E.M., Burbidge, G.R. and Hoyle, F. (1965). Astrophys. J. 142, 423.
Fusi-Pecci, F. and Renzini, A. (1975). Astron. Astrophys. 39, 413.
Giannone, P. (1967). Z. Astrophys. 65, 226.
Glasser, M.L. and Kaplan, J.I. (1975). Astrophys. J. 199, 208.
Goldreich, P. and Schubert, G. (1967). Astrophys. J. 150, 571.
Gross, P.G. (1973). Monthly Notices Roy. Astron. Soc. 164, 65.
Gurm, H.S. and Wentzel, D.G. (1967). Astrophys. J. 149, 139.
Howard, L.N., Moore, D.W., and Spiegel, E.A. (1967). Nature 214, 1297.
Iben, I. Jr. (1967). Ann. Rev. Astron. Astrophys. 5, 571.
Kippenhahn, R. (1970). Astron. Astrophys. 8, 50.
Kippenhahn, R. (1973). in Stellar Chromospheres, S.D. Jordan and E.H. Avrett, eds., NASA SP-317, Washington, p.265.
Kippenhahn, R. (1974). in Late Stages of Stellar Evolution, R.J. Tayler ed., Reidel Publ. Co., Boston, p. 20.
Levy, E.H. and Rose, W.K. (1974). Astrophys. J. 193, 419.
Lundquist, S. (1951). Phys. Rev. 83, 307.
Malkus, W.V.R. (1959). Astrophys. J. 130, 259.
Mengel, J.G. and Gross, P.G. (1976). Astrophys. Space Sci. 41, 407.
Mestel, L. and Weiss, N.O. (1974). in Magnetohydrodynamics, Geneva Observatory Publ., Sauverny, p. 141.
Michel, F.C. and Yahil, A. (1973). Astrophys. J. 179, 771.
Newell, E.B. (1973). Astrophys. J. Suppl. 26, 37.
Paczynski, B. (1974). Astrophys. J. 192, 483.
Parker, E.N. (1955). Astrophys. J. 121, 491.
Parker, E.N. (1974). Astrophys. Space Sci. 31, 261.
Rood, R.T. (1970). Astrophys. J. 161, 145.
Sarazin, C.L. and Bahcall, J.N. (1977). Astrophys. J. 216, L67.
Schwarzschild, M. and Härm, R. (1965). Astrophys. J. 142, 855.
Schwarzschild, M. and Härm, R. (1967). Astrophys. J. 150, 961 (SH).
Sweigart, A.V. (1974). Astrophys. J. 189, 289.
Sweigart, A.V. and Gross, P.G. (1976). Astrophys. J. Suppl. 32, 367.
Tapia, S. (1977). Astrophys. J. 212, L125.
Truran, J.W. and Iben, I., Jr. (1977). Astrophys. J. 216, 797.
Tuominen, I. and Musylev, V. (1974). Nauchnie Info. 33, 48.
Weidemann, V. (1977). Astron. Astrophys. 49, 411.
Weiss, N.O. (1971). Quart. J. Roy. Astron. Soc. 12, 432.
Wolff, R.J. and Wolff, S.C. (1976). Astrophys. J. 203, 171.
Woltjer, L. (1958). Proc. Nat. Acad. Sci. 44, 489.
Woltjer, L. (1972). Ann. Rev. Astron. Astrophys. 10, 129.
Wood, H.J., Weiss, W.W. and Jenkner, H. (1977). Bull. Am. Astron. Soc. 9, 434.

SURVEY OF SOME RECENT STELLAR-EVOLUTION CALCULATIONS

Allen V. Sweigart

NASA Goddard Space Flight Center

1. INTRODUCTION

A basic objective of stellar-evolution theory is to provide detailed quantitative information on the evolutionary characteristics of stars with different compositions and masses. For the evolutionary phases from the zero-age main sequence (ZAMS) through core-helium burning the important physical processes are sufficiently well understood to justify a systematic investigation, and, indeed, extensive grids of evolutionary sequences have been constructed for these phases by many researchers. However, several difficulties are sometimes encountered when using the available calculations. For example, differences in numerical techniques and physical assumptions can give rise to systematic differences among the results of various investigations. Furthermore, the available calculations do not always explore the full ranges of the input parameters, and in some cases they neglect physical effects that are now believed to be important.

In an attempt to overcome some of these limitations, I have undertaken a series of computational projects covering the main-sequence, red giant and horizontal-branch phases. The immediate aim has been to produce a self-consistent set of evolutionary sequences for wide ranges in the composition and mass. The long-term aim involves the detailed application of these results to observational data on cluster HR diagrams and the integrated properties of galaxies. Other persons who have contributed significantly to various aspects of this work include P. Demarque, J. Mengel and P. Gross.

A. G. Davis Philip and D. S. Hayes (eds.), The HR Diagram, 333-344.

TABLE I

MAIN-SEQUENCE-TURNOFF AND RED-GIANT SEQUENCES

Y	Z	$M/M_\odot$ (Turnoff)	$M/M_\odot$ (Red Giant)
0.10	0.00001	0.70 - 4.40	0.90
0.20	0.00001	0.70 - 3.50	0.70, 0.90
0.30	0.00001	0.55 - 3.50	0.70, 0.90
0.40	0.00001	0.55 - 3.50	0.70
0.10	0.0001	0.70 - 4.40	0.90
0.20	0.0001	0.70 - 3.50	0.70, 0.90, 1.40
0.30	0.0001	0.55 - 3.50	0.70, 0.90, 1.40
0.40	0.0001	0.55 - 3.50	0.70
0.20	0.0004	0.70 - 3.50	0.90
0.30	0.0004	0.55 - 3.50	0.70
0.10	0.001	0.70 - 4.40	0.90
0.20	0.001	0.70 - 3.50	0.70, 0.90
0.30	0.001	0.55 - 3.50	0.70, 0.90
0.40	0.001	0.55 - 3.50	0.70
0.20	0.004	0.70 - 4.40	0.90
0.30	0.004	0.70 - 3.50	0.70
0.10	0.01	0.90 - 4.40	0.90, 1.10
0.20	0.01	0.70 - 4.40	0.90, 1.10, 1.40, 1.75, 2.20
0.30	0.01	0.70 - 3.50	0.70, 0.90, 1.10, 1.40, 1.75, 2.20
0.40	0.01	0.55 - 3.50	0.70, 1.10
0.20	0.04	0.90 - 4.40	0.90, 1.75
0.30	0.04	0.70 - 4.40	0.70, 1.75
0.20	0.10	0.90 - 6.90	-
0.30	0.10	0.70 - 5.50	-

A listing of the computed sequences is provided in Tables I and II for all values of the helium abundance Y and the heavy-element abundance Z studied so far. The turnoff sequences extended from the ZAMS to the red giant branch except for a few sequences with high mass and low Z. The range in mass M given in column 3 of Table I was covered by approximately 10 turnoff sequences having roughly equally spaced values of log M. The red giant sequences listed in column 4 were continued to the onset of helium burning which in nearly all cases occurred under degenerate conditions. The quantity M_c in Table II denotes the mass within the helium core at the zero-age horizontal-branch (ZAHB) phase.

The stellar-evolution program developed by Sweigart (1972, 1973) from the Princeton program (Schwarzschild and Härm 1965) was used for all computations to insure the self-consistency of the results. A more complete description of this program can be found in Sweigart and Gross (1974, 1978). Modifications were

TABLE II

HORIZONTAL-BRANCH SEQUENCES

Y	Z	$M_c/M_\odot$	$M/M_\odot$
0.10	0.0001	0.475	0.50(0.04)0.90
0.20	0.0001	0.475	0.50(0.04)0.90
0.30	0.0001	0.475	0.50(0.04)0.90
0.40	0.0001	0.475	0.50(0.04)0.98
0.20	0.0001	0.525	0.54(0.04)0.90
0.30	0.0001	0.525	0.54(0.02)0.58(0.04)0.90
0.40	0.0001	0.525	0.54(0.04)1.02
0.30	0.001	0.425	0.46(0.04)0.90
0.40	0.001	0.425	0.46(0.04)0.90
0.10	0.001	0.475	0.50(0.04)0.90
0.20	0.001	0.475	0.50(0.04)0.90
0.30	0.001	0.475	0.50(0.02)0.70(0.04)0.94
0.40	0.001	0.475	0.50(0.04)1.02
0.10	0.001	0.525	0.54(0.04)0.90
0.20	0.001	0.525	0.54(0.04)0.90
0.30	0.001	0.525	0.54(0.04)0.90
0.40	0.001	0.525	0.54(0.04)1.06
0.30	0.001	0.575	0.58(0.04)1.06
0.30	0.01	0.475	0.50(0.02)0.58(0.04)0.94

made to this program during the course of these projects in order to incorporate those additional physical effects of particular importance for each evolutionary phase. For example, relativistic degeneracy and semiconvection were included in the red giant and horizontal-branch sequences, respectively. It is well-known that the CNO abundance can significantly influence the characteristics of the main-sequence turnoff, and consequently the approach to equilibrium of the ON cycle was followed during the turnoff calculations. The CNO abundance has been assumed here to scale with the Fe abundance, although there is observational evidence indicating that CNO and Fe may not be so tightly correlated.

The present results, consisting in total of over 100,000 stellar models, have a number of immediate astrophysical applications. From the fitting of theoretical isochrones with observed HR diagrams and from the relative red giant and horizontal-branch lifetimes one can in principle determine the ages and helium abundance of the globular clusters. Such information provides important constraints on cosmological models and has direct implications for the dynamical and chemical evolution of the Galaxy. The number distribution of stars along the giant branch can be compared with theoretical luminosity functions in an attempt to understand further the observed irregularities (Sandage *et al.*

1968; Sweigart 1978). The present red giant models can serve as the starting point for subsequent studies of the helium-core flash (Mengel 1978) and of the unusual C, N and s-process abundances found in some globular-cluster stars (Mengel and Sweigart 1978). Furthermore it will be possible to consider again in more detail some of the intriguing problems associated with the morphology of the horizontal branch.

In the following sections I will survey the various evolutionary calculations. For conciseness the discussion must, however, be limited to a few representative cases. A more thorough description of the available data is contained in the cited references.

2. TURNOFF SEQUENCES

The turnoff tracks for Z = 0.01 and Y = 0.20 and 0.30 are shown, respectively, in Figs. 1 and 2 (Mengel et al. 1978). In the higher mass sequences central hydrogen exhaustion occurs throughout a rather large region at the end of the main-sequence phase. This leads to an abrupt contraction of the star and hence to a blueward fluctuation in the track. Observationally this fluctuation would appear as a gap in a cluster HR diagram. Although not plotted in Figs. 1 and 2, a few intermediate mass tracks have been constructed in order to define the development of the gap more precisely. As expected, an increase in Y shifts

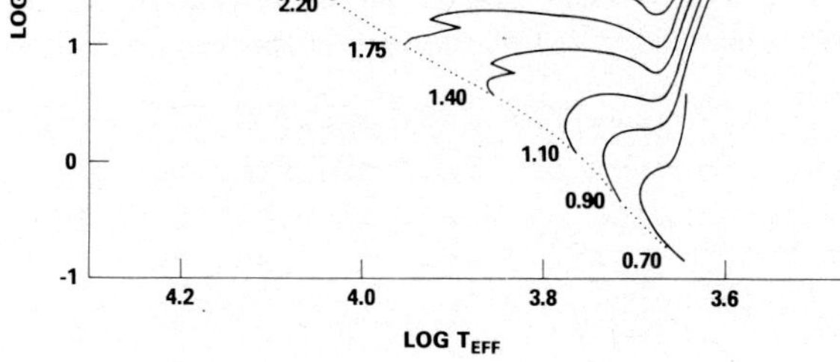

Fig. 1. Main-sequence-turnoff and red-giant tracks for (Y, Z) = (0.20, 0.01). The location of the ZAMS is indicated by the dotted line. Each track is labeled by the mass in solar units. The tracks for $0.90 \leq M \leq 2.20\ M_\odot$ have been evolved to the onset of helium burning.

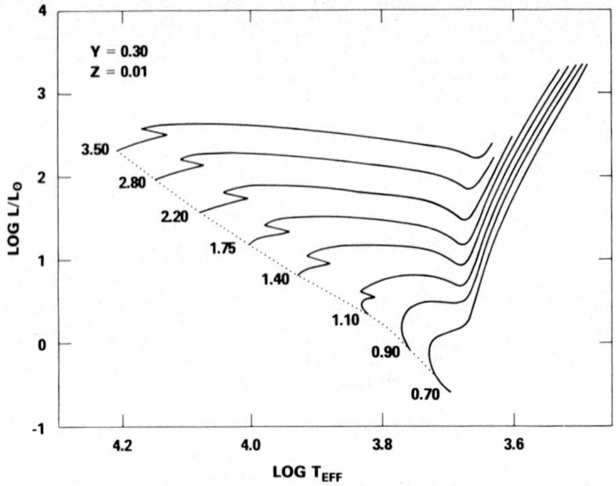

Fig. 2. Main-sequence-turnoff and red-giant tracks for (Y, Z) = (0.30, 0.01). The location of the ZAMS is indicated by the dotted line. Each track is labeled by the mass in solar units. The tracks for $0.70 \leq M \leq 2.20\ M_\odot$ have been evolved to the onset of helium burning.

---

a given track to larger values of log T_{eff} and Log L.

The dependence of the turnoff tracks on Z is illustrated in Fig. 3. Simoda and Iben (1968) have demonstrated that an increase in Z will reduce log T_{eff} and log L at the turnoff point. Such an effect is evident in Fig. 3 and is due primarily to the effects

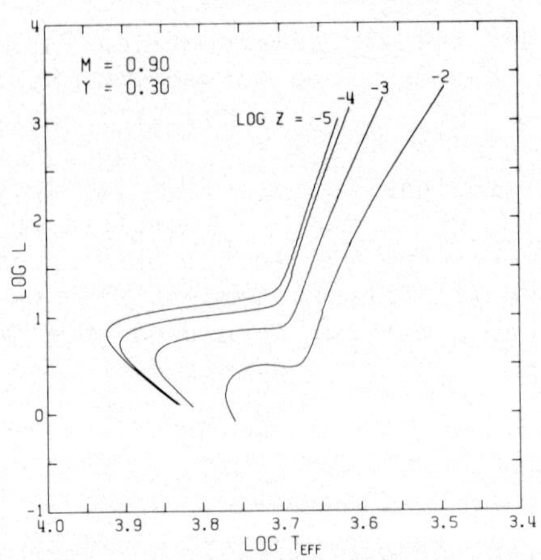

Fig. 3. Main-sequence-turnoff and red-giant tracks for (M, Y) = (0.90, 0.30). Each track is labeled by the value of log Z.

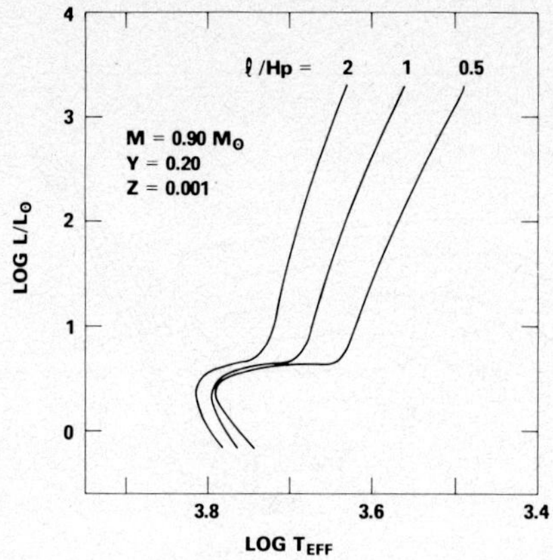

Fig. 4. Main-sequence-turnoff and red-giant tracks for (M, Y, Z) = (0.90, 0.20, 0.001). Each track is labeled by the value of the mixing-length ratio $\alpha = \ell/H_p$.

of the CNO elements on the hydrogen-burning rate and the opacity (Renzini 1977). Any variations in the CNO abundance relative to the Fe abundance could therefore significantly influence the turnoff characteristics.

Since the lower mass sequences have a convective envelope, their tracks are affected by the value adopted for the ratio α of the mixing length ℓ to the pressure-scale height H_p (Sweigart 1978). All of the present sequences use $\alpha = 1$ except for the 2 cases plotted in Fig. 4. In this figure log T_{eff} at the ZAMS and along the red-giant branch is approximately proportional to log α. Not surprisingly, log L is rather insensitive to the value of α. Around the turnoff point the dependence of log T_{eff} on log α is quite nonlinear, since the envelope convection has almost disappeared in the $\alpha = 0.5$ and 1 cases while it is still present in the $\alpha = 2$ case. Sequences with a deeper convective envelope would be expected to show a smoother variation of log T_{eff} at turnoff with log α. The caution which must be exercised when ages are obtained by fitting isochrones to cluster HR diagrams is indicated in Fig. 4. If the fit is based essentially on the turnoff color and if the correct α has not been used, the derived ages can be rather uncertain, a point emphasized by Iben and Rood (1970).

3. RED-GIANT SEQUENCES

Some of the red-giant tracks have already been displayed in Figs. 1 to 4. From these tracks and the others listed in Table I the following conclusions can be drawn (Rood 1972; Sweigart and

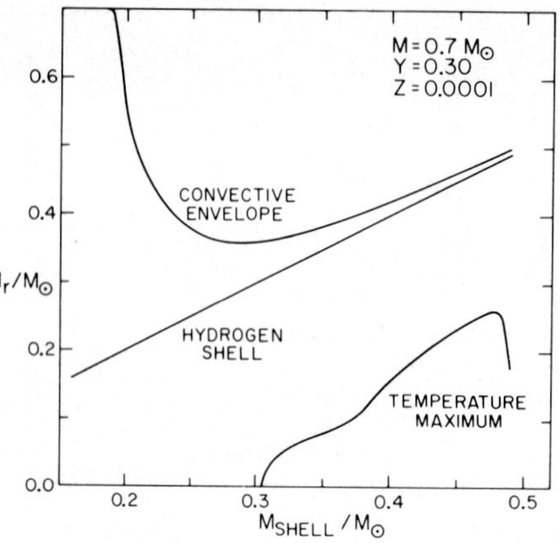

Fig. 5. Variations in the locations of the convective envelope, the hydrogen shell and the core-temperature maximum along a red-giant sequence for (M, Y, Z) = (0.70, 0.30, 0.0001). The evolution between the end of the main-sequence phase and the onset of the helium-core flash is shown.

Gross 1978; Sweigart 1978). First, there is a blueward shift of the red giant branch with increasing Y, M or α or decreasing Z. Second, an increase in Y or a decrease in Z lowers the surface luminosity at the tip of the red-giant branch and reduces the maximum stellar mass at which the helium-core flash still takes place.

How the edge of the convective envelope and the point of maximum temperature within the degenerate helium core vary along a typical red giant sequence is illustrated in Fig. 5. The abscissa gives the location of the center of the hydrogen shell, defined as the point where the hydrogen abundance is one half of its envelope value. The location of the hydrogen shell is thus represented by a straight line. The temperature inversion produced by neutrino emission becomes progressively more extensive as the mass within the core grows. Shortly before the helium-core flash the temperature maximum suddenly shifts inward. The flash point can be quite off center, especially for small values of Y and M.

Of particular interest is the behavior of the convective envelope. Initially the convective envelope moves inward in mass until deepest penetration is reached on the lower subgiant branch. At that time the convective envelope extends into the region that was partially depleted of hydrogen during the main-sequence phase. A discontinuity in the hydrogen abundance is therefore created at the point of deepest penetration. The convective envelope then retreats outward due to the approach of the hydrogen shell. When

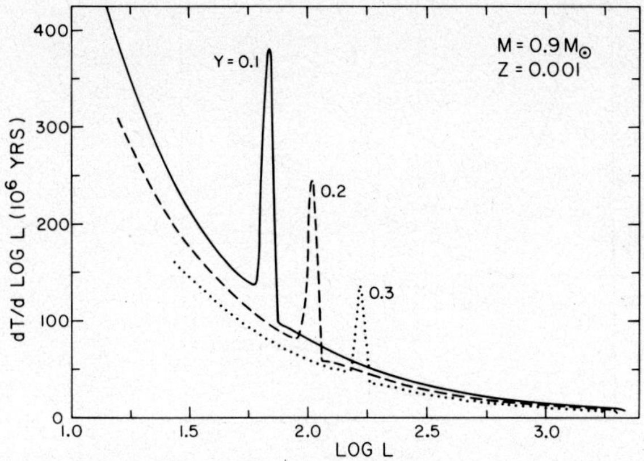

Fig. 6. Theoretical red-giant luminosity functions for (M, Z) = (0.90, 0.001) and Y = 0.10, 0.20 and 0.30. The ordinate gives the time T spent per unit interval in log L in units of 10^6 yrs.

---

the hydrogen shell later encounters this discontinuity, a readjustment of the interior structure takes place, leading to a hesitation in the rate of evolution up the giant branch and often to a temporary drop in the surface luminosity (Rood 1972). The result can be a rather pronounced peak in the luminosity function as is demonstrated by Figs. 6 and 7 (Sweigart 1978). In general the prominence of this peak is increased and the luminosity at which it occurs is lowered by smaller values of Y and larger values of Z. Such composition changes tend to produce a deeper

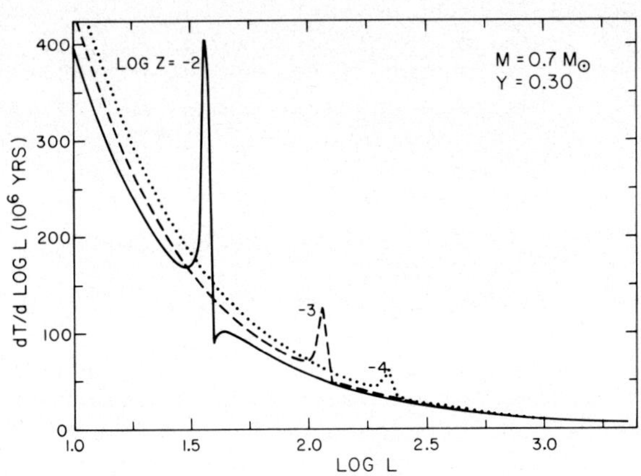

Fig. 7. Theoretical red-giant luminosity functions for (M, Y) = (0.70, 0.30) and log Z = -2, -3 and -4. The ordinate gives the time T spent per unit interval in log L in units of 10^6 yrs.

penetration of the convective envelope on the subgiant branch and thus a more pronounced hydrogen discontinuity.

A major reason for constructing red-giant sequences is to determine the core mass M_c at the time of the helium-core flash. The precise value of M_c strongly affects the morphology of the horizontal-branch tracks as well as the horizontal-branch lifetimes. The present calculations give

$$M_c/M_\odot = 0.474 - 0.24(Y - 0.30) - 0.007\,(\log Z + 3) - 0.038(M - 0.80) \quad (1)$$

in the neighborhood of the case (M, Y, Z) = (0.80, 0.30, 0.001). Eqn. (1) assumes the neutrino emission rates of Beaudet et al. (1967). If neutrino emission is omitted, M_c decreases by 0.03 $M_\odot$, while the additional neutrino emission due to neutral current interactions raises M_c by 0.007 $M_\odot$ (Sweigart and Gross 1978).

4. HORIZONTAL-BRANCH SEQUENCES

The interior structure and observable properties of horizontal-branch models are substantially altered by convective overshooting and semiconvection within the helium core (Schwarzschild 1970). During the initial horizontal-branch evolution the convective core propagates outward in mass because of convective overshooting. This growth in the convective core is driven fundamentally by the increase in the opacity and hence the radiative gradient caused by the gradual depletion of the central helium abundance (Castellani et al. 1971a, b). Tentative estimates for the rate of propagation suggest that convective neutrality will, to a good approximation, be maintained at the convective-core edge (Renzini 1977). When the central helium abundance falls below about 0.7, a semiconvective zone develops in the layers surrounding the convective core. The subsequent changes in the composition distribution are shown in Fig. 1 of Sweigart and Gross (1974). The effects of semiconvection on the evolution have been examined in a series of papers (Demarque and Mengel 1972; Sweigart and Demarque 1972; Sweigart and Gross 1974, 1976). The results show that semiconvection can almost double the horizontal-branch lifetime and track length when compared with calculations in which neither convective overshooting nor semiconvection is permitted.

Some typical horizontal-branch tracks computed with semiconvection (Sweigart and Gross 1976) are presented in Fig. 8. These tracks extend from the ZAHB to the initial helium-shell burning stage and are similar to those obtained by Gingold (1976). The 0.50 $M_\odot$ track evolves vertically in the HR diagram and would not be expected to return to the asymptotic branch. The 0.54 $M_\odot$ track

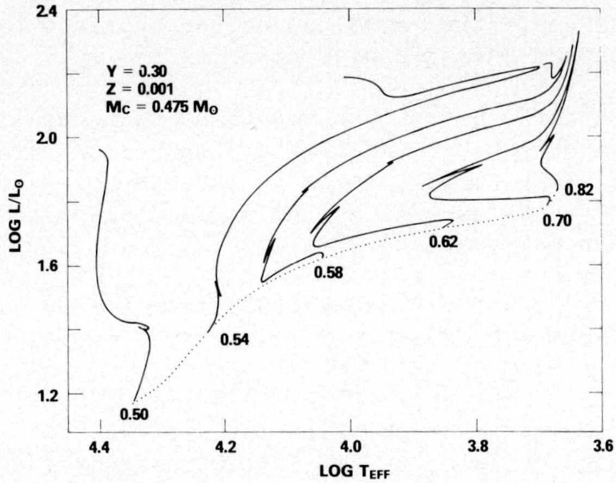

Fig. 8. Horizontal-branch and post-horizontal-branch tracks for $(Y, Z, M_c) = (0.30, 0.001, 0.475)$. The location of the ZAHB is indicated by the dotted line. Each track is labeled by the mass in solar units.

---

makes an excursion into the region of the BL Her stars in a manner similar to the 0.51 $M_\odot$ track of Sweigart et al. (1974). At higher masses the tracks all evolve back to the asymptotic branch. The spike visible at the end of the main horizontal-branch phase of each track is due to a rapid growth in the extent of the convective core and appears to be caused by an instability in the composition distribution within the core (Sweigart and Demarque 1973). The same effect has been reported by Gingold (1976) and Taam et al. (1976).

The author wishes to express his appreciation to G. Lasher, whose helpful contributions made much of this research possible, and to S. Sofia for constructive comments on the manuscript. The support of grant AST72-04418 A04 from the National Science Foundation is gratefully acknowledged.

REFERENCES

Beaudet, G., Petrosian, V. and Salpeter, E.E. (1967). Astrophys. J. 150, 979.
Castellani, V., Giannone, P. and Renzini, A. (1971a). Astrophys. Space Sci. 10, 340.
Castellani, V., Giannone, P. and Renzini, A. (1971b). Astrophys. Space Sci. 10, 355.
Demarque, P. and Mengel, J.G. (1972). Astrophys. J. 171, 583.
Gingold, R.A. (1976). Astrophys. J. 204, 116.

Iben, I., Jr. and Rood, R.T. (1970). Astrophys. J. 159, 605.
Mengel, J.G. (1978). in preparation.
Mengel, J.G. and Sweigart, A.V. (1978). in preparation.
Mengel, J.G., Sweigart, A.V., Demarque, P. and Gross, P.G. (1978). in preparation.
Renzini, A. (1977). Lectures given at the 7th Advanced Course on Advanced Stages in Stellar Evolution, Saas-Fee, Switzerland.
Rood, R.T. (1972). Astrophys. J. 177, 681.
Sandage, A., Katem, B. and Kristian, J. (1968). Astrophys. J. Letters 153, L129.
Schwarzschild, M. (1970). Quart. J. Roy. Astron. Soc. 11, 12.
Schwarzschild, M. and Härm, R. (1965). Astrophys. J. 142, 855.
Simoda, M. and Iben, I., Jr. (1968). Astrophys. J. 152, 509.
Sweigart, A.V. (1972). Bull. Am. Astron. Soc. 4, 203.
Sweigart, A.V. (1973). Astron. Astrophys. 24, 459.
Sweigart, A.V. (1978). in preparation.
Sweigart, A.V. and Demarque, P. (1972). Astron. Astrophys. 20, 445.
Sweigart, A.V. and Demarque, P. (1973). In IAU Colloquium No. 21, Variable Stars in Globular Clusters and in Related Systems, J.D. Fernie, ed., D. Reidel, Dordrecht, p. 221.
Sweigart, A.V. and Gross, P.G. (1974). Astrophys. J. 190, 101.
Sweigart, A.V. and Gross, P.G. (1976). Astrophys. J. Suppl. 32, 367.
Sweigart, A.V. and Gross, P.G. (1978). Astrophys. J. Suppl. in press.
Sweigart, A.V., Mengel, J.G. and Demarque, P. (1974). Astron. Astrophys. 30, 13.
Taam, R.E., Kraft, R.P. and Suntzeff, N. (1976). Astrophys. J. 207, 201.

DISCUSSION

COX: Can you tell me what your best Y value is for the RR Lyrae stars?

SWEIGART: The value of Y can be estimated from the ratio R of the number of horizontal-branch stars to the number of red giant stars, since the corresponding ratio of theoretical lifetimes is a sensitive function of Y. The original number counts gave a value of 1 for R, implying a Y value of about 0.3. This Y value was determined from a theoretical calibration in which the effects of semiconvection were neglected. It is now well-known that semiconvection can substantially increase the horizontal-branch lifetime. The resulting change in the calibration between R and Y reduces the Y value for R = 1 to 0.12. Renzini (1977) has recently re-examined the number counts in more detail and has found that a significant number of faint horizontal-branch stars were omitted in the original counts. According to Renzini (1977), the average value of R for a number of globular clusters is about 1.6. For this R value one obtains $Y \sim 0.22$, when the calibration including semiconvection is used. This value for the helium abundance represents the best estimate currently available from the R method.

COX: You know, of course, that pulsation theory gives a value of about 0.3. Do you have any comment?

SWEIGART: All I can say is that the value based upon the number counts is about 0.20 - 0.25. I do not think that 0.3 is possible. However, a detailed calibration of the R method using the latest horizontal-branch and red giant sequences has not yet been made, and thus the value I have given is a preliminary one.

ROOD: In reply to Cox's suggestion that RR Lyrae theory requires $Y \sim 0.28$ to 0.30, I think most of the indications, at least from the linear theory, are that $Y \sim 0.20$ to 0.25. Iben and Tuggle's blue edge calculations are consistent with this.

CASTELLANI: What is the dependence of "RG clumps" on the assumed value of the mixing length?

SWEIGART: The peak in the red-giant luminosity function does depend somewhat on the assumed value for the ratio, α, of the mixing length to the pressure-scale height. An increase in α reduces the size of the peak and raises the luminosity L at which it occurs. Although detailed luminosity functions have not yet been computed for $\alpha \neq 1$, this dependence appears to be mild. For example, in the case (M, Y, Z) = (0.90, 0.20, 0.001), a change in α from 1 to 2 increases log L at the peak by only 0.02.

CASTELLANI: I have one more comment on the problem of the RR Lyrae stars raised by Dr. Cox. I know that in decreasing the helium content the width of the instability strip is expected to decrease. Nevertheless I wish to emphasize that, if one assumes a constant width of the strip (i.e. the observed one) and looks at the fitting of the observed period-frequency histograms then available pulsational theories, when connected with evolutionary prescriptions, clearly indicate a helium content on the order of 0.23 - 0.25. On the contrary, no agreement can be found with the assumption that Y = 0.30 for the large majority of galactic globular clusters.

FLOWER: I have one comment to make concerning the proper choice of the value of the mixing length. The evolutionary tracks (Mengel, et. al., 1978) used to construct isochrones were computed using a value of 1.0 for the mixing length. Since, for the present generation of evolutionary codes, a value closer to 1.5 or 1.6 seems to be necessary to (1) reproduce the surface temperature of the present day Sun; (2) reproduce the location of the red giant branch on the HR diagram for old galactic clusters; and (3) reproduce the location of red giants on the HR diagram in young populous clusters in the LMC, the isochrones must be corrected to reflect these "calibrations" of the mixing length.

SWEIGART: I agree that a value closer to 1.5 gives a better fit with observations. It is particularly important to be aware of the dependence of the sequences on the mixing length when using the present results.

THE EVOLUTION OF THE INTEGRATED COLORS OF GLOBULAR CLUSTERS AND ELLIPTICAL GALAXIES

R. B. Ciardullo and P. Demarque

Yale University Observatory

I. INTRODUCTION

The purpose of this paper is to show the first results of a program designed to construct population models of stellar systems on the basis of theoretical data of stellar evolution. Synthetic HR diagrams were obtained using the isochrones and luminosity functions of Ciardullo and Demarque (1977). This last work is based on the stellar models of the Yale - IBM research group (Mengel, Sweigart, Demarque and Gross 1978; Sweigart and Gross 1976; 1978). The conversion from the theoretical HR diagram to observable parameters is made with the help of theoretical model stellar atmospheres principally due to Kurucz (1976), Relyea (1976) and Bell and Gustafsson (1975). The program provides a synthetic C-M diagram and integrated properties for the system given the following parameters: the chemical composition in terms of the (Y,Z) pair, the age, the total number of stars, the initial mass function (IMF) and the law of mass loss from stars in the system. In this brief paper, we shall concentrate on the integrated properties of theoretical star clusters and then remark on some interesting consequences for the color evolution of elliptical galaxies.

II. MODELS OF GLOBULAR CLUSTERS

We have considered two extreme chemical compositions, one corresponding to a very metal-poor stellar population with $(Y,Z) = (0.20, 10^{-4})$, the other characteristic of the metal-rich component of the galactic halo with $(Y,Z) = (0.30, 10^{-2})$. This last metallicity could also be considered representative of many of the stars in the galactic disk, except for some possible differences in the value of Y (Demarque and McClure 1977a, b). The

A. G. Davis Philip and D. S. Hayes (eds.), The HR Diagram, 345-348.

IMF was taken to be the Salpeter (1955) function in the case illustrated in the figures. A standard law of mass loss due to a stellar wind, which is most effective at the tip of the giant branch (Heasley and Mengel 1971; Fusi-Pecci and Renzini 1975) was adopted. In addition we assumed a Gaussian dispersion in mass loss in order to obtain a small mass spread on the horizontal branch. The total number of stars was varied from 10^5 to 10^6. Fig. 1 shows the evolution in the two-color diagram of a metal poor system of 10^6 stars. Fig. 2 illustrates the same result for the metal rich cluster.

The necessity for including mass loss in the models is demonstrated in Figs. 1 and 2. It is well known that without mass loss the extremely luminous asymptotic giant branch stars dominate the light and render the system too red. In the present work, mass loss also affects the color evolution in an important way through its effect on horizontal branch morphology.

Racine's (1973) calibration of the intrinsic colors of galactic globular clusters is drawn on Figs. 1 and 2. On this curve the clusters form a sequence with the most metal poor systems occupying the blue end, the most metal rich lying at the red end. This is verified in a general way by our models. However, age effects are also striking and may be the cause for the intrinsic scatter noted by Racine (1973) in the relation between $(B-V)_o$ and metallicity. The loop described as a function of age in the two-color plane is the result of the inclusion of horizontal-branch evolution.

The globular clusters in M31 have been found by van den Bergh (1969) to have a brighter M_V (by 0.5 mag.) and a redder $(B-V)_o$

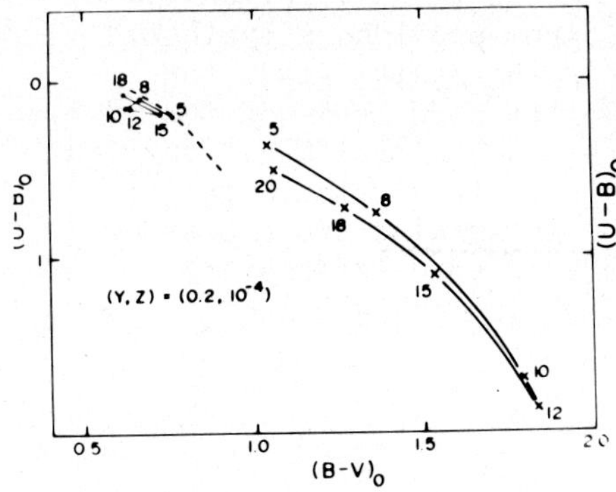

Fig. 1. Colors for metal-poor system as a function of age (in Gyrs) with mass loss (dots) and without mass loss (crosses). The dashed line is Racine's (1973) calibration.

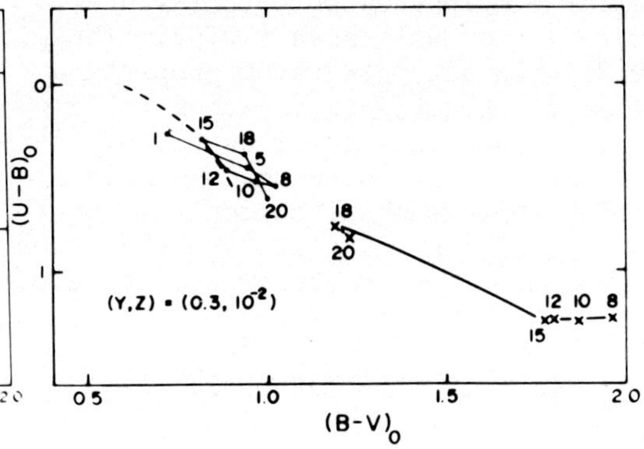

Fig. 2. Same as Fig. 1 for the metal-rich system.

(by 0.1 mag.) than their galactic counterparts. Van den Bergh (1969) then concluded that the M31 clusters are the more metal rich and the more massive. The same data also seem consistent with the alternative interpretation that the globular clusters in M31 are 3 Gyrs younger than those in our Galaxy.

Statistical fluctuations in the numbers of stars in rapid but luminous phases of evolution can be the source of larger uncertainties in color. Some of the model clusters for 10^5 stars confirm this. Fig. 3 shows an extreme example of this effect which confirms some earlier discussions by Racine (1975) of the clusters NGC 8749, NGC 7492 and Pal 13. It seems possible that the differences between the brightest red giants in the Fornax galaxy and its associated globular clusters (Hodge 1961) is due to a similar statistical effect.

III. COMMENT ON THE COLOR EVOLUTION OF ELLIPTICAL GALAXIES

The models for metal-rich systems presented here are also directly applicable to elliptical galaxies. Fig. 2 illustrates the interesting result that the integrated $(B-V)_o$ color does not necessarily increase monotonically toward the red as a function of time as found in current models of galaxies (Tinsley 1977; Tinsley and Gunn 1976; Prather 1976). In fact inclusion of the evolution of the horizontal-branch results in a loop in the two-color plane and in the possibility of blueward evolution with time. In fact in the age interval of 8 to 15 Gyrs, we find that:

$$\frac{\partial(B-V)}{\partial t} = -0.03\ \mathrm{Gyr}^{-1} \qquad (1)$$

Equation (1) is consistent with the observations of the color evolution of elliptical galaxies by Crane (1975), recently rediscussed by Oke and Wilkinson (1977). A more reliable estimate of equation (1) will require a better quantitative understanding of the role of mass loss in stellar evolution and of the theory of the horizontal branch. Financial support of this research by grant AST 72-04418 from the National Science Foundation is gratefully acknowledged.

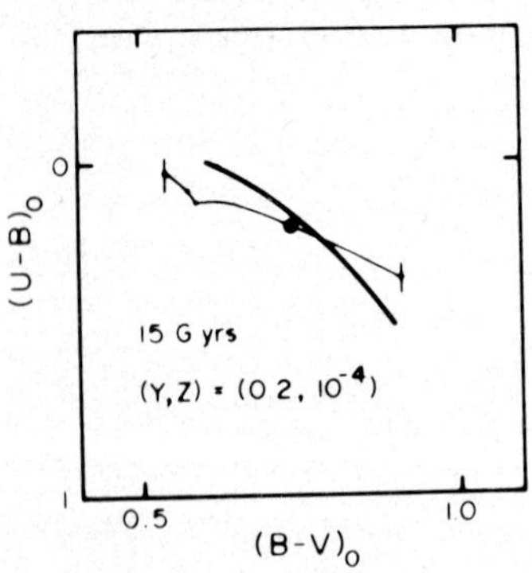

Fig. 3. Range of colors for a model system containing 10^5 stars.

REFERENCES

Bell, R.A. and Gustafsson, B. (1975). In Multicolor Photometry and the Theoretical HR Diagram, A.G.D. Philip and D.S. Hayes, eds., Dudley Obs. Report No. 9, p. 319.
Ciardullo, R.B. and Demarque, P. (1977). Trans. Yale Univ. Obs. Vol. 33, 34, 35.
Crane, P. (1975). Astrophys. J. Letters 198, L9.
Demarque, P. and McClure, R.D. (1977a). Astrophys. J. 213, 716.
Demarque, P. and McClure, R.D. (1977b). Yale Conf. on The Evolution of Galaxies and Stellar Populations, B.M. Tinsley and R.B. Larson, eds., Yale Univ. Obs., New Haven, p. 199.
Fusi-Pecci, F. and Renzini, A. (1975). Astron. Astrophys. 39, 413.
Heasley, J.N. and Mengel, J.G. (1972). Observatory 92, 93.
Hodge, P.W. (1961). Astron. J. 66, 83.
Kurucz, R.L. (1976). unpublished.
Mengel, J.G., Sweigart, A.V., Demarque, P. and Gross, P.G. (1978). to be published.
Oke, J.B. and Wilkinson, A. (1977). to be published.
Prather, M.J. (1976). Ph.D. dissertation, Yale University.
Racine, R. (1973). Astron. J. 78, 180.
Racine, R. (1975). Astron. J. 80, 1031.
Relyea, L. (1976). Ph.D. dissertation, SUNY Albany.
Salpeter, E.E. (1955). Astrophys. J. 121, 161.
Sweigart, A.V. and Gross, P.G. (1976). Astrophys. J. Suppl. 32, 367.
Sweigart, A.V. and Gross, P.G. (1978). to be published.
Tinsley, B.M. (1977). to be published.
Tinsley, B.M. and Gunn, J.E. (1976). Astrophys. J. 203, 53.
van den Bergh, S. (1969). Astrophys. J. Suppl. 19, 145.

DISCUSSION

WEIDEMANN: Did you include differential mass loss besides the Fusi-Pecci/Renzini-type mass loss on the first giant branch in order to produce the horizontal branch?

DEMARQUE: A Gaussian distribution in the mass loss from individual stars was also included to obtain the derived spread in mass on the horizontal branch.

EVOLUTION OF MASSIVE STARS THROUGH LATE EVOLUTIONARY PHASES

Susan A. Lamb

University of California, Los Angeles

1. INTRODUCTION

The theoretical study of the evolution of massive stars has considerable relevance to many branches of observational astronomy. As an example, one may mention, first, the direct application of theoretical models in interpreting the observed properties of supergiants, and second, the use of calculations of the nucleosynthesis taking place in these stars, together with observed supergiant mass loss rates and supernova calculations, to estimate the rate and type of nucleosynthetic enrichment of the interstellar medium by supergiants.

In this brief review of the evolution of massive stars, I shall emphasize theoretical calculations of the later evolutionary phases, namely those beyond core helium burning, and their observational implications, although I shall make some reference to the earlier stages where necessary.

In Section 2 I discuss nucleosynthesis in massive stars and its observable consequences. The interpretation of the ratio of blue to red supergiants is considered in Section 3, and in Section 4, I briefly discuss the observed minimum absolute magnitude of red supergiants, and the possibility that the highest mass Cepheids observed are massive stars in late evolutionary phases. Section 5 contains a summary.

A. G. Davis Philip and D. S. Hayes (eds.), The HR Diagram, 349–356.

2. NUCLEOSYNTHESIS IN MASSIVE STARS

2.1 Theoretical Calculations

The most extensive nonexplosive nucleosynthesis takes place in those stars that are sufficiently massive to burn carbon quiescently. The lower mass limit to this class of stars is approximately 9 or 10 $M_\odot$; the lack of precision reflects the uncertainty in the rate of the carbon burning reaction. The products of carbon burning are later reprocessed through a series of nuclear burning phases which eventually produce the "iron peak" elements, but none of these nuclear reactions ignite in a sufficiently degenerate environment with enough energy release to explode the star. Quasi-static evolution is terminated when the core of the star becomes unstable due to the production of electron-positron pairs, or photodissociation of the 'iron peak' nuclei into α-particles. The former instability sets in prior to the latter only in very massive stars ($M \gtrsim 80$-$100\ M_\odot$, see Arnett, 1972) and, therefore, stars with initial masses in the range $9 - 10 < M/M_\odot \lesssim 80$-$100$ progress through the full range of possible nuclear burning stages. As stars in this mass range are very much more numerous than those with $M>80M_\odot$, and potentially produce a wider range of nuclei in their interiors, they have been the subject of most nucleosynthetic studies.

Many authors have investigated the detailed structural evolution and the associated chemical evolution of massive stars through the phases of core hydrogen and core helium burning, but only relatively recently has there been detailed investigation of the later nuclear burning phases, that is, from core carbon burning onwards. Many of the calculations have been for stellar cores devoid of envelopes, where the focus of interest has been the nucleosynthesis that takes place in the deep interior. Thus Rakavy *et al.* (1967) and Ikeuchi *et al.* (1971, 1972) have evolved a series of carbon-oxygen cores from before carbon ignition to the onset of instability, while Sugimoto (1970) has followed helium cores from core helium burning through core oxygen burning; these latter models have been carried through to the onset of instability by Sugimoto and Nomoto (1974, and references therein). Arnett (1977, and references therein) has made an extensive study of helium cores, evolving them through Si-burning. Ignoring the hydrogen envelopes of massive stars is a reasonable way to explore the detailed nucleosynthesis in the core, assuming that envelope convection does not penetrate through the base of the hydrogen envelope, but no information on observable surface parameters is obtained.

Early investigations of the late evolutionary stages of complete massive stars were made by Hayashi and Cameron (1962) and by Stothers and Chin (1969). More recently, Endal (1975)

has evolved a 15 $M_{\odot}$ star from the main sequence to neon ignition in the core, while Lamb et al. (1976) have evolved 15 $M_{\odot}$ and 25 $M_{\odot}$ stars from the main sequence through core carbon burning. The latter work includes evolutionary tracks in the theoretical Hertzsprung-Russell diagram.

The results of nucleosynthesis are sensitive to the physical conditions in the star, that is, the run of temperature and density with radius, and the details of energy transport. In the cores of massive stars these quantities are determined, from core carbon burning onwards, predominantly by the loss of energy due to neutrino emission. 'Pair', 'plasma', and 'photo' neutrino processes dominate (see Beaudet et al., 1967) whereas bremstrahlung neutrinos (see Festa and Ruderman, 1969) make only a small contribution. The final distribution of nuclei through the stellar core depends sensitively on the extent in mass of the convective core formed during each successive nuclear burning phase. A sequence of convective regions diminishing in mass as evolution progresses is required to produce a layering of regions with different mean molecular weights, that is, a wide variety of nuclei, in the core. The size of the convective core is determined by the rates and location of neutrino energy losses (Arnett, 1972, and Lamb et al., 1976). As the central temperature of the star increases, these energy loss rates approach the nuclear energy generation rates, ensuring smaller convective cores. The calculations of Arnett (1977, and references therein), which do include neutrino energy losses, yield a stratified stellar core with the unburned helium surrounding successive layers composed largely of carbon, oxygen, neon, magnesium, silicon, and iron, as one proceeds inward. The precise mass and composition of each layer depends on the mass of the star; tables of the final core composition as a function of mass are given by Arnett (1978).

Although the major burning stages of stellar evolution only produce nuclei with $A \lesssim 60$, nuclei of much higher atomic weight can be produced during these phases by neutron captures on the lighter nuclei. Interesting quantities of neutrons are produced in the core during the later part of core helium burning, and the resulting nucleosynthesis has been investigated by Couch et al. (1974) and Lamb et al. (1977). It should be noted that the nuclei produced in this way are contained wholly within the helium convective core. Many will not survive the later evolution but will act as seed nuclei for further processing. A very detailed calculation of the composition of a massive star nearing the end of its quasi-static evolution has not yet been performed.

2.2 Observational Implications

A surface composition displaying products of nuclear burning

can occur in supergiants if these products are transported upwards through the envelope to the surface or if the outer layers are shed, thereby revealing inner regions which have undergone nuclear processing. Early-type supergiants do not possess convective envelopes. Thus stars of this class which display evolved surface compositions are most likely to have undergone considerable mass loss (see Walborn, 1976, and Dearborn and Eggleton, 1977) or to have experienced some mixing process such as meridianal mixing. On the other hand, the rapid drop in surface temperature which occurs as a supergiant crosses the Hertzsprung gap towards the red is accompanied by the development of a convective envelope. This dredges up material which has been partially burned during core hydrogen burning earlier, and deposited outside the retreating convective core. The ratio of nitrogen to hydrogen in the envelope is raised to about five times its initial value by this process (see Lamb et al., 1976, for details). At no time during the major fraction of the life of these massive stars does it appear that the envelope convection penetrates through the hydrogen burning shell. Thus the products of helium and later burning stages are not expected to be brought to the stellar surface by convective transport. Mass loss of the entire hydrogen and some of the helium envelope is required to reveal these burning products.

If supergiants end their quasi-static evolution as supernovae, it may be possible to detect the products of nuclear processing in young supernova remnants which have not yet become mixed with large masses of interstellar medium. The supernova remnants Cas A, Pup A, and Kepler are possible examples. Studies of Cas A by Peimbert and van den Bergh (1971), Peimbert (1971), and Chevalier and Kirshner (1978) reveal quite different products of nuclear processing in different parts of the remnant. An interpretation of these abundances in terms of the evolutionary history of the progenitor has been suggested by Lamb (1978). In general, young supernova remnants may display the nucleosynthetic products of several stages of core nucleosynthesis, if the progenitor has undergone substantial mass loss prior to the supernova explosion.

3. THE RATIO OF BLUE TO RED SUPERGIANTS

The observed ratio of blue to red supergiants is of considerable interest, as theoretical calculations have shown that this ratio is very sensitive to the neutrino energy loss rate. This has led several authors (Stothers, 1969, and references therein) to attempt to confirm or refute the existence of the universal Fermi interaction, by a detailed comparison of observation and theory. However, other uncertainties in the calculations are known to effect this ratio, for example, the choice of convective

stability criterion (see Chiosi and Summa, 1970) and the amount of mass loss during the early evolution (Hartwick, 1967, de Loore et al., 1977, and Chiosi et al., 1977). Further, it has also been pointed out that interpreting the ratio of blue to red supergiants as a measure of the ratio of evolutionary timescales of single massive stars through different evolutionary phases is erroneous. Thus, for example, Choisi and Summa (1970) demonstrated that the presence of binary systems among massive stars increases the observed blue to red ratio by an unknown amount, while Ferrari et al. (1970) argued that the sample of red supergiants used previously is contaminated by lower mass stars, which have evolved into the red supergiant region, thereby decreasing the blue to red ratio. The spread in the time of formation of stars in any stellar association, or collection of associations, used to determine this ratio will necessarily introduce the latter uncertainty (Lamb et al., 1976). Thus it would appear that the observed ratio of blue to red supergiants is a poor tool for probing basic physics, but is, nevertheless, an observational quantity which must eventually be given a full theoretical explanation.

4. RED SUPERGIANTS AND MASSIVE CEPHEIDS

Observations of stellar associations and galactic clusters (cf. Humphreys, 1970, and this volume, p. 263) indicate that red supergiants have a minimum $M_{bol} \sim -9$. This bolometric magnitude corresponds to the luminosity expected for a star of mass slightly greater than 25 $M_{\odot}$, during its later evolution, and the question then arises of the fate of the still more massive stars. One possibility for preventing the most massive stars from reaching the red supergiant region is to invoke substantial mass loss during the early evolution, that is, during core hydrogen and core helium burning, when the stars are O and B supergiants. Hartwick (1967) was the first to demonstrate that substantial mass loss can take the evolutionary tracks back towards the blue. On the other hand, the possibility that stars of about 25 $M_{\odot}$ and greater may not evolve into the red supergiant region, even without mass loss, until very late in their evolution, if at all, has emerged from recent calculations by Lamb et al. (1976). They find that a 25 $M_{\odot}$ star stays to the blue of $\log_{10}(T_s) \sim 3.9$ throughout its evolution up to at least the end of core carbon burning, that is, to very near the end of its observable life. By contrast, they find that a 15 $M_{\odot}$ star evolves across the Hertzsprung gap, and burns carbon as a red supergiant with $\log_{10}(T_s) \sim 4.5$. The different surface temperatures of these two stars reflect substantially different envelope structures, the 25 $M_{\odot}$ star having no convective envelope.

Turning now to a consideration of long period Cepheids, we

note that the longest period Cepheids observed by Gascoigne (1969) have periods ∿ 130 days, which indicate stars of about 20 $M_{\odot}$. It has been suggested by Lamb et al. (1976) that these Cepheids are in the core carbon burning phase of evolution, and occur at the intersection of the instability strip and the core carbon burning band. However, the evolution after the carbon burning phase is so rapid that the stellar envelope has no time to adjust to changes taking place in the core. Thus, if a star becomes a Cepheid during core carbon burning it will stay one until its demise. Chiosi et al. (1977) have also pointed out that stars which become Cepheids need not have had an initial mass of 20 $M_{\odot}$ or less, if appreciable mass loss occurred during their early evolution.

5. SUMMARY

In this brief review of some aspects of the late evolution of massive stars I have discussed, 1) the theoretical calculations of the evolution, placing particular emphasis on the factors influencing nucleosynthesis, 2) possibilities for observing the products of this nucleosynthesis, at the surface of stars, and in young supernova remnants, 3) the interpretation of the ratio of blue to red supergiants, 4) the observed maximum luminosity limit for red supergiants, and 5) the possibility that the longest period Cepheids are massive stars in late evolutionary phases.

I wish to thank Dave Arnett, Icko Iben, Jim Truran, and Roger Ulrich for interesting and helpful discussions on different aspects of the material presented in this paper, and the Aspen Center for Physics for hospitality.

REFERENCES

Arnett, W.D. (1972). Astrophys. J. 176, 699.
Arnett, W.D. (1977). Astrophys. J. Suppl. 35, 127.
Arnett, W.D. (1978). Astrophys. J. (in press).
Beaudet, G., Petrosian, V. and Salpeter, E.E. (1967). Astrophys. J. 150, 979.
Chevalier, R.A. and Kirschner, R.P. (1978). Astrophys. J. (in press).
Chiosi, C., Nasi, E. and Sreenivasan, S.R. (1977). Astron. Astrophys. (in press).
Chiosi, C. and Summa, C. (1970). Astrophys. Space Sci. 8, 478.
Couch, R.G., Schmeidekamp, A.B. and Arnett, W.D. (1974). Astrophys. J. 190, 95.
Endal, A.S. (1975). Astrophys. J. 197, 405.
Ferrari, A., Galeotti, P., Silvestro, G. and Trussoni, E. (1970). Astrophys. Space Sci. 9, 181.

Festa, G.G. and Ruderman, M.A. (1969). Phys. Rev. 180, 1227.
Gascoigne, S.C.B. (1969). Mon. Not. Roy. Astron. Soc. 146, 1.
Hartwick, F.D.A. (1967). Astrophys. J. 150, 953.
Hayashi, C. and Cameron, R.C. (1962). Astrophys. J. 136, 166.
Humphreys, R.M. (1970). Astrophys. Letters 6, 1.
Ikeuchi, S., Nakazawa, K., Murai, T., Hoshi, R. and Hayashi, C. (1971). Prog. Theor. Phys. Kyoto 46, 1713.
________ (1972). ibid. 48, 1870.
Lamb, S.A. (1978). Astrophys. J. (in press).
Lamb, S.A., Howard, W.M., Truran, J.W. and Iben, I. (1977). Astrophys. J. 217, 213.
Lamb, S.A., Iben, I. and Howard, W.M. (1976). Astrophys. J. 207, 209.
de Loore, C., De Grève, J.P. and Lamers, H.J.G.L.M. (1977). (preprint).
Peimbert, M. (1971). Astrophys. J. 170, 261.
Peimbert, M. and van den Bergh, S. (1971). Astrophys. J. 167, 223.
Rakavy, G., Shaviv, G. and Zinamon, Z. (1967). Astrophys. J. 150, 131.
Stothers, R. (1969). Astrophys. J. 155, 935.
Stothers, R. and Chin C.-W. (1969). Astrophys. J. 158, 1039.
Sugimoto, D. (1970). Prog. Theor. Phys. Kyoto 44, 599.
Sugimoto, D. and Nomoto, K. (1974). In IAU Symposium 66, Late Stages of Stellar Evolution, R.J. Tayler and J. Hesser, eds., Reidel, Dordrecht, p. 105.
Walborn, N. (1976). Astrophys. J. 205, 419.

DISCUSSION

TAYLER: I hope you'll excuse me while I congratulate the last two speakers on their results and also make a historical reminiscense. About four years before Henry Norris Russell died I was evolving stars - I haven't, since then, evolved any stars - and I evolved a massive star towards the onset of hydrogen exhaustion, using a desk calculator. I calculated that it would take me 20 months, working 6 hours a day, five days a week to evolve a one solar mass star - and that is why I gave up stellar evolution at that moment. I think that the results shown by Allen Sweigart and by Susan Lamb have shown us how far things have come in these last 20 - odd years.

BIDELMAN: Could you say a word about the evolution of stars of somewhat smaller mass, say 5 to 10 $M_{\odot}$?

LAMB: Stars in this mass range are thought to end their evolution in carbon-detonation supernova, although if all of these stars exploded in this way it would lead to an overproduction of carbon. Various possible mechanisms for avoiding the explosion or reducing the mass range of stars involved in this type of supernova have been investigated, for example, the Urca process and substantial mass loss. The end points of evolution of these stars are not well studied partly because the evolutionary calculations are severely complicated by the occurence of helium shell flashes. These latter have been extensively studied by Iben.

COX: What is the maximum mass that you believe has a blue loop?

LAMB: I think that it would be somewhere between 10 and 13 $M_{\odot}$, perhaps. But that's an uncertain area.

THE STRUCTURE AND EVOLUTION OF MASSIVE STARS

C. Chiosi

Instituto di Astronomia, Universita di Padova

1. INTRODUCTION

This review is restricted to the most recent studies of the structure of stars in the approximate range from 10 to 100 $M_\odot$, during the core H- and He-burning phases. Other recent major reviews on this subject are by Dallaporta (1971), Massevich and Tutukov (1974) and Iben (1974). The lower limit was chosen to be just above the transition from degeneracy to non-degeneracy in the core at carbon ignition (Schwarzschild and Härm 1958). The upper limit is very uncertain. The canonical value of about 60 $M_\odot$ for Pop I composition was set by Schwarzschild and Harm (1959), using linear pulsation theory. More recent non-linear dynamical calculations lift the limit above 100 $M_\odot$, and also show that mass loss by vibrational instability occurs at such a rate that stars in the range from 100 to 200 $M_\odot$ can survive for a time comparable to the total main sequence lifetime (Appenzeller 1970a, b, Ziebarth 1970, Talbot 1971a,b and Papaloizou and Taylor 1974).

2. SEMICONVECTION

Our understanding of massive stars in core H- and He-burning is still affected by the long-standing problem of the mixing of convectively unstable layers at the boundary of the convective core (semiconvection).

Excellent reviews are available on the basic ideas of semiconvection (Dallaporta 1971, Massevich and Tutukov 1974, and Iben 1974) and on the variety of structures of the intermediate layers which have been suggested since its discovery (Stothers 1970). We will review only the controversy over which criterion

A. G. Davis Philip and D. S. Hayes (eds.), The HR Diagram, 357-368. All Rights Reserved.

to use for stability against convection in regions with a gradient of molecular weight. Two approaches have been suggested. Schwarzschild and Härm suggested that the composition in these layers is modified such that the temperature gradient is equalized to the adiabatic value to maintain a neutrality condition ($\nabla_T = \nabla_{Ta}$, hereafter called Eqn. 1). In the second case, due originally to Ledoux (1947), the density gradient is equalized to the adiabatic value ($\nabla_\rho = \nabla_{\rho a}$, hereafter called Eqn. 2).

It is evident that whereas Eqn. 1 implies a layer which is dynamically stable, Eqn. 2 implies a layer on the verge of dynamical instability. It is therefore likely that complete mixing could take place.

The basic mechanism for mixing is unknown in both cases. Kato (1966) pointed out that Eqn. 2 must be applied in a medium with a gradient of molecular weight when thermal dissipation is neglected. If, however, it is included, one ends up with Eqn. 2. In Kato's picture the mixing mechanism is thought to rely on the vibrational instability of the gravity modes of non-radial oscillations. Since the instability sets in as soon as $\nabla_T > \nabla_{Ta}$, mixing would take place in such a way that Eqn. 1 is satisfied. But Kato's analysis is a local one, based only on the intermediate layers themselves. This approximation is very drastic, and it is valid only for eigenfunctions with very large amplitudes in the destabilizing region and small amplitudes everywhere else. Moreover the onset of vibrational instability is known to involve the whole star, so a careful analysis is necessary. This problem was investigated by Gabriel (1969) and Aure (1971) using approximate stellar models and eigenfunctions, and no instability was found. Gabriel and Noels (1976) reconsidered this problem, analyzing the vibrational instability of gravity modes for a 30 $M_\odot$ stars, when no prescription for semiconvection is adopted in the intermediate layers. The result is that a large number (increasing as evolution proceeds) of eigenfunctions, with large values of the surface spherical harmonic ℓ, are trapped in the region with a gradient of molecular weight. Most of the eigenfunctions are vibrationally unstable with a time scale on the order of 10^3 to 10^4 years. The idea is that these unstable gravity modes are able to reduce the temperature gradient to a value very close to the adiabatic one.

This result, if confirmed by non-linear analysis, would allow us to disregard models with layers obeying Eqn. 2 in favor of those obeying Eqn. 1.

3. EVOLUTIONARY RESULTS

Numerous calculations have shown that semiconvective mixing does not play a significant role (whatever the adopted stability

criterion) on the main sequence. However, H shell burning and especially He core burning are, on the contrary, completely dominated both by any small difference resulting from the choice of stability criterion and by secondary details of the input physics (opacity, nuclear reaction rates, chemical composition, and mixing length ratio).

3.1 Helium Ignition

Because of the large body of literature on this subject, only a few references will be quoted here. For further details see Stothers and Chin (1975, 1976). The lower limit of the mass, for Pop I composition, at which the choice of stability criterion produces a significant difference in the evolutionary behavior was estimated by Barbaro, et al. (1972) to be around 13 $M_\odot$. The adoption of Eqn. 2 in computing stellar models results in the devlopment of a shallow gradient of chemical composition. The onset of an H-burning shell does not modify this structure, and no matter where core He-burning is initiated, whether at high or low T_{eff}, thermal equilibrium is first partially achieved in the red supergiant phase. Further evolution either takes place entirely at low T_{eff}, or with more or less wide loops in the HR diagram. Several causes of the loops are discussed next.

The adoption of Eqn. 1 leads to a more intriguing situation. The onset of an H-burning shell causes the development of a fully convective shell in the region of intermediate instability, which drastically changes the H-profile throught the zone. Models of this type were first computed by Chiosi and Summa (1970) and by Simpson (1971), who also pointed out the main reasons for this transition from semiconvection into full convection.

In the mass range from 13 to 40 $M_\odot$ (these values are only estimates, as they depend upon several parameters, mostly the composition), due to the occurrence of the fully convective zone, or more precisely, the plateau-shaped H-profile, core He-burning takes place either entirely in the blue supergiant region or partly in the blue and partly in the red supergiant region. This is no longer true, though for different causes, for both lower and higher masses. In fact, models around 11 $M_\odot$, which use Eqn. 1 and which have a plateau-shaped H-profile, start core He-burning as red supergiants and perform a loop in the HR diagram (Barbaro, et al. 1972). Two sequences at 12 $M_\odot$, by Robertson (1972), are exceptions, having the entire He-burning phase in the red supergiant region without a loop. On the other hand, stars with mass greater than about 40 $M_\odot$ spend either their entire He-burning phase, or a time comparable to the total time of He-burning as red supergiants (Barbaro, et al. 1971, Varshansky and Tutukov 1973, and Stothers and Chin 1976). Barbaro, et al. (1971) suggested a possible qualitative interpretation of this behavior for both types

of models. In studying the influence of the H-profile on the core He-burning phase, they introduced two characteristic time scales which describe the pre-He-ignition phase: the time T_{sh} at which the H-burning shell advances through the H-profile, and the time T_d of diffusion of radiation from the shell to the surface. When $T_{sh} < T_d$ (low H-content) the models are not in thermal equilibrium because the outer layers cannot readjust to the rapid changes in the interior. If the H-content in the shell is high then T_{sh} becomes greater than T_d, and thermal equilibrium can be restored. Moreover, T_d becomes quite small in the presence of extended convective zones in the outer envelope. It is also easy to understand the relation between T_{sh} and the H-profile resulting from the choice of different stability criteria. Models in the intermediate mass range ($13 < M/M_{\odot} < 40$) using Eqn. 1 have the merger of the H-burning shell with the inner edge of the H-plateau, formed by the convective shell, occurring at high T_{eff}, so that $T_{sh} > T_d$ and thermal equilibrium is restored. At lower mass ($M < 13 M_{\odot}$), however, as the inner extension of the convective shell is smaller, an extended zone with low H content is left between the He-core and the H-plateau; the models reach the Hayashi line before the H-burning shell has reached the H-plateau. In the higher mass ($M > 40 M_{\odot}$) range, even though T_{sh} is quite soon equal to or greater than T_d, the additional effect of the high L/M makes the models red (Stothers and Chin 1976). Models constructed with Eqn. 2, owing to their shallow H-profile ($T_{sh} < T_d$), must undergo a preliminary red phase before partially restoring thermal equilibrium.

Subsequent evolutionary behavior, after the star has reached the Hayashi line in the early phases of core He-burning, is not well understood. In particular, whether these stars become blue supergiants while performing a loop in the HR diagram, or whether they remain red supergiants until core He-depletion, is not yet well established.

3.2 Loop Formation

A general view of the variety of models obtained by many authors gives the impression of contradictory and rather erratic results. It was not until the work of Ziolkowski (1972), though the result had been predicted by Paczynski (1970), that the crucial effect of small changes of the depth of the convective envelope at the tip of the red giant branch on the loop formation was shown. The critical point is whether the H-burning shell is able or not to approach, before the core He-depletion stage, the discontinuity in the composition profile previously produced by the penetration of the outer convection zone. The numerical computations show that, once the H-burning shell passes the composition discontinuity, the models rapidly shift into the region of high T_{eff}, producing a loop.

A deeper analysis of this problem shows that the loop phenomenon is ultimately related to the composition profile throughout the star. It was shown by Lauterborn, et al. (1971a, b), Kozlowski (1971) and Fricke and Strittmatter (1972) that thermally stable models in the core He-burning phase with a rectangular H-profile can exist only in the region of high T_{eff}. In this context it is evident that any mixing that contributes to the formation of a discontinuous H-profile, due to convection either in internal layers or in the outer envelope, plays a very important role. Models reaching the Hayashi line with a shallow H-profile must stay there until the H-burning shell reaches the discontinuity in composition produced by the outer convection zone. A significant fraction of the core He can be burnt during this stage. Some models may even have the entire core He-burning as red supergiants. On the other hand, those models which suffered an intermediate homogenization while crossing the HR diagram prior to core He-ignition, quite soon build up a quasi-rectangular profile and go back into the region of high T_{eff}. A negligible fraction of the core He is burnt in this case. It is also easy to see that a number of parameters, such as the opacity source, the nuclear reaction rates and the mixing length ratio may significantly affect this picture (Stothers and Chin 1975, 1976).

3.3. Stationary He-burning

Models without a pre-He-burning red phase (Chiosi and Summa 1970, Simpson 1971, Stothers and Chin 1976), or models that did not undergo any significant He-depletion before the loop formation (Chiosi and Summa 1970, case A, Barbaro, et al. 1972), have their major stage of core He-burning as blue supergiants, with the remainder as red supergiants. The two subphases are separated by an intermediate yellow phase with a thermal time scale. Over an extended mass range Lauterborn, et al. (1971a, b), Fricke and Strittmatter (1972), and Barbaro, et al. (1973) suggested that the two stages could be described in terms of the fractional mass in the He core, q_{He}, and the central He-content Y_c. The parameters q_{He} and Y_c are found to control the blue and red stages, respectively. The existence of the thermally unstable yellow region is also confirmed by the sequences of equilibrium models by Barbaro, et al. (1973) and Lauterborn, et al. (1971b). The ratio $(T_B/T_R)_{He}$ of the lifetimes of the blue and red stages is not uniquely correlated with the stellar mass, but is also affected by a number of other factors. It increases with Z, Y, ε_{C+DC} and decreases with $\varepsilon_{3\alpha}$. On the other hand, models whose lifetime in the red supergiant region is long enough to significantly deplete the core He-content, and models which spend their entire core He-burning lifetime as red supergiants, do not fit this picture. In particular, the former exhaust the remaining core He as blue supergiants only (Ziolkowski 1972, Stothers and Chin 1975). In any case, the lifetime ratios $(T_B/T_R)_{He}$ are always quite small.

3.4 Multiple Solutions

The possibility that multiple solutions may exist for models of the same mass and distribution of composition was first suggested by Paczynski (1970). His suggestion was confirmed by Lauterborn, et al. (1971a, b), Kozlowski (1971), and Fricke and Strittmatter (1972). They showed the existence of three different models (blue, yellow and red) of a star with a given mass and composition profile in a certain range of the mass of the He core.

The same authors demonstrated the occurrence of secular instabilities, associated with the yellow solutions, during the loop phase. These apparent violations of the Vogt-Russell theorem were discussed by Kähler (1972) and further investigated by Paczynski (1972), who developed the concept of the similarity between a sequence of static models and the linear series of Poincaré. The relationship between multiple solutions, the sign of the Henyey determinant, and thermal stability was also clarified. Moreover, the possibility of multiple branches in the linear series also accounts for the multiplicity in the solutions found by numerical computations. The most salient advantage of these studies is that they not only give a systematic interpretation of the basic evolutionary features, but also foresee the response to variations in the input parameters.

4. CARSON'S OPACITIES

The results we have just outlined are mostly based on the use of the radiative opacities calculated by Cox and Stewart (1965, 1970) using the hydrogenic model of the atom. On the other hand, the new radiative opacities of Carson (1974) are based on the hot "Thomas Fermi" model of the atom for all elements heavier than He. The largest difference between the new and old opacities is the bump (absent in the old ones) due to the ultimate ionization of the CNO group of elements at moderate temperatures and low densities. Evolutionary models by Stothers (1977), for the mass range from 7 to 60 $M_\odot$ and Pop I composition, show that the new feature of Carson's opacity becomes important for masses greater than 10 $M_\odot$. As a consequence of it, a fraction of the envelope, growing with mass and proceeding evolution, becomes convective even at high T_{eff}. This result broadens the main sequence band. The main features of semiconvection and its dependence upon the stability criteria are still preserved under the new opacities. However, for masses above some value, about 20 - 30 $M_\odot$, the intermdiate instability is suppressed when the outer layers expand to very large radii, or when the outer convection zone penetrates inwards very deeply. Moreover, the present computations seem to indicate that stars of the highest mass (around 60 $M_\odot$) could spend a large fraction of the core H-burning phase itself as red supergiants. According to this interpretation the upper main

sequence could extend across the whole HR diagram. In addition, the results indicate that no static red supergiant models can exist for masses greater than 30 $M_\odot$. Stothers (1977) speculates that, as a plausible consequence of it, mass loss in the red region would occur.

The advantages and improvements in the comparison with observations have been extensively discussed by Stothers (1976b) and references quoted therein. However, a number of serious difficulties are still present, such as the observed rarity of very luminous M supergiants, and the presence of intermediate spectral type stars (B2-A5), which do not have a theoretical counterpart in the proposed scenario (Stothers 1977).

5. MASS LOSS

In the spectra of OB supergiants, the presence of displaced ultraviolet resonance lines, similar effects in the visible region, Hα emission, and radial velocity gradients are commonly interpreted as an indicator of mass outflow (Hutchings 1976, Conti 1976, and Snow and Morton 1976). The mass loss rates are estimated to be from 10^{-7} to 10^{-5} $M_\odot$/yr. A further indication of mass loss in OB supergiants is provided indirectly by the upper limit of the luminosity of hot supergiants, which decreases from $M_{bol}=-10^m.5$ at $\log T_{eff}=4.5$ to $M_{bol}=-8^m.5$ at $\log T_{eff}=4.3$ (Hutchings 1976). Moreover, anomalies in the surface abundances of some OB stars, which appear to be carbon or nitrogen rich (OBC, OBN), seem also to imply that mass loss occurs (Dearborn and Eggleton 1977). The observational evidence of mass loss from red giants and supergiants is discussed by several authors (Weyman 1963 and Reimers 1975). The mass loss rate ranges from 10^{-9} to 10^{-5} $M_\odot$/yr. with an uncertainty on the order of a factor of two either way, according to Reimers (1975). For red supergiants with masses greater than 3 $M_\odot$, Reimers (1975) also suggests either an additional previous phase of mass loss, or an increase in the rate by a factor of three. No accurate estimate exists for the most luminous red supergiants.

Theoretical investigations of mass outflow from red supergiants offer two possible pictures, either radiation pressure on grains (Lucy 1976), or stellar wind from a hot corona due to non-thermal processes in the convective envelope (Fusi-Pecci and Renzini 1975, and references quoted therein).

Evolutionary sequences incorporating the mass loss phenomena, for the mass range 20 - 100 $M_\odot$ during the H and He core burning phases, have been calculated by Chiosi, et al. (1977) and de Loore, et al. (1977a, b). Because of several uncertainties in our theoretical understanding of mass loss, both of these attempts use one or two parameters to reproduce the observational data. Core H-burning sequences with mass loss evolve at lower luminosity and T_{eff}, thus broadening the main sequence. However, if the mass loss rate

is high, the most massive stars (60 - 100 $M_{\odot}$, the limits depending upon the average mass loss rate) invert their path in the HR diagram long before core H-exhaustion and shrink towards the main sequence, eventually crossing it. This behavior leads to a plausible interpretation of the upper luminosity boundary of OB supergiants if the mass loss rate ranges from $2x10^{-6}$ $M_{\odot}$/yr., for an initial mass of 20 $M_{\odot}$, to $2x10^{-5}$ $M_{\odot}$/yr., for an initial mass of 100 $M_{\odot}$. Semiconvective instability is suppressed during the entire core H-burning phase. At the onset of an H-burning shell, a convective layer appears above it, as is usual when the Schwarzschild criterion is used. However, its extension in mass is much smaller than in the case of conservative evolution. Stars in the upper range of mass, roughly from 60 to 80 $M_{\odot}$ with a sufficiently high mass loss rate, may lose as much as 40 to 50% of their mass during the main sequence phase. Layers of the original convective core are brought to the surface, with the result that the surface He abundance is increased by almost a factor of two at the end of core H-burning, and the H abundance is correspondingly decreased. Since CNO processed material is brought to the surface, enhancements of ^{14}N and ^{13}C relative to ^{12}C and ^{17}O relative to ^{18}O are expected. Following Dearborn and Eggleton (1977), mass loss and these anomalies in the surface composition may be clues to the interpretation of OBN stars. On the basis of the quite low N(H)/N(He) ratios (1-2) predicted theoretically for the last stages of core H-burning in very massive stars (roughly 80 and 100 $M_{\odot}$), de Loore, et al. (1977b) suggested that single WN7-WN8 stars were still burning H in their cores. An alternative suggestion was made by Chiosi, et al. (1977), who preferred the idea that WN7-WN8 stars are in advanced stages of core He burning. In this case the WN7's and WN8's would be the result of mass loss in both the blue and red stages. Thus, an attempt was made by Chiosi, et al. (1977) to follow evolutionary sequences, with mass loss given by the acoustic flux mechanism, through both the blue and red stages and up to the intermediate core He-burning phase. As a result of this study we may suggest the following scenario. Stars with an initial mass smaller than $\sim$15 $M_{\odot}$ reach the Hayashi line and undergo moderate mass loss by a wind driven by the acoustic flux. Their evolution is similar to the classic scheme. Stars with an initial mass in the range from $\sim$15 to $\sim$35 $M_{\odot}$ reach the Hayashi line but suffer a significant mass loss so that their subsequent evolution is greatly modified. In fact, the core He-burning takes place after the red phase, while the star is slowly moving towards higher T_{eff}'s. The area in the HR diagram which is occupied is similar to the one resulting from the use of the Ledoux criterion, but covers a broader strip, giving better agreement with the observations.

Stars with higher initial masses ($M > 40$-50 $M_{\odot}$) do not reach the Hayashi line because of their high mass loss rate. Nevertheless, their subsequent evolution is similar to that of

intermediate-mass stars. Within this framework, Chiosi, et al. (1977) suggested that single WN7 and WN8 stars originate from stars of the highest mass and are in the late stages of core He-burning. Moreover, single WN's and WC's should have progenitors in the mass range from 20 to 40 $M_\odot$ (or, better, from 30 to 50 $M_\odot$) which reached the Hayashi line but also suffered extensive mass loss. WN's and WC's, then, may be in advanced stages of central He burning. The numerical results seem also to indicate that stars in the latter mass range may lose their entire H-rich envelope, before the completion of core He-burning, as blue stars. Then the WC's should correspond to that fraction of the core He-burning phase which occurs after the removal of the H-rich envelope and before He-exhaustion. However, because of the many uncertainties prsent in this picture, more work is necessary to achieve a deeper insight into the problem.

REFERENCES

Appenzeller, J. (1970a). Astron. Astrophys. **5**, 355.
Appenzeller, J. (1970b). Astron. Astrophys. **9**, 216.
Aure, J.L. (1971). Astron. Astrophys. **11**, 345.
Barbaro, G., Bertelli, G., Chiosi, C. and Nasi, E. (1973). Astron. Astrophys. **29**, 185.
Barbaro, G., Chiosi, C. and Nobili, L. (1971). In Colloquium on Supergiant Stars, M. Hack, ed., Trieste Obs., p. 334.
Barbaro, G., Chiosi, C. and Nobili, L. (1972). Astron. Astrophys. **18**, 186.
Carson, T.R. (1974). unpublished.
Castor, J.I., Abott, D.C. and Klein, R.J. (1975). Astrophys. J. **195**, 157.
Chiosi, C. and Summa, C. (1970). Astrophys. Sp. Sci. **8**, 478.
Chiosi, C., Nasi, E. and Sreenivasan, S.R. (1977). Astron. Astrophys., in press.
Conti, P.S. (1976). Mem. Soc. Roy. Sci. Liege 6° Serie **9**, 193.
Cox, A.N. and Stewart, J.N. (1965). Astrophys. J. Suppl. **11**, 22.
Cox, A.N. and Stewart, J.N. (1970). Astrophys. J. **19**, 243.
Dallaporta, N. (1971). In Colloquium on Supergiant Stars, M. Hack, ed., Trieste Obs., p. 250.
Dearborn, D.S.P. and Eggleton, P.P. (1977). Astrophys. J. **213**, 448.
de Loore, C., De Greve, J.P. and Lamers, H.J.G.L.M. (1977a). preprint.

de Loore, G., De Greve, J.P. and Voubereven, D. (1977b). preprint.
Fricke, K.J. and Strittmatter, P.A. (1972). Mon. Not. R. Astron. Soc. 156, 129.
Fusi-Pecci, F. and Renzini, A. (1975). Astron. Astrophys. 39, 413.
Gabriel, M. (1969). Astron. Astrophys. 1, 321.
Gabriel, M. and Noels, A. (1976). Astron. Astrophys. 53, 149.
Hutchings, J.B. (1976). Astrophys. J. 203, 438.
Iben, I. (1974). Ann. Rev. Astron. Astrophys. 12, 215.
Kahler, H. (1972). Astron. Astrophys. 20, 105.
Kato, S. (1966). Publ. Astron. Soc. Japan 18, 374.
Kozlowski, M. (1971). Astrophys. Letters 9, 65.
Lauterborn, D., Refsdal, S. and Weigert, A. (1971a). Astron. Astrophys. 10, 97.
Lauterborn, D., Refsdal, S. and Weigert, A. (1971b). Astron. Astrophys. 13, 119.
Ledoux, P. (1974). Astrophys. J. 105, 305.
Lucy, L.B. (1976). Astrophys. J. 205, 482.
Lucy, L.B. and Solomon, P.M. (1970. Astrophys. J. 159, 879.
Massevich, A.G. and Tutukov, A.V. (1973). In IAU Symposium No. 66, Late Stages of Stellar Evolution, R.J. Tayler and J.E. Hesser, eds., D. Reidel, Dordrecht, p. 73.
Paczynski, B. (1970). Acta Astron. 22, 327.
Paczynski, B. (1972). Acta Astron. 22, 163.
Papaloizou, J.C.B. and Tayler, R.J. (1974). In IAU Symposium No. 59, Stellar Instability and Evolution, P. Ledoux, A. Noels and A.W. Rodgers, eds., D. Reidel, Dordrecht, p. 77.
Reimers, D. (1975). Mem. Soc. Roy. Liege 6° Serie 8, 369.
Robertson, J.W. (1972). Astrophys. J. 177, 473.
Simpson, E. (1971). Astrophys. J. 165, 295.
Snow, T.P. and Morton, D.C. (1976). Astrophys. J. Suppl. 32, 429.
Stothers, R. (1970). Mon. Not. R. Astron. Soc. 151, 65.
Stothers, R. (1972). Astrophys. J. 175, 431.
Stothers, R. and Chin, C.W. (1975). Astrophys. J. 198, 407.
Stothers, R. and Chin, C.W. (1976a). Astrophys. J. 204, 472.
Stothers, R. and Chin, C.W. (1976b). Astrophys. J. 209, 800.
Stothers, R. (1977). Astrophys. J. 211, 189.
Schwarzschild, M. and Härm, R. (1958). Astrophys. J. 128, 348.

Schwarzschild, M. and Härm, R. (1959). Astrophys. J. 129, 637.
Talbot, R.J. (1971a). Astrophys. J. 163, 17.
Talbot, R.J. (1971b). Astrophys. J. 165, 121.
Varshavsky, V.I. and Tutukov, A.V. (1973). Nauch. Inf. 26, 35.
Weymann, R. (1963). Ann. Rev. Astron. Astrophys. Vol. 1, L. Goldberg, ed., Annual Reviews, Palo Alto, p. 97.
Ziebarth, K. (1970). Astrophys. J. 162, 947.
Ziolkowski, J. (1972). Acta Astron. 22, 327.

DISCUSSION

KÄHLER: I have a comment and a question. First the comment: You discussed the first loop which depends on details of the chemical profile. We have investigated the second loop for a star with $9M_\odot$. It turns out that the occurrence of the second loop depends critically on the mass of the burnt-out C-core but is nearly independent of the chemical profile. My question: Do you prefer the new opacities or the old ones?

CHIOSI: The results I have just described refer only to massive stars ($M > 10\ M_\odot$) and in the central He-burning phase. It is generally accepted that the chemical profile throughout the models is the leading parameter for the occurrence of a loop, although other factors may strongly affect its features. Moreover, as far as I know, in this range of masses, phases subsequent to the central He exhaustion occur only along the Hayashi line, so a second loop does not occur. However, I agree with you that in the case of lower mass stars the leading parameter for the loop formation in the double shell-burning phase is the mass of the carbon core. As far as your second point is concerned, I don't have an easy answer. The results by Stothers, using the new radiative opacities, give a rather embarassing picture for the evolution of most massive stars even in the core H-burning phase. Very recently, Stothers has discussed possible improvements achieved by the use of the new radiative opacities. However, the new opacities also give a number of serious problems when the theoretical results are compared with observations. For instance, they do not account for the occurrence of intermediate spectral type supergiants and the lack of very luminous red supergiants (mass-loss in the case is absolutely necessary). My personal opinion is still in favor of the old radiative opacities. In any case more work is necessary to cast light on this problem.

FLOWER: I would like to comment on the Carson opacities. The group calculating stellar opacities at Los Alamos under the direction of Al Mertz has not been able to reproduce these opacities using the same atomic model as Carson. Therefore, evolutionary calculations using these opacities may be very suspect.

CONTI: Stothers' models using Carson's opacities do have what I consider a serious problem. The temperature of the most massive, O-type stars is predicted not to be greater than 35000 K or so. This is in conflict with what I feel is the well determined temperature scale based upon the observed ionization equilibrium of helium. We observe normal O-type stars with temperatures up to 50000 K. This is in conflict with Stothers' models using Carson's opacities.

CHIOSI: I agree.

EVOLUTION OF MASSIVE STARS: WHEN DOES AN Of STAR BECOME A WR STAR?

Peter S. Conti*

Joint Institute for Laboratory Astrophysics, University of Colorado and National Bureau of Standards

ABSTRACT

Detailed spectroscopic work has begun for several stars that have been classified WN 7. All show absorption lines in the upper Balmer series and in ionized helium. The absolute magnitudes, established from cluster membership, are relatively brighter than those of most other WR stars. In this paper the outflow velocity near the stellar photosphere is shown to be similar to that in Of stars. These WN 7 stars show somewhat larger emission line widths and probably have less hydrogen in their envelopes, but otherwise have similar luminosities and temperatures as Of stars.

The overall properties of WR stars are well known (e.g., Smith 1973). The stars are hydrogen poor, helium rich, and characterized by strong emission lines which dominate the optical spectrum. The two classes, WN and WC, are thought to be chemically distinct stars in which the products of hydrogen burning CNO reactions (WN) or helium burning (WC) are brought to the surface (Paczynski 1973). In principle, WR objects can be produced by stripping away the hydrogen rich layers of massive stars, either by mass exchange in a binary (Paczynski 1973), or by winds from single or binary stars (Conti 1976). Of stars are those O-type objects in which the stellar wind is sufficiently strong that op-

*Visiting Astronomer, Cerro Tololo Inter-American Observatory, operated by Association of Universities for Research in Astronomy, Inc., under contract with the National Science Foundation.

A. G. Davis Philip and D. S. Hayes (eds.), The HR Diagram, 369–374. All Rights Reserved.

tical emission lines are seen (Conti and Leep 1974). The observed mass loss rates range from 10^{-5} to $10^{-6} M_{\odot} yr^{-1}$ (Snow and Morton 1976; Conti and Frost 1977). A mass loss rate of a few times $10^{-6} M_{\odot} yr^{-1}$ is sufficient to considerably alter the evolution of the star while it is still burning hydrogen in the core, and subsequently, as discussed independently by Chiosi et al. (1977) and de Loore et al. (1977). In particular, the general lack of later type supergiants, compared to blue supergiants, can be understood in terms of substantial mass losses in the previous evolutionary phases.

There is a distinct set of WN stars that, unlike most WR stars, contain appreciable amounts of hydrogen. These stars have been called "transition" objects (Conti 1976) because of their spectroscopic similarities to Of stars. Their emission line widths and strengths are less than those of most other WR stars. In some cases absorption lines in the upper Balmer series are observed (Niemela 1973). Although the term "transition WR" implies only spectroscopic similarities to Of stars, it may also have an evolutionary connotation: The transition stars may be descendant from Of stars and evolving to WR stars. Chiosi et al. (1977) and de Loore et al. (1977) are able to place these "transition" stars in an evolutionary sequence descendant from massive Of stars with sufficiently high mass loss rates. For the former authors, the stars are burning helium in their cores; for the latter, hydrogen core burning still operates. The location of these objects in the theoretical HR diagram is, however, similar in both cases.

The mass loss rate of a star is established by the envelope. The most important parameters are the velocity gradient and its terminal value, and the wind density. The extent, composition, and radiative temperature are important but only indirectly affect the determination of the mass loss rate. It appears likely that most spectroscopic differences between O, Of and WR stars can be completely accounted for by variations of these envelope parameters. Any other physical differences between Of and WR stars may then be differences of interior structure, hydrogen or helium burning, and stellar mass.

Three "transition" WN stars and the related object HD93129, type O3f, will be discussed briefly here. This work is part of a more detailed study of these stars (Conti, Niemela and Walborn 1978). The WN stars, HD 92740, HD 93131, HD 93162 and HD 93129 are members of the Carina association (Walborn 1973) and have relatively bright M_v, between $-6\overset{m}{.}5$ and $-7\overset{m}{.}0$. HD 92740 appears to be a single line spectroscopic binary (Niemela 1973) with emission and absorption lines in phase. The period has now been firmly established at about 80 days (Conti, Niemela and Walborn 1978). Although the mass function is 1.83, the secondary has not been detected as yet.

The star HD 93131 shows no evidence for periodic variable radial velocity. The star HD 93162 shows some scatter in radial velocity but at the writing of this paper all the data have not yet been reduced. HD 93129 shows no significant velocity variation. For these three stars, then, it is possible to assume for the present they are single and average velocities for all lines measured. The results are plotted in Fig. 1. The velocity of the Carina association has not been firmly established but the few single O-star members measured so far by Conti, Leep and Lorre (1977) have velocities in the range -5 to -20 km s^{-1}.

It can be seen from Fig. 1 that the absorption and emission line velocities of HD 93129, the O3f star, are slightly negatively displaced with respect to the association. There is no convincing "Balmer progression" up to $H\gamma$. $H\beta$ is weak on the plates used and both it and $H\alpha$, which is in emission, are distorted by nebular emission. HD 93162 and HD 93131 have a very strong Balmer progression and all absorption lines are violet displaced with respect to the (narrow) emission lines of N III, N IV (4057 Å only), and Si IV. These emission lines are at roughly the same velocity as those in HD 93129. The Balmer progression is stronger in HD 93131 than in HD 93162, which is also the ordering of emission line strengths.

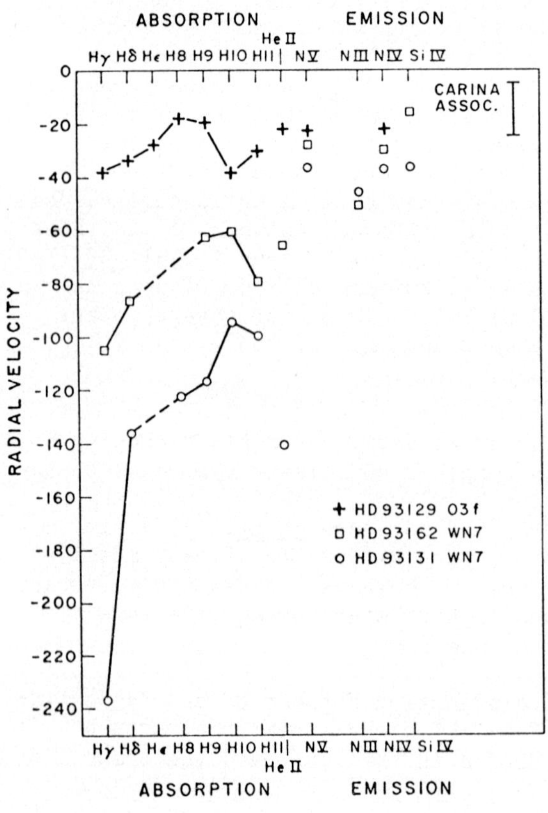

Fig. 1. Absorption and emission lines as indicated for two WN7 stars and HD 93129. There is no emission associated with the upper Balmer lines but in the two WN7 stars emission begins to appear at about Hε and lower in the series. By Hγ emission dominates the line profile. There is evidence of a strong Balmer progression (increasing outward velocity with decreasing quantum number) in the two WN stars. There is little or no progression in HD 93129. The emission and N V absorption velocities in all three stars are close to that of the Carina association. This figure gives a quantitative estimate of the beginnings of the outflow of the envelope in the three stars.

Under the usual assumption that velocity increases outwards, it must be that the narrow emission lines are formed in a level similar to, or lower than, that of the Balmer absorption lines. The N V absorption lines, of high excitation and ionization, are also formed at this low level. Balmer progressions have been found previously in Of stars (Hutchings 1968; Conti, Garmany and Hutchings 1977; Leep 1978), and tentatively by Niemela (1977) for two of these WN stars. A line-by-line comparison of the actual radial velocities shows at least some Of stars have outflow velocities intermediate between those in HD 93162 and HD 93131.

The broad emission lines, such as 4686 Å He II, 3478, 3483 Å N IV and 4603, 4618 Å N V, are shifted redwards of the other emission lines in this figure. This suggests they are formed in a relatively large extended region around each star; the net red shift is then a result of scattering in an expanding envelope -- the so-called Auer-van Blerkom effect (Mihalas 1974). This effect is well known in double-lined WR binaries where the gamma velocities of the broad emission from the WR star and the absorption lines from the O star frequently differ. An analogous effect is also seen in certain Of stars (Bohannan and Conti 1976; Massey and Conti 1977).

A further inference from Fig. 1 is that the presence of absorption lines in a WR star is no guarantee that the star is double. Statements of WR duplicity including such stars may well overestimate the actual fraction. Very preliminary results on emission line widths by Leep (1978) show that the lines in even these WN stars are somewhat wider than those strongest lined Of objects. The wind densities are probably different, even though the velocity gradients may be similar in some cases. It appears that the H/He ratio is different between Of and these WN stars. The normal ratio of 10 may not be true for all Of stars (e.g., Conti and Frost 1977) but probably no Of star has a value as low as 2 as found by Smith (1973) for several other late-type WN stars. An abundance analysis of these transition objects is currently under way.

Masses unfortunately cannot be derived for stars showing no velocity variations and even single lined binary systems such as HD 92740 can add little to our knowledge. From evolutionary tracks of Chiosi _et al_. (1977) and de Loore _et al_. (1977) it appears likely that transition stars will be relatively massive objects of about 40 solar masses. These values certainly overlap those of Of stars. The transition WN stars occupy the same region of the HR diagram as Of stars.

The remaining physical distinction between Of and these transition stars may be one of H/He ratio and interior structure. If the Of stars are burning hydrogen in the core, and the transition

and WR stars helium in the core, the differences might be understood.

This research has been supported by the National Science Foundation under grant AST76-20842 through the Univ. of Colorado.

REFERENCES

Bohannan, B. and Conti, P.S. (1976). Astrophys. J. 204, 797.
Chiosi, C., Nasi, E. and Sreenivasan, S.R. (1977). Astron. and Astrophys. in press.
Conti, P.S. (1976). Mem. Soc. Roy. des Sci. de Liege, 6e Serie, Tome IX, p. 193.
Conti, P.S. and Frost, S.A. (1977). Astrophys. J. 212, 728.
Conti, P.S., Garmany, C.D. and Hutchings, J.B. (1977). Astrophys. J. 215, 561.
Conti, P.S., Leep, E.M. and Lorre, J.J. (1977). Astrophys. J. 214, 759.
Conti, P.S. and Leep, E.M. (1974). Astrophys. J. 193, 113.
Conti, P.S., Niemela, V.S. and Walborn, N.R. (1978). in preparation.
Hutchings, J.B. (1968). Mon. Not. R. Astr. Soc. 141, 329.
Leep, E.M. (1978). Astrophys. J. in press.
de Loore, C., de Greve, J.P. and Lamers, H.J.G.L.M. (1977). Astron. and Astrophys. 61, 251.
Massey, P. and Conti, P.S. (1977). Astrophys. J. in press.
Mihalas, D. (1974). Astron. J. 79, 1111.
Niemela, V.S. (1973). Publ. Astron. Soc. Pacific 85, 220.
Niemela, V.S. (1976). In Colloques International du C.N.R.S., #250 (Nice Symposium) R. Cayrel and M. Steinberg, eds., CNRS, Paris, p. 467.
Paczynski, B. (1973). In Wolf-Rayet and High Temperature Stars, M.K.V. Bappu and J. Sahade, eds., Reidel, Dordrecht, p. 143.
Smith, L.F. (1973). In Wolf-Rayet and High Temperature Stars, M.K.V. Bappu and J. Sahade, eds., Reidel, Dordrecht, p. 15.
Snow, T.P., Jr. and Morton, D.C. (1976). Astrophys. J. Suppl. 32, 429.
Walborn, N.R. (1973). Astrophys. J. 179, 517.

DISCUSSION

PARTHASARATHY: Spectral lines of the secondary components in some of these systems are seen only at certain phases and also show variations in intensity with orbital phase. Radial velocities of the secondary component obtained from these lines may not represent the true orbital motion.

CONTI: I don't think this is now the case for any of these Of systems. There is a new orbit of Plaskett's star by Cowley and Hutchings (Astrophys. J. 1977) and I think their evidence about the secondary is reasonably compelling. Dr. Nancy Morrison of JILA has many high dispersion spectrograms of HD 47129 and 29CMa and is presently studying them. The secondaries are visible.

BIDELMAN: What is your explanation for the WC stars?

CONTI: It could be either of the following ways to bring the products of helium burning to the surface: (1) evolution from a WN star, sufficient mass loss occurs only after an O star has commenced helium burning; or (2) perhaps only OC stars can become WC stars directly: the carbon anomaly then would already exist.

COX: How is it that we know that the Wolf-Rayet stars are helium burning?

CONTI: There are two arguments, as outlined by Paczynski at IAU Symposium #49: (1) The WR stars generally do not show evidence of hydrogen in their spectra; helium dominates. Since little or no hydrogen remains, hydrogen burning is not possible. (2) The luminosity, compared to the mass, is consistent with that for pure helium stars burning helium in their cores.

PISMIS: Are there sufficient data to know whether there are differences in the kinematic characteristics of the spectral groups you discussed as well as the distribution in the Galaxy, I mean for the groups of normal O, Of and WR stars?

CONTI: All stars belong to the extreme Pop. I in terms of their kinematic properties. There is an absence of WR stars towards the galactic anti-center. probably a result of fewer O stars being present. There are, of course, some nuclei of planetary nebulae with spectra which are Of or WR type. These phenomena are therefore not restricted to Pop. I objects.

PARTHASARATHY: What are the reasons for the under-luminosity of the massive secondary components in systems like HD 47129, UWCMa and others?

CONTI: I don't really have a detailed answer for this. For close binaries such as these we will undoubtedly have to consider mass loss from the system, along with mass exchange, as affecting the stellar evolution. I would think a very non-equilibrium situation exists for the stellar structure configurations.

NUCLEOSYNTHESIS IN MASSIVE POPULATION II STARS

David S. P. Dearborn

Steward Observatory, U. Arizona

Virginia Trimble

Astronomy Program, U. Maryland

1. INTRODUCTION

According to the conventional wisdom, the first stars formed in our Galaxy, including the massive ones responsible for most heavy element synthesis, must have had metal abundances much lower than solar. They may also have had somewhat lower helium abundances, corresponding to an enrichment $\Delta Y/\Delta Z = 2.7 \pm 1.0$ over the history of the Galaxy, a larger ratio than is readily accounted for by standard (Arnett-type) supernovae (Hacyan et al. 1977) or by mass shed from lower mass stars (Gingold 1977). But early nucleosynthesis must have occurred in stars with Pop II compositions. Several authors (Ezer and Cameron 1969; Cary 1974; Chiosi and Nasi 1974; Trimble et al. 1973) have modelled and evolved such stars, but none of the studies followed the stars far enough from the main sequence to be able to say anything very quantitative about their contribution to galactic chemical evolution. We address here the question of the effects of initial composition upon helium and heavy element production in massive stars.

2. THE CALCULATIONS

The stellar evolution code developed by Eggleton (1973) has been modified (Dearborn and Eggleton 1977) to apply to massive stars, incorporating the standard physics (nuclear reaction rates from Fowler et al. 1975; neutrino loss rates from Beaudet et al. 1967; opacities from Cox and Stewart 1970). Convective mixing was treated as a diffusion problem. By adjusting the rate of diffusion

A. G. Davis Philip and D. S. Hayes (eds.), The HR Diagram, 375-378.

it was possible to examine the effects of semi convection. We evolved stars of 12 and 32 $M_\odot$ with three compositions (X=0.7, Z=0.02; X=0.7, Z=0.0004; X=0.78, Z=0.0004) representing Pop I and Pop II with high and low initial helium. Calculations were stopped early in carbon-burning. We can, therefore, say nothing about relative amounts of heavy elements, except C and O. The C/O ratio is quite sensitive to diffusion time scale, because every He nucleus that leaks into the CO core turns a previously formed C into an O. Thus the fact that we see lots of carbon in the Universe limits the amount by which semi convection can erode the helium supply prior to expulsion by a supernova event. Diffusion time scales probably cannot be much shorter than those used here (about 10^{-6} of the nuclear time scale). Thus we believe that the net change in masses of helium and heavy elements in our stars during their evolution should be realistic estimates of $\Delta Y/\Delta Z$ expelled by them as SN's.

The models and their net production of He and "metals" (mostly C and O in our models, but appreciable fractions of the core will burn to Ne, Mg, Si, Fe, etc. in late, very short-lived stages) are listed in Table I. The last column for each model is the expelled $\Delta Y/\Delta Z$ if a 1.4 $M_\odot$ remnant, made of heavy elements is left.

TABLE I

NET HELIUM AND HEAVY ELEMENT PRODUCTION BY STARS AS A FUNCTION OF MASS AND COMPOSITION. ALL MASS QUANTITIES ARE IN $M_\odot$.

M	X	Z	ΔY	ΔZ	ΔY/ΔZ
12	0.7	0.02	0.9	1.8	2.24
12	0.7	0.0004	0.8	2.2	1.08
12	0.78	0.0004	1.0	1.9	1.96
32	0.7	0.02	2.0	9.3	0.26
32	0.7	0.0004	1.5	10.0	0.18
32	0.78	0.0004	3.4	8.3	0.49

3. IMPLICATIONS FOR GALACTIC CHEMICAL EVOLUTION

We explore here the consequences of varying two quantities that enter into calculations of galactic chemical evolution. These are initial stellar composition, as discussed above, and mass loss early in the lives of massive stars, as discussed by Dearborn and Eggleton (1977) and Dearborn et al. (1976a, 1976b). Our models are exceedingly simple. We assume that stars from 6 to 100 $M_\odot$ are formed according to a Salpeter birthrate function with $\alpha = 2.35$. All stars more massive than 6 $M_\odot$ give rise to supernovae that blow off everything but a 1.4 $M_\odot$ core. Mass loss, if it is important at all, acts to reduce the effective stellar masses that do the nucleosynthesis so as to steepen α to 3.49 or 4.97 above 15 $M_\odot$. This reduces the amount of mass in 100 $M_\odot$ stars relative to that

in 15 $M_\odot$ stars by factors of 10 or 100. Finally, the Galaxy is assumed to be capable of making the heavy elements we see, so that the only interesting number for each model is $\Delta Y/\Delta Z$. Since stars below 6 $M_\odot$ are more likely to make He than heavy elements, our ratios are lower limits to what would actually be produced by a generation of stars with these compositions and mass losses. We give in Table II the values of $\Delta Y/\Delta Z$ for 9 models, have three initial compositions and three values of α for M greater than 15$M_\odot$. It is clear that initial composition can vary the production ratio by 50% either way and that significant mass loss can increase $\Delta Y/\Delta Z$ to values large enough to match any of the observations.

TABLE II

NET $\Delta Y/\Delta Z$ FOR A GENERATION OF STARS AS A FUNCTION OF COMPOSITION AND IMPORTANCE OF MASS LOSS.

X/Z	no mass loss	moderate	extreme
0.7/0.02	0.44	2.87	3.34
0.7/0.0004	0.32	2.02	2.30
0.78/0.0004	0.60	2.82	3.27

ACKNOWLEDGEMENTS
We are indebted to Dr. Peter Eggleton for use of his stellar evolution program. D.S.P.D. is a Visiting Astronomer at Kitt Peak National Observatory and is grateful to them for hospitality.

REFERENCES
Beaudet, G., Petrosian, V. and Salpeter, E. (1967). Astrophys. J. **150**, 979.
Cary, N. (1974). Astrophys. Space Sci. **31**, 3.
Chiosi, C. and Nasi, E. (1974). Astron. Astrophys. **35**, 81.
Cox, A.N. and Stewart, J. (1970) Astrophys. J. Suppl. **11**,1.
Dearborn, D. and Eggleton, P.P. (1977). Astrophys. J. **312**, 177&448.
Dearborn, D., Eggleton, P. and Schramm, D. (1976a). Astrophys. J. **203**, 455.
Dearborn, D., Kozlowski, M. and Schramm, D. (1976b). Nature **261**,210
Eggleton, P.P. (1973). Monthly Notices Roy. Astron. Soc. **163**, 279.
Ezer, D. and Cameron, A. (1971). Astrophys. Space Sci. **14**, 399.
Fowler, W.A., Caughlan, G. and Zimmerman, B. (1975). Ann. Rev. Astron. Astrophys. **13**, 69.
Gingold, R. (1977). Monthly Notices Roy. Astron. Soc. **178**, 569.
Hacyan, S., Dultzin-Hacyan, D., Torres-Peimbert, S. and Peimbert, M. (1976). Rev. Mex. Astron. Astrof. **1**, 355.
Trimble, V., Paczyński, B. and Zimmerman, B. (1973). Astron. Astrophys. **25**, 35.

DISCUSSION

TINSLEY: How much does your mass loss change the mass function of stars between formation and death? Is it enough to affect the estimates Arnett makes of the relative amounts of different "Z" elements shed per generation of stars?

TRIMBLE: Our "moderate" mass loss essentially eliminates all stars more massive than 30 $M_{\odot}$, in accordance with the observed absence of very massive M supergiants. This would affect Arnett's nucleosynthesis, but not disasterously, in the sense that he can make all the elements we need using masses up to about 22 $M_{\odot}$. Our "extreme" mass loss eliminates most stars above 15 or 20 $M_{\odot}$ and is probably not in the realm of practical politics for making the heavy elements.

CAYREL de STROBEL: Your results for $\Delta Y/\Delta Z$ are very nice, because they resemble closely our observational results about a simultaneous variation of the helium content with the metal content (see papers in Part VII) for older disk stars. The fact that you get these results theoretically and we get them observationally gives us confidence that they are real.

CHIOSI: I would like to know more about how mass loss is included in your computations. I wonder whether your evolutionary models for this range of mass, with the mass loss rates you seem to adopt, are able to reproduce the observed features of the HR diagram of supergiant stars.

I have the feeling that the results by Eggleton and Dearborn (1977) are obtained by using mass-loss rates too high for initial masses of $\sim$ 30 $M_{\odot}$ on the main sequence, in that they are not supported by the observational information on mass flow in OB stars.

TRIMBLE: This problem has enough free parameters in initial mass function, initial composition, choice of mixing length, and treatment of semi convection, as well as rate and duration of mass loss, that it should be possible to fit an elephant, let alone any observed HR diagram one wants.

We, in fact, put all the mass loss during hydrogen burning, so that the denuded star merely mocks one of lower initial mass in its nucleosynthetic properties. The total amount of mass lost in the "extreme" case is consistent with the work of Eggleton and Dearborn (1977) and may well be too high. Our "intermediate" case is perhaps more realistic.

THE COLORS OF G AND K GIANT STARS

R. A. Bell

Astronomy Program, University of Maryland

B. Gustafsson

Astronomical Observatory, Uppsala University

The material on which this talk is based is contained in a paper submitted to Astronomy and Astrophysics. The summary of this paper is given below.

SUMMARY

In previous papers, we have discussed the calculation of model atmospheres for metal-deficient and solar-abundance giant stars and have presented a grid of such models. We have now carried out synthetic spectrum calculations at a resolution of 0.1 Å between 3000 and 7200 Å for most of the models of this grid. We have also computed synthetic spectra at wavelengths up to 12000 Å for some of the grid models. The calculations allow for the presence of thousands of atomic and molecular lines. Examples of these spectra are shown in this paper.

We show a comparison of observed and computed spectra for short wavelength intervals for the Sun, α Boo, γ Tau and the G8 III-IV stars ϕ^2 Ori (HD Nos. 124897, 27371 and 37160). Low dispersion synthetic spectra are used to illustrate spectral classification criteria. We have compared observed and computed spectral scans for a number of K giants with different metal abundances.

We have convolved the synthetic spectra with the sensitivity frunctions of various photometric systems in order to obtain theoretical colors as well as bolometric corrections. The zero

A. G. Davis Philip and D. S. Hayes (eds.), The HR Diagram, 379–381.

points of the colors have primarily been found by adopting certain fundamental parameters for the star ϕ^2 Ori. Some colors of Sirius, Vega, Procyon and the solar-type star HR 483 (HD Nos. 48915, 172167, 61421 and 10307) have also been used, in combination with their corresponding models, to check the zero points. The color grid is given in an accompanying paper submitted to Astronomy and Astrophysics.

We have studied the influence of terrestrial atmospheric and interstellar extinction on some of the colors. The influence of varying the abundances of particular elements, such as nitrogen, has also been studied, as have the effects of changes in the Doppler broadening velocity and in the treatment of damping.

We have compared the grid of colors with the observed colors of a sample of giant stars with relatively well-known fundamental parameters. The overall agreement was found to be good, although certain systematic discrepancies occur, especially for the ultraviolet colors and the CN indices. These discrepancies are interpreted in terms of incomplete blocking. In the first case this is probably the consequence of not including weak spectral lines in the computations or possibly of assuming LTE, and the second is probably the result of nitrogen enrichment when CNO processed material is mixed to the surface during the subgiant phase.

We have also used the theoretical colors for analysis of the properties and possibilities of the different photometric systems. Using the locations of the models and corresponding stars in certain two-color diagrams as well as a more general approach, where the effects of observational errors are transformed to effects on the fundamental parameters deduced for the stars, we find quantitative measures of the power of various color systems. This study also leads to the disclosure of certain new properties of some systems, which should be exploited in future investigations. The computed colors will be applied to the analysis of observations of globular cluster stars in forthcoming papers.

This work was supported by the U. S. National Science Foundation and the Swedish National Science Research Council.

DISCUSSION

KODAIRA: You have mentioned a certain discrepancy between your model prediction for α Boo and the observations in the near IR region. Is α Boo a special case or have you the same problem for other red giants? If this is the case, what reasons do you suspect?

BELL: I don't think we have any troubles in the IR but we are examining this region. Our problems occur in the UV and we have

examined various possible causes.

KODAIRA: How large is the atmospheric extent in your plane-parallel atmosphere with the lowest surface gravity log g = 0.75?

BELL: The temperature errors which are caused by our neglect of sphericity effects are thought to be smaller than 100K. Details are given in Gustafsson, et al. (Astr. Astrophys. **42**, 407).

BUSCOMBE: Have you tested your models with very luminous G supergiants? Can you discern the effects of the stellar wind which Cox believes should deplete atmospheric hydrogen?

BELL: We have not yet computed models for G supergiants, although we plan to do so. We will have to compute opacity distribution functions for a higher Doppler broadening velocity in order to do this. I would think that the effects suggested by Cox, at least for the most extreme depletion of hydrogen, would cause the spectra to appear super-metal-rich. The color-temperature relation would be substantially altered. I'd be surprised if this is consistent with the observations.

COX: The calculation of high luminosity models with high helium is exactly what is needed to check whether a large helium enrichment really exists. We must remember that helium is neutral and there still is plenty of hydrogen. Do you plan to do such models?

BELL: Possibly.

PEL: In our comparison of the Walraven photometry with the computed fluxes of Kurucz we find discrepanices in the UV similar to the ones you mentioned, but surprisingly enough the agreement improves when going from main-sequence stars to supergiants. The latter fit quite well even at 3250 Å. Do you find this also in your computations?

BELL: I don't think so. However our low gravity models are cooler than the Kurucz models which you refer to.

PEL: But at the lower temperatures of your models you do not see that the excess UV flux decreases towards lower gravity?

BELL: I don't think so. However, the UV flux is quite sensitive to the adopted Doppler broadening velocity. Lines of OH, as well as atomic lines, might affect the flux at the wavelengths which you mention.

STENCEL: Could you briefly compare your models with the grid of cool model atmospheres published by H.R. Johnson?

BELL: Johnson has shown that the opacity sampling method does yield a model similar to the corresponding model in our grid. I have not done detailed comparison with his models. The model used for α Tau by Lambert and Luck, which I think was computed by Dragon of Indiana, does have lower pressures near the surface than does our corresponding model. This does have some effect on the emergent spectrum, for example in the TiO bands.

MICROTURBULENCE AS A THIRD DIMENSION IN THE G-K GIANT REGION OF THE HR DIAGRAM

R. Foy

Paris-Meudon Observatory

The structure of the HR diagram in the G-K giant region is complex, because of the funneling effect and of the intersection of evolutionary tracks for different stellar masses. Therefore it is impossible to derive the age of a cool giant from its location in the HR diagram alone. To remove this indeterminacy a third dimension is needed, and we suggest microturbulence as the appropriate observational parameter.

Three points have to be discussed before supporting this idea:

1) microturbulence is not a spurious effect due to departures from LTE,
2) microturbulence is related to physical parameters of stellar atmospheres, and,
3) microturbulence is a function of stellar evolution.

1. MICROTURBULENCE AND NON-LTE EFFECTS

If microturbulence is a spurious parameter resulting from non-LTE effects in stellar detailed analyses, it is expected to increase with decreasing atmospheric gaseous pressure or with decreasing gravity. Some star to star comparisons do not confirm this relation. For instance ε Vir (HD113226) and HD 192947 have the same effective temperature T_{eff}, gravity g and metal content. Their saturated spectral lines have to be equally sensitive to non-LTE effects. On the contrary, ε Vir and HD 192947 have markedly different microturbulent velocities: 1.8 and 0.8 km/s respectively. Hence, for such stars as ε Vir, departures from LTE cannot dominate in the interpretation of the line saturation level.

A. G. Davis Philip and D. S. Hayes (eds.), The HR Diagram, 383-386. All Rights Reserved.

2. MICROTURBULENCE VERSUS ATMOSPHERIC PHYSICAL PARAMETERS

We have searched for relations between ξ and T_{eff}, g and metal content. For this purpose we have reanalyzed a set of 98 G-K giants and supergiants for which reliable equivalent width data are available (22 from Helfer and Wallerstein (1964), 19 from Luck (private communication), 12 from Oinas (private communication), 11 from Pilachowski (private communication), and the 34 remaining from our group in Meudon: Cayrel, G. and R., Foy, Kovacs, Pasinetti, and Spite, F. and M.

Here all stars have been analyzed in the same way, so that if systematic errors occur, they do not affect our actual results, since we are interested in relative effects only. The effective temperatures have been deduced from the (R-I) versus T_{eff} relation by Johnson (1966), and the gravities from the condition of ionization equilibrium applied to iron for the assumed values of T_{eff}, i.e.: weak lines of both Fe I and Fe II must lead to the same iron abundance. The abscissae of the curves of growth have been computed using model atmospheres rescaled from those of Gustafsson et al. (1975). Iron abundances and microturbulent velocities have been derived from the Fe I curves-of-growth. We have found reliable results for 84 stars out of 98.

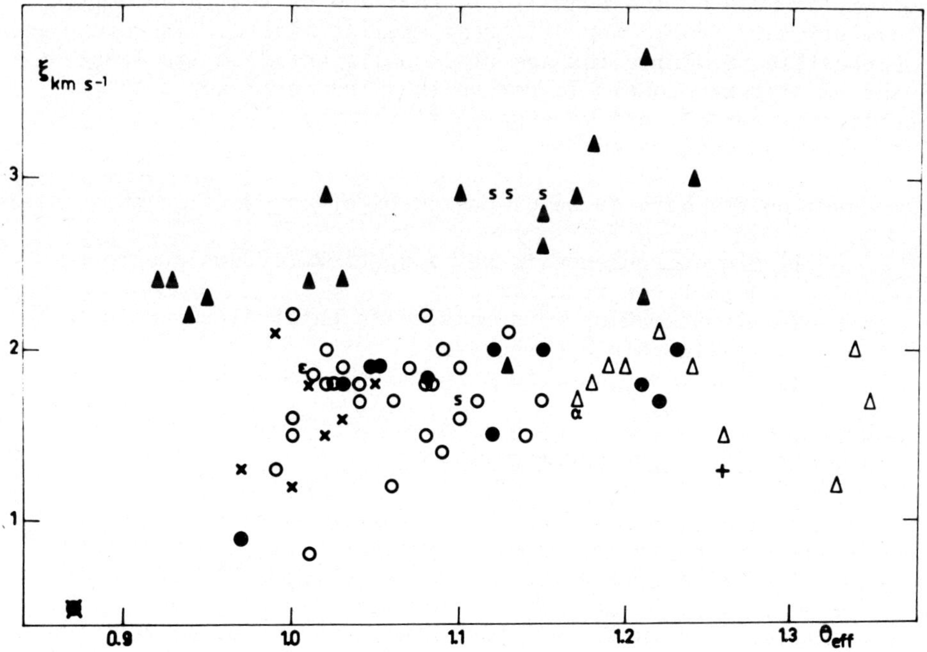

Fig. 1. Microturbulence ξ as a function of Θ_{eff}. Symbols; supergiants = ▲. giants = log g > 3.0 = X; 2.5 < log g < 3.0 =0; 2.0 ≤ log g < 2.5 = ●; 1.0 ≤ log g < 2.0 = Δ; log g < 1.0 = +; very metal-poor stars = S.

The interpretation of the results has demonstrated the following points:

1) ξ does not correlate with Fe/H;

2) ξ correlates with $\theta_{eff} = 5040/T_{eff}$ and g (Fig. 1). For supergiants there is a trend of increasing ξ with increasing θ_{eff}. A similar trend occurs for giants, except in the coolest part of the diagram. A main feature of Fig. 1 is also the dependence with g. In spite of the spread of the relation, one sees that ξ is a function of T_{eff} and g. Thus we confirm the preliminary results given in Foy (1976, 1977) for a smaller sample of stars.

This θ_{eff}/ξ relation does not fit the relation: $\xi \approx T^{1.1}g^{-0.2}$ predicted by Reimers (1973, 1975) from a compilation of ξ measurements in literature, and from a relation between ξ and M_V derived from the H and K emission lines.

3. VARIATION OF ξ WITH EVOLUTION

Note that during the evolution of a star near and along the first asymptotic branch both T_{eff} and g decrease. It is clear on a log g/θ_{eff} diagram that the observed log g/θ_{eff} relation of Fig. 1 fits the theoretical evolutionary tracks in the vicinity of the Hayashi boundary. Furthermore it appears that ξ is lower at the base than at the tip of the asymptotic branch. Thus microturbulence is related to stellar evolution.

The low-ξ giant HD 192947 markedly departs from the mean ξ/log g relation. It appears that it is evolving not in the hydrogen burning shell, as most of the analyzed stars, but in the helium burning core phase. The same conclusion occurs for the few low-ξ stars for which an evolutionary state can be determined, so that ξ could the the third parameter needed for this part of the HR diagram.

An improvement of this study could be an analysis of a set of stars as numerous as the present one, but based on spectra obtained with a same spectrograph, in order to minimize the scatter of the correlations.

REFERENCES

Foy, R. (1976). In Physique des Mouvements dans les atmospheres stellaires, R. Cayrel, ed., Nice, p. 433.

Foy, R. (1977). In Highlights of Astronomy, E. Muller, ed., 4, 115.

Gustafsson, B., Bell, R.A., Erikson, K and Nordlund, A. (1975). Astron. Astrophys. **42**, 407.

Helfer, H.L. and Wallerstein, G. (1964). Astrophys. J. Suppl. **9**, 81.

Johnson, H.L. (1966). In Ann. Rev. Astron. Astrophys. Vol. **4**, Annual Reviews, Palo Alto, p. 193.

Reimers, D. (1973). Astron. Astrophys. 24, 79.
Reimers, D. (1975). In Physique des Mouvements dans les atmospheres stellaires, R. Cayrel, ed., Nice, p. 421.

DISCUSSION

BELL: Firstly, how have you checked that your material is homogeneous? Different observers draw continua differently.

Secondly, have you compared your results with those of Gustafsson, Kjaergaard and Andersen (Astr. Astrophys. 34, 99)? Their very narrow band photometry using two bands, one containing lines sensitive to abundance and the other containing lines sensitive to microturbulence, allows them to obtain very precise measures of both quantities.

FOY: I have rescaled the equivalent width sets, using stars studied in common, generally ε Vir.

Too few stars are in both the list by Gustafsson et al. and my list to allow this comparison. I would mention that the gravity and microturbulent velocity ranges in the sample by Gustafsson et al. is much smaller than the present one. This would make such a comparison more difficult.

STENCEL: I am pleased to see your correlation between microturbulence and gravity for G and K stars since it agrees with my own work on this from the Ca II lines. However, we should be perhaps more cautious in deducing log g from the ionization equilibrium of iron in K stars due to departures from LTE recently demonstrated by Ramsey (1977 Astrophys. J.).

Do you have any results on the variation of macroturbulence with temperature and gravity?

FOY: I think it is possible that departures from LTE cause decreasing ξ in the coolest part of the ξ versus θ_{eff} diagram for giants. But I do not think that NLTE effects give a significant bias for most of my sample. Indeed Oinas (1974) had mentioned problems with the ionization equilibrium for K giants. But Perrin, et al. (1975) have removed these difficulties and Oinas (1977) now thinks that problems occur only after spectral type K3III.

I am just now beginning to study macroturbulence.

CAYREL de STROBEL: I have a comment on Oinas' unknown absorber: It is difficult to imagine that the departure from ionization equilibrium is due to an unknown absorber in late K stars. For us, it is easier to attribute this departure to effects of NLTE.

SPHERICITY EFFECTS OF EXTENDED ATMOSPHERES OF LATE-TYPE GIANTS IN THE HR DIAGRAM

Keiichi Kodaira

University of Tokyo

1. THEORETICAL AND EMPIRICAL HR DIAGRAMS

In the late phases of stellar evolution, evolutionary tracks of stars with different masses come together along the Hayashi line in the HR diagram. The theoretical HR diagram (log L, log T_{eff}) is accordingly partially degenerate in the domain of late-type giants and supergiants, with respect to the third parameter, the stellar mass M. The stellar radius, R, being determined by log L and log T_{eff}, the mass determines the surface gravity log g at the radius R. These parameters enable us to transform a point in the theoretical HR diagram to the corresponding point in the empirical HR diagram M_V, (R-I) or spectral type. This transformation is conventionally carried out within the framework of the plane-parallel approximation in stellar atmospheres, and the parameters for the abscissa of the empirical HR diagram are dependant upon T_{eff} and log g alone, irrespective of the mass itself. In this case, the parameter M indirectly affects the observable quantities through log g, but the effects of a variation by Δlog g= $\pm$0.5, corresponding to Δlog M=$\pm$0.5, are almost insignificant (cf. Tsuji 1976). The transformation between the theoretical and the empirical HR diagram is, therefore, almost one-to-one, within the framework of the plane-parallel approximation. Late-type giants and supergiants, however, have moderately extended atmospheres in general (cf. Schmid-Burgk and Scholz 1975), and their photometric colors and spectra are expected to be influenced by the sphericity of the atmospheric structure. Consequently, in comparing empirical HR diagrams with theoretical ones, it is important to know how atmospheric sphericity affects the transformation in the degenerate domains of the theoretical diagram. The sphericity can be represented, for instance, by the atmospheric extent η relative to the

A. G. Davis Philip and D. S. Hayes (eds.), The HR Diagram, 387–390. All Rights Reserved.

stellar radius;

$$\eta = \frac{(\text{geometrical thickness between } \tau=10^{-3} \text{ and } 1)}{(\text{stellar radius defined at } \tau=1)} ,$$

where τ is the radial optical depth for the Rossland mean opacity.

2. SPHERICITY EFFECTS IN THE HR DIAGRAM

Spherical model atmospheres (static, non-grey, LTE, solar composition) were constructed for late-type giants and supergiants, by a revised method analogous to that proposed by Hundt, Kodaira, Schmid-Burgk and Scholz (1975). The line blocking and blanketing effects of the CO, CN, OH, H_2O and TiO molecules were taken into account by an approximation which is realistic enough to yield the magnitude of the sphericity effects; details of the model calculations will be published elsewhere (Watanabe and Kodaira 1977). Model parameters were chosen roughly along theoretical evolutionary tracks computed or suggested by different authors [Refsdal and Weigert (1970), Robertson (1972), Paczynski (1970, 1971); for massive stars, see the references in Stothers (1972)], and are given in Table I. The relative extents, η, in the table were determined from the spherical model atmospheres. The models in the A series have the same values of T_{eff} and log g. Consequently, they would have the same colors and spectrum if the plane-parallel atmospheric model are applied. The models in the series B, C, D, or E, have common values of L and T_{eff}, but differing values of η corresponding to M. By examining the spherical model atmospheres in the A series, we found that the sphericity effect primarily appears as a dilution of the radiation field, leading to a general cooling of the outer layers. This cooling, however,

TABLE I. PARAMETERS FOR SPHERICAL MODEL ATMOSPHERES

Model		log $L/L_\odot$	log $R/R_\odot$	$m/m_\odot$	log g	log T_{eff}	η(%)	M_V	(R-I)	(J-K)
	1	5.207	3.12	20			5.0	$-4^{m}.45$	$2^{m}.16$	$1^{m}.05$
A	2	4.384	2.71	3	-0.50	3.505	12.0	-1.98	2.24	1.04
	3	3.906	2.47	1			23.0	-0.93	2.27	1.04
	1			15	0.05		3.0	-7.28	0.91	0.87
B	2	4.822	2.78	4.2	-0.50	3.580	12.0	-6.97	1.00	0.85
	3			1.5	-0.95		54.0	-5.19	1.34	0.89
	1			3	1.24		1.3	-0.50	1.55	0.96
C	2	2.800	1.84	1	0.76	3.544	4.6	-0.17	1.62	0.96
	3			0.3	0.24		17.4	+0.60	1.77	0.97
	1			10	0.71		1.6	-1.35	2.02	1.00
D	2	3.700	2.36	3	0.18	3.505	5.5	-0.87	2.13	1.03
	3			1	-0.29		16.6	-0.34	2.22	1.01
	1			15	-0.25		2.6	-3.16	2.21	0.93
E	2	4.600	2.93	7	-0.58	3.447	6.1	-2.93	2.26	1.00
	3			3	-0.95		17.1	-2.39	2.38	0.99

enhances the formation of the temperature-sensitive molecules such as TiO and H_2O, and the blanketing effects of these molecules further modify the atmospheric structure in a complicated manner. Among the models in series B through E, a decrease in the stellar mass leads to a lowering of the surface gravity but, contrary to the plane-parallel cases, enhances the formation of TiO and H_2O, owing to the temperature decrease caused by the increased dilution effect in the extended atmospheres.

The distribution of emergent energy flux is slightly shifted towards longer wavelength, reflecting the general cooling of the surface layers. The most striking effect is the enhancement of the molecular absorption bands of TiO: Strengths of TiO bands increase with the increasing atmospheric extension, and strongly decrease the emerging flux in the photometric passband V. The effect of the enhanced formation of H_2O is partially cancelled by the general shift of the continuous energy distribution, and the colors in the near-infrared region are less affected by the sphericity effects than the colors in the red to blue region. The resulting effects in the absolute visual magnitude M_V and colors (R-I) and (J-K) are given in the last three columns of Table I. Since no special effort was made to bring the photometric colors precisely into the Johnson system, the absolute values are subject to an uncertainty of about $\pm 0\overset{m}{.}2$, but the relative values, which are important in estimating the sphericity effects, are estimated to have an accuracy of about $\pm 0\overset{m}{.}02$. The values of M_V also should be interpreted differentially. According to the present result, when L and T_{eff} are fixed, a decrease of stellar mass by $\Delta \log M \simeq 1$ causes lowering of the absolute visual magnitude by $\Delta M_V \simeq 1\overset{m}{.}$ and reddening of $\Delta(R-I) \simeq 0\overset{m}{.}2$ which approximately corresponds to the difference for one spectral sub-class in the MK classification (cf. Lee 1970).

REFERENCES

Hundt, E., Kodaira, K., Schmid-Burgk, J. and Scholz, M. (1975). Astron. Astrophys. **41**, 37.
Lee, Th. A. (1970). Astrophys. J. **162**, 217.
Paczynski, B. (1970). Acta Astron. **20**, 47.
Paczynski, B. (1971). Acta Astron. **21**, 4.
Refsdal, S. and Weigert, A. (1970). Astron. Astrophys. **6**, 426.
Robertson, J.W. (1972). Astrophys. J. **173**, 631.
Schmid-Burgk, J. and Scholz, M. (1975). Astron. Astrophys. **41**, 41.
Stothers, R. (1972). Astrophys. J. **175**, 431.
Tsuji, T. (1976). Publ. Astron. Soc. Japan, **28**, 543.
Watanabe, T. and Kodaira, K. (1977). submitted to Publ. Astron. Soc. Japan.

DISCUSSION

FOY: I have two comments. Firstly, I think that the sphericity could be related to the microturbulence variations which I have described this morning if microturbulence is induced by overshooting. Indeed, the sphericity obviously depends on the stellar radius, and the overshooting scale height is a function of the thickness of the convective zone and consequently of the stellar radius, too.

Secondly, I wish to mention a new work by Labeyrie and his group and myself. New speckle interferometer observations at Palomar in narrow spectral bands show abrupt variations in the diameter of Mira and R Leonis as a function of wavelength. The diameter is at least twice as large in strong TiO bands as outside them. From Tsuji's models we have interpreted these variations to be due to the large opacity of TiO. In the strong TiO features the optically thick region of the atmosphere extends several astronomical units outwards from the continuum formation layer (up to $\tau_{cont} \simeq 10^{-5}$). Since the stellar radius in the continuum is smaller than one astronomical unit, the plane parallel approximation is surely not valid in the intense TiO bands. This supports your arguments.

KODAIRA: I agree with you on the first point. As for the second point, I am afraid that you may have to invoke dynamical extended atmospheres in order to interpret the large radii observed. I hope that you can obtain a qualitative insight into sphericity effects from our model calculations.

SUMMARY

Louis C. Green

Haverford College

The Symposium opened with a morning devoted to the accomplishments and character of Henry Norris Russell, as one of the two originators of the Hertzsprung-Russell diagram, on the hundredth anniversary of his birth. The papers presented at this session, together with some shorter remarks made at other times during the Symposium, are to appear in a separate volume as Dudley Observatory Report. No. 13. Suffice it to say here that the picture which emerged was that of a fine man of enormous energy, prodigious memory, committed to his scientific work but with many interests, always polite and pleasant to those around him but somewhat remote.

1. FUNDAMENTAL ASEPCTS OF THE HR DIAGRAM

At the first scientific session of the Symposium, W.W. Morgan described his extension of the original MK system. Between O9 and G0 there are 100 to 150 cells in a two-dimensional array of luminosity class against spectral type, each cell being defined by the lists of standard stars given by Morgan and his co-workers from time to time. A classification defined in this way has the great advantage of being free of theoretical constructs. It may also be extended by the insertion in the existing cells of one or more vectors to represent other properties of normal and peculiar spectra --both those we already know and those which may be discovered. The vectors are themselves to be defined by standard stars, but are regarded as secondary aspects or perturbations of the spectrum-luminosity class array. Three questions illustrate problems of current interest: 1.) Does the HR diagram have any strong line, normal--that is, Hyades-like--stars in the cell G0 III? 2.) The

A. G. Davis Philip and D. S. Hayes (eds.), The HR Diagram, 391-403. All Rights Reserved.

cell A0 V contains A0 Va and A0 Vb stars with narrow and broad H lines respectively; the latter occur in clusters. Are there any A0 Vb stars not in clusters? 3.) The cells B7 III to B9 III contain giants with relatively narrow, stellar K lines and broad HeI lines with rotational velocities less than 100 km/sec. Are there any stars in these cells for which the K and HeI lines are different?

Since the present work is based on the spectral range 3500-4700 Å, Morgan emphasized the desirability of constructing classification systems independently of the MK system but based on similar principles in the far UV, green-red, and infrared, as well as for weak-lined stars in the photographic region. When these systems are sufficiently complete, a comparison of the four systems with each other and with the MK system could be most instructive.

N.R. Walborn described an empirical luminosity classification for early O stars to serve as an extension of the MK system, which does not include types earlier than O9. This extension is based on the suggestion that the negative luminosity effect of the weakening of the absorption lines of HeII 4686 Å and NIII 4634-4640-4642 Å in the O9 to B0 stars is in fact a filling in by emission, which with greater luminosity leads to these same lines appearing in emission in the Of stars. Application of this idea to the earlier O's leads to luminosity classes in agreement with those found from other, less sensitive, spectroscopic criteria. The Of stars no longer appear peculiar but simply as early supergiants. From known distances, $M_V = -7\overset{m}{.}0$ for the Of's versus $-5\overset{m}{.}5$ for the hottest main sequence O's.

With normal O stars defined as above, several groups show peculiarities related to the presence of extended envelopes. The Onfp stars have broad absorption lines and HeII 4686 Å as a centrally reversed emission line. The Of?p stars have strong CIII 4647-4650-4651 Å in emission. The emission features are sharp relative to the absorption lines, suggesting a shell source. The OIafpe stars are very luminous supergiants with a P Cygni spectrum. The WN stars are characterized by broad emission bands. Other early type stars show chemical anomalies, which in some cases appear to be related to binary character. Possibly all OBN stars and no OBC stars are members of short-period binaries. The OBN's may result from nitrogen enrichment and the OBC's from nitrogen deficiency but the relative importance of mass transfer, mixing, and mass loss in these phenomena is unclear. There is also the group of helium rich stars near spectrum-luminosity class B2V which show enhanced and, in many cases, periodically variable HeI. Finally, Walborn described the deficiency in the strength of the metal lines in the OB stars of the SMC.

P.C. Keenan presented an HR diagram for G, K and M stars designed to bring out the fine structure in the luminosity-spectrum plane by

plotting only those stars with the best established positions. The luminosity of the GB is known to 0.2 mag. and the dispersion is 0.4 mag. The ridge of the GB is defined by luminosity class IIIab, stars on the upper side are called IIIa and those on the lower IIIb. The three classes yield separate bands in his HR diagram. There is a scarcity of the most luminous K supergiants, which is not an artifact of the classification scheme; from G7 to K1, class II bright giants are extremely rare.

P.C. Boeshaar reported that the coolest halo subdwarfs are not subluminous and that very cool M subdwarfs can be conveniently identified by the enhanced CaOH 5530 Å band and NaI D-lines.

One might be excused for approaching the next three papers of this session with some skepticism but the quality of the results in each case were impressive. I. Appenzeller and H. Zekl described a completely automated procedure designed to yield spectroscopic parallaxes in which spectra at 115 Å/mm are scanned, approximate spectral type derived, and line widths, depths, and ratios are compared to those in standard stars. Preliminary results yield parallaxes as accurate as conventional visual classification. Th. Schmidt-Kaler reported on a completely computerized process for deriving two-dimensional classifications from objective prism spectra. Since the resolution of such spectra depends on the seeing, he employs equivalent widths, which are little affected. Plate calibration is obtained from the numerous stellar spectra themselves. Classification of B0 to F2 stars at 570 Å/mm to the accuracy of the MK system can be obtained in 30 seconds. Finally M. Parthasarathy showed slides which demonstrated convincingly that spectra taken at the ultra-low dispersion of 10,000 Å/mm can be used effectively for reconnaisance, for example on objects such as the LMC. The resulting microspectra are about 250 microns in length along the spectrum, and spectral types can be assigned to an accuracy within half a spectral class.

V.M. Blanco, A.A. Hoag and M.F. McCarthy have found that the ratio of the number of carbon to late-type M stars per unit area is a useful indicator of stellar populations at various points both in the nucleus of the Galaxy, where it is low, and the central regions of the LMC and SMC, where it is high.

From BVRI photoelectric photometry and Van Vleck Observatory parallaxes, A.R. Upgren has constructed enriched main sequences for K and M spectral types for the Hyades and field stars in the solar neighborhood. The dispersion of the latter group about the main sequence is about ± 0.4 mags. after removal of the dispersion from parallax errors.

A. Heck reported on absolute magnitudes for various ranges of spectral types as determined by statistical parallaxes. S.V.M. Clube and J.A. Dawe used statistical parallax procedures to invest-

igate the motions and luminosities of RR Lyrae variables in the solar neighborhood as a function of the period and Preston's ΔS = 10 (spectral type from H - spectral type from Ca). They found a low and a high velocity group and identified the latter with the RR Lyrae's in the Galactic center and halo and in the SMC and LMC. This led to corrections to the distances to these objects. R.E. Stencel pointed out that weak emission lines of iron-peak elements in the wings of H and K in late giants and supergiants can be used to determine M_V as well as the run of the temperature and non-thermal line broadening as a function of atmospheric height. The latter information can be used to study mass loss.

W.J. Luyten presented a set of informative plots of reduced proper motion, H, against color. The former, first used by Hertzsprung for stars with unknown parallaxes, is found from an expression identical with that for M using m and the parallax, p, except that to find H, p is replaced by μ. On average, H therefore tends to increase when M does. Plots based on proper motion surveys to some limit are more representative of the stars in a given volume than those based on some limiting magnitude. The four plots presented were for 5000 degenerate dwarfs, 5000 doubles with common proper motion, stars of very low luminosity and more than 10^5 stars of all types together.

D.S. Hayes discussed very clearly the transformation, freed as far as possible from theoretical constructs, from the observed magnitudes and spectral types or color indices to M_{bol} and log T_{eff} by way of absolute energy distributions and angular diameters.

2. HR DIAGRAMS: SOLAR NEIGHBORHOOD

W. Gliese gave some interesting statistics on our knowledge of stars nearer than 22 parsecs. 1500 stars in this volume have trigonometric parallaxes and of these about 1000 have M_V known to 0.5 mags. Only 600 have MK spectral types, but (B-V) photometry has been done on 1300. There are 53 degenerate dwarfs in the (B-V, M_V) diagram. The percentage of dMe stars increases in the later spectral types, many are flare stars. D.C. Barry reported that although the solar spectrum defines type G2 V, its color and line strengths are close to G3 and G4 stars. N. Houk and R. Fesen showed an interesting HR diagram based on the quality 1 and 2 normal stars of Volume 1 of the Michigan Catalogue of Two-Dimensional Spectral Types for the Henry Draper Stars. The use of appropriate symbols made immediately clear the approximate number of stars assigned to a given spectral type and luminosity type.

3. HR DIAGRAM, SUBLUMINOUS STARS

A number of interesting but somewhat separate points were made

in this session. J. Greenstein reporting on his large, significant program had found the halo G and K dwarfs to be as much as 1.0 mag. brighter than those of the old disk population, but with no difference between the M's. Composition differences in the M's affect their spectra little, but Pop. II sdM stars are recognizable by the strength of MgH and CaH compared to TiO. Halo stars fainter than visual absolute magnitude $15^{m}_{\cdot}5$ and old disk stars fainter than $18^{m}_{\cdot}5$ are uncommon. Dwarf Me's were found above but in the scatter of the main sequence.

Large composition variations are found among the degenerates. Very few red degenerate dwarfs have been found, very possibly as the result of envelope convection which leads to rapid cooling and short lifetimes. The lowest temperatures found are 4500 to 5000 K. Bolometric luminosities run from 10^{-4} $L_\odot$ to 10 $L_\odot$. P.R. Wesselius reported that hot subdwarf O's can be 5 mags. brighter at 1550 Å than in the V band.

H.L. Shipman pointed out that for degenerates, colors derived from different model atmosphere calculations are in relatively good agreement. These fluxes together with known parallaxes yield average radii of 0.013 $R_\odot$ for 95 hydrogen-rich, and 0.011 $R_\odot$ for 13 helium-rich, white dwarfs. These values are not significantly different and, when corrected for estimated observational selection, yield 0.011 $R_\odot$. This radius corresponds to a mass of 0.7 $M_\odot$. V. Weidemann emphasized the constant average radius, $\log R/R_\odot = -1.90 \pm 0.10$, from hot to cool DA's. The corresponding range of masses is (0.55 ± 0.2) $M_\odot$. Using surface gravities, the mass range is 0.40 to 0.65 $M_\odot$. Mass loss in the red giant stage is therefore substantial. There was disagreement in regard to the existence of more than one cooling sequence. S. Tapia found evidence for two sequences among the DA's but Weidemann did not. On the basis of an astrometry and photometry program at the Naval Observatory, C.C. Dahn and R.S. Harrington also found no support for more than one sequence. They also did not find convincing evidence for subluminous stars between the subdwarfs and the degenerates.

4. HR DIAGRAMS: CLUSTERS

A. Gutiérrez-Moreno and H. Moreno employed a photometric quantity, ΔD, which measures emission or absorption in the Balmer continuum, to confirm emission in circumstellar envelopes in pre-main sequence stars in the Orion and Upper Scorpius associations. W. Herbst and R.J. Havlen reported that there are gaps in the color magnitude diagram for the Ara OB1 association, just on the red side of the turn-up points. These gaps measure 0.07 mag. in (B-V) and 0.25 mag. in (U-B).

D.L. Crawford gave a systematic review of the wide variety of

"HR diagrams" which are employed for intercomparison of magnitudes, spectra, temperatures, and numerous color indices. R.F. Garrison discussed the overlaping of the main sequence of 6 clusters, the Hyades, Alpha Persei, II Sco., NGC 6231, NGC 2244 and NGC 2264, to obtain a composite main sequence extending from O4 V to K4. This overlapped sequence can be used for absolute magnitude calibration.

The recent suggested change in the distance to the Hyades from 40 to 46 pc. implies an increase in absolute magnitude of the ZAMS, which could formally be obtained by taking Z in the range from 0.06 to 0.10. Rather than accept such high metallicities, P.J. Flower has discussed the influence on the comparison of the color-magnitude diagram with the theoretical ZAMS of reasonable changes in the parameters of the latter due to uncertainties in the opacity, Z-value, stellar rotation, and mixing length, and changes in the color-magnitude diagram due to evolution. He concluded that a distance of 43 pc. is more nearly correct.

H.A. Abt and H. Levato from a study of 455 stars in 12 open clusters found that Ap stars first appear after about 10^6 yrs and increase thereafter to the frequency among field stars in about 10^8 yrs. The percentage of Am stars does not depend on cluster age. The rotational velocities of Ap and Am stars decrease inversely as the square and fourth root of their age respectively. Half of the stars in the Orion Nebula cluster (age $10^{5.7}$ yrs) have very broad H line wings, but only 2 percent in the Orion association ($10^{6.7}$ yrs). C.A. Pilachowski and W.K. Bonsack have made a survey of clump giants in open clusters, looking for peculiar stars and evolutionary effects. Indices have been developed which, when quantified, will yield overall cluster abundances of CH, CN, and Fe group elements. The "barium index" can be used to estimate the mean mass of a clump giant. A Ba star in the clump of NGC 2420 has been tentatively identified.

R.C. Henry and J.E. Hesser reported on their K-line photometry of A stars in Coma and NGC 6475. In the latter cluster, 2/3 of the 31 stars studied are Am stars.

From a study of several intermediate to old galactic clusters, R.D. McClure finds He and/or CNO abundances are independent of Fe abundance, with the differences possibly related to position with respect to the galactic plane and to age. In the open clusters, one finds much the same abundance ratios in all cluster members. Norris suggested that the difference between globular and open clusters is that the globulars are massive enough to hold whatever is made but is lost or blown out of open clusters.

R.P. Kraft opened his excellent discussion of non-homogeneity of the metals in globular clusters with a qualitative explanation due to A. Renzini of the effects of Z_{CNO} and Z_{Fe} on model calculations. Near the turn-off point from the main sequence for stars

between 0.6 and 1.0 $M_\odot$, the CNO cycle becomes dominant over the pp-chain with the result that the rate of energy generation increases linearly with Z_{CNO}, H exhaustion occurs earlier, and turn-off occurs sooner and at a lower luminosity. In two temperature ranges, the opacity depends heavily on Z. At 4000 K or below, the principal contributor is H^-, with the electrons determined primarily by Z_{Fe}. Near 10^6 K, the opacity depends on the b-f transitions of C, N, and O, that is on Z_{CNO}. Thus near turn-off, for surface temperatures greater than 4000 K, evolution depends on Z_{CNO} whereas for surface temperatures of 4000 K or less subgiant, giant branch, and red giant top move to the right with increasing Z_{CNO} and Z_{Fe}, but much more rapidly with Z_{Fe}.

ω Cen and M22 have been found to contain stars with enhanced Ba, and possibly Sr, along the red edge of the giant branch. This result implies a slow neutron flux but the mechanism is unclear. From a study of 23 giants in M22 and two other clusters with $<[Fe/H]>=-1.7$ to -2.0, Mallia finds $[Ba/Fe]=0.0$, in contrast to the aging effect found by Peterson for field halo giants.

Since Fe-peak and α-process elements are not produced in low mass stars, the great width of the giant branch of ω Cen indicates primordial variations in abundance but Norris and Bessell found a correlation of CaII (K) and CN bands suggesting that mixing also plays some role, possibly by way of effects of CN on the atmospheric opacity, with resulting influences on the intensity of strong resonance lines.

The overabundance of C in the CH stars of ω Cen and M22 can be accounted for as a result of the mixing of triple-α products to the surface at He-core flash or He-shell flashes. The N overabundance in M92 and the C^{12}/C^{13} ratio close to the equilibrium value follow from CNO-processing and mixing. The fact that metal-rich globular clusters show wide variations in the strength of CN while metal-poor clusters for the most part do not can also be accounted for by CNO-processing and mixing. Exceptions among the metal-poor clusters can be explained by assuming a second ascent of the giant branch following He shell-flashs for ω Cen and M22 or primordial N overabundance for M10 and NGC 7006, with their CN-strong stars. Again CNO-processing and mixing can account for the weakness of the g-band of CH on the asymptotic giant branch as compared with the subgiant branch. If future observations can reach stars fainter than those for which, according to theory, mixing first occurs, we can separate the effects of primordial abundance from those resulting from mixing.

M. and F. Spite reported on Ba differences in halo dwarfs and giants. G. Wallerstein and C. Pilachowski measured CO and OI in globular clusters with different horizontal-branch morphologies and found C and O enhanced relative to Fe in clusters with red horizontal branches. From a study of halo globular clusters, R. Canterna and

R.A. Schommer found no [Fe/H] gradient beyond 20 kpc but enhanced CN appears in some cases. They suggested that age and/or primordial He abundance are responsible for horizontal-branch morphologies. D. Butler, R.A. Bell, R.J. Dickens and E. Epps reported that the range of values of [Fe/H] for the RR Lyrae variables in ω Cen is sufficient to explain the great width of the giant branch. Both metal-rich ([Fe/H] > -1.0) and metal-poor ([Fe/H] < - 1.0) groups appear, with average periods of 0.57 and 0.66 days respectively. Superposition of evolutionary tracks supports composition-dependent mass loss, with the metal-rich group becoming 0.1 $M_\odot$ less massive than the metal-poor one but not more than 0.1 mag. fainter. J. Norris described his work done with M.S. Bessell and K.C. Freeman on ω Cen and 47 Tuc. The latter cluster shows the same anomalous bimodal distribution of CN in the giant, asymptotic, and horizontal branches. 47 Tuc appears to have no CH or Ba stars and no RR Lyrae's but TiO is normal. DDO photometry shows that the M stars of ω Cen and 47 Tuc are not alike. The giant branches of ω Cen and 47 Tuc are clearly not similar. They find that at $M_V \approx -1^m$ stars on the red side of the giant branch of ω Cen have much stronger CN than those on the blue. T. Lloyd Evans finds the giant branch of ω Cen much narrower in the (I, V-I) diagram than in (V, B-V) with a number of members on the red side of the giant branch. These stars show strong BaII and, where the temperature allows, strong CN. A high primoridal metal content and mixing are suggested. Dr. R.J. Tayler argued that the peculiarities of ω Cen must originate in unusual primordial abundances--possibly its great mass enabled it to retain more gas--rather than in mixing because otherwise the stars would have to know they were in ω Cen and required to give a broad giant branch.

6. HR DIAGRAMS: HORIZONTAL-BRANCH STARS

A.G. Davis Philip finds from a review of Strömgren four-color photometry of HB stars that A-F spectral types and luminosity classes III-V can be transformed to theoretical HR diagrams, which for both Pop I and II agree with the theoretical predictions. Future studies of blue HB stars offer the possibility of comparing the dispersion in the giant and horizontal branches and of investigating the role of primordial abundances and mixing. V. Castellani finds that the position of the hot HB stars in the HR diagram is largely independent of the original chemical composition but raises the question of the Draco dwarf galaxy which has a red HB cluster but $Z=3\times10^{-5}$, a value so low that theoretically it would not be possible for the HB to evolve to the red side of the instability strip for $Y \gtrsim 0.20$. F. Fusi-Pecci and A. Renzini in a very clear paper on HB morphologies and dispersions investigated the effects of mass loss and initial rotation, and the identity of the "second parameter" beyond the abundance of the elements of low ionization potential. They find that if the cores of evolving Pop II stars keep most of their original angular momentum, the spread in the

latter can explain the HB dispersion. From 30,000 VBLUW measurements on 170 Cepheids and 100 RR Lyrae Stars, J.W. Pel and J. Lub find that: a) the short period, Pop I Cepheids have very similar compositions; b) the Cepheid strip is wider in the HR diagram at higher luminosities; c) the blue Cepheids of small amplitude are probably pulsating in the first harmonic; d) the double-mode Cepheids have compositions of Pop I and are probably related to fundamental-overtone transitions; e) the RR Lyrae stars have a wide range of Z; f) members of group ab typically have radial amplitudes of 14 percent and those of group c about 5 percent; g) among the field stars one finds groups with properties, such as average periods of $0\overset{d}{.}55$ and $0\overset{d}{.}65$, like those identified by Oosterhoff in his globular clusters of type I and II; h) for group ab, $Y=0.28 \pm 0.02$. On the assumption of $Y=0.75$ in Cepheid convection zones, involving 10^{-4} or less of the star's mass, and possibly originating in a stellar wind much richer in H than He, A.N. Cox and S.W. Hodson have computed ratios of the periods of the second overtone and fundamental, positions of bumps in the light and velocity curves, and periods at various points in the instability strip, all in agreement with observations. L.A. Balona reported a study of a small sample of RR Lyrae stars which yielded an absolute visual magnitude of -0.5 and supported the idea that group c variables pulsate in the first harmonic.

7. GALAXIES AND OTHER SUBJECTS

Beatrice M. Tinsley gave a superb talk on HR diagrams of galaxies. Her paper must be read to appreciate the many interesting and significant points she makes. Extreme types of stellar population are those rich in young stars, Pop I, and those deficient in young stars, similar to Baade's Pop II, but metal rich. Differing mixtures of the extreme and intermediate populations account for the properties of the sequence of galaxies. Low mass stars radiate most of their energy as late red giants. Bright stars are the principal source of the integrated light. The star formation rate, SFR, is indicated by the relative number of upper main sequence and turn-off stars. In the integrated light, the contribution of the young component from the upper main sequence stars and their descendent supergiants indicate the SFR; and the contribution of the turn-off stars and their descendent red giants indicate the total number of stars formed. Color corrected normal galaxies define a narrow band in the UBV plane, of which the lower part contains the central bulges of spirals, ellipticals, and S0's. Using the local initial mass function, solar composition, and declining SFR, Larson and Tinsley found for all models a narrow locus for normal giant galaxies with ages from 5 to 20×10^9 yrs. Galaxy formation from primordial gas clouds may account for the observed morphology, mass, and population. Ellipticals and the nuclear bulges of spirals formed in a few free fall times. Disks form slowly, and the lower the SFR the flatter they are. The SFR also

depends on the rate of accretion of ambient gas, sudden brief increases in accretion rate can yield bursts of star formation. Intracluster gas creates van den Bergh's anemic spirals and S0's by sweeping out the interstellar matter.

D. Crampton pointed out the similarity of the spectra and luminosities of OB stars in the LMC to those in the Galaxy.

R.M. Humphreys reported that the supergiants in the Galaxy and the LMC are very similar with M_{bol} between -10^m and -12^m for O stars but an upper limit of $-9\overset{m}{.}5$ for other spectral types. S. van den Bergh presented arguments which suggest that Fe/CNO is lower in disk than in halo stars. R.A. Schommer, E.W. Olszewski and W.E. Kunkel described the color-magnitude diagram for the Ursa Minor dwarf galaxy, which indicates $[Fe/H] = -2.1$ and exhibits a HB which is heavily populated on the blue side of the RR Lyrae gap in contrast to other halo objects and dwarf galaxies. B. Hauck described the use of the Geneva photometric system to detect Am and Ap stars and to obtain an estimate of their chemical composition.

M. Golay reported photometric parallaxes for several galactic clusters. D.S. Evans described a remarkably close relationship between a stellar visual surface brightness parameter and Johnson (V-R) color, which is largely independent of interstellar extinction and luminosity class but lacks a clear physical foundation. It has been used to calculate angular diameters, and changes therein, from photometric observations. If linear diameters, or variations thereof, are known as for Cepheids, white dwarfs, or eclipsing binaries, distances can be found. At the least, this relationship appears to be very useful for reconnaisance. E.E. Mendoza V described the use of narrow, intermediate, and broad band photoelectric photometry to place A stars precisely in the HR diagram.

V. Straižys described the effective, seven-color, intermediate band, photometric system of the Vilnius Observatory. After appropriate calibrations, the intersection of five lines based on the seven-color measurements yields the reddening-free location of a normal star in the log T_{eff}-log g or the M_V-spectral type diagrams, and deviations from a close intersection can be immediately interpreted by a trained investigator as indicating subdwarf, metal-deficient giant, Am or Ap star, white dwarf, or double star. Discussions are underway in regard to some unifying of the Geneva and Vilnius systems, and D.S. Evans suggested that any resulting system be referred to as the "Genius System".

8. THEORETICAL PAPERS

H. Kähler in an interesting paper pointed out that the mathematical arguments which have been used to "prove" the Vogt-Russell theorem--that a unique solution of the equations of stellar

structure exists for given mass and chemical composition--apply to systems of linear differential equations with appropriate boundary conditions, whereas the equations of stellar structure are highly non-linear. Indeed, counter examples to the theorem are familiar. A "local" formulation--small changes in the parameters lead uniquely to small changes in the model--can be established for the neighborhood, in the mathematical sense, of a stable model, and some statements can be made about the number and stability of equilibrium models in general.

M.N. Perrin and G. Cayrel de Strobel find that stars in the same six-dimensional box, 0.01 mag. on a side in each of the 6 measures or combinations of the Geneva photometry, can differ substantially in M_{bol}. In a separate paper, these two authors report that the interpretation of an observational log T_{eff}-M_{bol} diagram derived for nearby F, G, and K stars using the theoretical diagrams of Hejlesen and those of Ciardullo and Demarque yields similar ages and compositions.

P.G. Gross discussed the possible contribution of magnetic mixing to surface chemical composition. Magnetic fields arising from flux conservation or the dynamo action of turbulence and rotation could create magnetic bubbles, buoyant because of the magnetic pressure forces. Detachment could result from the buoyancy or the aerodynamic drag of differential rotation. The bubbles would be expected to rise 1 or 2 pressure scale heights, which under appropriate conditions could carry processed nuclear matter from the He flash region to the bottom of the outer convective envelope, with the concomitant downward motion of the H leading to the possibility of s-process element formation.

A.V. Sweigart reviewed certain results of his large computational program covering about 100,000 stellar models using uniform numerical procedures and similar input physics. In a number of cases, his conclusions confirm in a particularly clear way already known or suspected results. For higher mass, central H exhaustion occurs in a larger volume and is followed by sudden contraction and evolution to the left. Near turn-off, log T_{eff} and log L increase for increasing Y and decreasing Z, primarily Z_{CNO}. The RGB moves to the left with decreasing Z or increasing mass, Y, or ratio of mixing length to pressure scale height. The luminosity at the RGB tip and the maximum stellar mass for a He-core flash decrease with increasing Y or decreasing Z. The maximum extent of the temperature inversion from neutrino losses occurs shortly before He-core flash and increases with core mass. On the SGB, when the H-burning shell moves up from the region partly H exhausted during the MS stage to the region mixed by the envelope convection, it encounters a higher H concentration and pauses in its evolution, yielding a larger concentration of stars at that point and therefore a peak in the luminosity function. The duration of this pause and the luminosity at which it occurs decrease with smaller Y and

larger Z. Semi convection can increase HB life-times by a factor of almost two.

R.B. Ciardullo and P. Demarque have computed CM diagrams, luminosity functions, integrated colors, etc. based on extensive stellar models as a function of time for several chemical compositions, converted to observables using model stellar atmospheres, and analyzed the effects of such parameters as initial mass function and mass loss.

S.A. Lamb gave a particularly clear review of the evolution of massive stars beyond core He burning. Stars of more than about 10 $M_\odot$ can burn C quiescently with the iron peak elements as the eventual product. Except for the few stars of more than 80 $M_\odot$, quasistatic evolution ends in the photodissocation of the iron peak elements, primarily into α-particles. The products of nucleosynthesis are dependent on the temperature, density, and the extent of convective mixing at various depths, all of which at this stage for these massive stars are largely determined by the energy loss from neutrinos arising from pari-, plasma-, or photo-interactions. The heavier products tend to occur in the deeper layers. Products appear at the surface if convective envelopes carry them there or mass loss reveals the inner layers as for the late-and early-type supergiants respectively. Envelope convection does not penetrate the H burning shell so that the products of He and later burning stage are expected only if mass loss occurs. The products of core nucleosynthesis may appear in supernova remnants if the explosion was preceded by sufficient mass loss.

C. Chiosi gave a very complete and interesting review of the structure and evolution of massive stars, starting with the role of semiconvection, the relation of the Schwarzschild and Ledoux conditions for convective stability in regions with a gradient of molecular weight, Kato's discussion on the local basis and the need for a satisfactory treatment of non-local convection. Semi-convection is not a major consideration in the main sequence phase but is a dominant consideration in H shell, or largely He core, burning, since the chemical profile and the resulting changes in the depth of the convective envelope at the upper end of the red supergiant branch influence looping in the HR diagram. Mass loss for stars of less than 15 $M_\odot$ is moderate and their evolution is not greatly influenced. For more massive stars, mass loss can reach substantial fractions of the initial mass with resulting changes in the He abundance and increases in N^{14}, C^{13}/C^{12}, and O^{17}/O^{18}.

It became apparent in the course of discussion that the new Carson opacities, which for stars of more than 10 $M_\odot$ lead to a broadened main sequence and modified evolutionary tracks, are not yet generally accepted. The detailed calculations have not been

published and some attempts to reproduce specific values have not been successful.

P.S. Conti reported mass loss rates of the order of 10^{-6}, 10^{-5} and 10^{-4} $M_{\odot}$/yr. for O, Of, and WR stars respectively but found no gross differences in the velocity structure of the envelopes. He suggested that if the mass is sufficient, evolution proceeds from O to Of stars and thence to WR stars.

D.S.P. Dearborn and V. Trimble evolved stars of 12 and 32 $M_{\odot}$ of Pop I composition and Pop II metal composition but two different Y's to early carbon burning. Assuming a Salpeter birthrate function and various mass loss rates, they find lower limits to the ratio of the enrichment of He to that of the metals in the galaxy, from the time of the first stars, which can match any value derived from the observations.

Using their own model atmospheres, R.A. Bell and B. Gustafsson have computed synthetic spectra and colors of G and K giants for different metal abundances. Included in the computations are some 50,000 atomic and molecular lines. The effects of terrestrial atmospheric and interstellar extinction and changes in the Doppler broadening velocity and the damping were studied. Agreement of the computed and observed colors was good except for the ultraviolet and for the CN indices. These differences can be traced to the omission of weak lines and failure to include N enrichment from CNO processed matter respectively.

R. Foy suggested that microturbulence could be used as a third dimension to yield the age of a G-K giant in the HR diagram. He finds that microturbulence is not a result of departures from LTE but is related to stellar atmosphere parameters and is a function of stellar evolution.

K. Kodaira has found that the effect on the transformation from the theoretical to the empirical HR diagram of considering the sphericity of the extended atmospheres of late type giants, with the accompanying decrease in g and increase in the dilution of the radiation field, cools the outer layers and enhances the TiO and H_2O, with the blanketing effects leading to changes in M_V and color.

In summary one can say that Symposium 80 was most pleasant and informative and that the participants and attenders owe substantial thanks to A.G. Davis Philip and the Scientific and Local Organizing Committees.

CONTINUATION OF DISCUSSION

BOESHAAR DISCUSSION, Continued from page 22

the stars in my sample appear to exhibit a negative luminosity effect in the Na 'D' lines. Mould's model atmosphere analysis suggests that this effect is due to pressure broadening. I tend to agree.

LIEBERT: George Wallerstein has likened the search for a true K, M abundance subdwarf to the search for the 'Holy Grail'. But we must remember here that, even though the high velocity M dwarfs are easily selected out in surveys based on proper motion or reduced proper motion, these stars must be found within a much more restricted space volume about the Sun than the F and G stars. You have only a handul of high velocity stars, I believe, in your sample and there aren't a great many more M stars known with either high velocity or these spectral peculiarities. Yet only a small fraction of halo stars would be expected to have [Fe/H] ~ -2. I wonder if we should expect to have any really extreme cases in our current sample.

BOESHAAR: Very few. I agree.

BLANCO, HOAG AND McCARTHY DISCUSSION, Continued from page 38

a group of low dispersion spectroscopists that 'the best auxiliary equipment for a low dispersion spectral study is a fine slit spectrograph with high dispersion'. We look to slit spectroscopy for further illumination of our results.

FLOWER DISCUSSION, Continued from page 154

0.10. So, you can change both Z and Y, and change the modulus.

KÄHLER: You mentioned the uncertainty of the opacities. In addition one should consider the uncertainty of the burning rates. There might be undetected resonances in the CNO-cycle which could be important.

WEIDEMANN: Are you aware of our paper (Koester and Weidemann, *Astron. Astrophys.* **25**, 437, 1973) in which we discussed the chemical composition and the location of the Hyades main sequence in the color magnitude and mass-luminosity diagrams? We used older models, by Copeland et al. (1970). How would your results change, had you used those models?

FLOWER: Yes, I am but their models only go up to about $2.5M_{\odot}$ so I cannot compare them with my upper main sequence.

COX: Can you tell us whether the ratio of mixing length to scale height affects the results you have given?

FLOWER: Since my main sequence models are hotter than about log $T_{eff} = 4.0$, the models are radiative. This is, of course, an advantage, because comparisons with the upper main sequence minimizes the uncertainties in the mixing length theory, as well as, uncertainties in line blanketing and, possibly, in abundance effects on temperature scales.

ABT AND LEVATO DISCUSSION, Contined from page 156

is unusually good for detecting faint peculiarities. Hartoog did not have a similar advantage. Otherwise the results would probably have been the same.

MORGAN: What is the metallic line type of those broad-lined Am stars in the Orion association?

ABT: They're early A-stars like Sirius.

HIRSHFELD, McCLURE AND TWAROG DISCUSSION, Continued from page 166

factor, not age. This, of course, still permits the original picture that the heavy element abundance greatly increased with time during the earlier phases of the Galaxy, but in such a way as to produce the present position-dependent abundances.

DEMING: I notice that you adopt an approximately solar metallicity for M67. However the CN strength of M67 is fairly high, about halfway between the Hyades and the average value for the solar neighborhood. Are you prepared to claim that M67 may be somewhat CNO-rich?

McCLURE: No. I'm sure that the difference in inferred abundance on the basis of UV excess and on CN strength is less than the errors of observation.

A. G. Davis Philip and D. S. Hayes (eds.), The HR Diagram, 405–406.

NORRIS DISCUSSION, Continued from page 204

You see the width in (B–V) because it reflects the blanketing due to the Fe peak metals that cover the spectrum. But if you had the corresponding variations in the CNO group that were primordial, the question arises whether you would get much spread in the giant branch.

NORRIS: In our (Bessell and I) investigation in the CN-strong stars in ω Cen we found a deep depression at 4200 Å. There also seems to be a depression stretching blueward of 4000 Å, due to some unknown molecule. We thought that we could explain a large part by some blocking in the B-band, while the V-band does not contain CN or anything like that.

DICKENS: The feature at 4215 Å seems enhanced in one of your spectra of a CN-strong star in 47 Tuc. Whilst this could perhaps be caused by the CN itself in this star, it would be interesting to investigate the possibility of Sr enhancements in the stars which show CN enhancement, particularly on the AGB, by looking at other lines of the Sr multiplet. Some stars in ω Cen showed simultaneous enhancements of N (H-burning) and Sr and Ba (S-process) which are difficult to understand theoretically.

NORRIS: Our computations suggest that the feature at 4215 Å increases because of CN enhancement. The high dispersion spectra (obtained with the Anglo Australian Telescope) which we have in this region run from 4120 – 4600 Å, so that we do not have the possibility of commenting on the Sr II line at 4077 Å.

PHILIP DISCUSSION, Continued from page 220

constant $g\theta^4$ yielding constant M/L. You have models subject to only small differences depending on helium content. Consistent with the cosmologically produced helium wouldn't you derive a reliable mass? Your stars would help determine the helium core mass of a red giant of the old halo population.

PHILIP: Yes. In a paper in preparation concerning the field and globular cluster HB stars, masses will be calculated. However, I have not yet made these calculations.

RENZINI DISCUSSION, Continued from page 228

KRAFT: There are other ways to get the spread on the HB, but if (and that's a big 'if') there are primordial abundance variations in globular clusters, then they will *directly* cause a spread on the HB without appeal to rotation, although rotation certainly should be considered.

LUB AND PEL DISCUSSION, Continued from page 236

COX: Let me be sure that you have not solved the mass anomaly for the bump or double mode Cepheids.

PEL: Yes, what I said refers only to the general mass-problem with all Cepheids. Our data do not change the mass-discrepancies of bump-Cepheids and double-mode Cepheids.

MADORE: There does exist one other independent and totally fundamental method for determining the masses of Cepheids: the use of binary systems. Purely photometric means exist for discovering Cepheid binaries (Madore, B. F., *Mon. Not. R. Astron. Soc.*, 1976) and one of the brightest candidates, SU Cygni, has been found to be a spectroscopic binary. A preliminary orbit has been derived and a lower limit to the Cepheid mass has been obtained. This lower limit is already consistent with the evolutionary masses and is higher than the anomalously low pulsation mass.

POSTER SESSION

The Banquet on November 4, 1977.

A REVISED CLASSIFICATION SYSTEM IN THE RED FOR S STARS

Thomas B. Ake

Warner and Swasey Observatory,
Case Western Reserve University

ABSTRACT

A classification scheme in the red spectral region for the S stars is described. For those stars exhibiting both ZrO and TiO bands, a modified version of the Keenan (1954) system is used for the assignment of temperature subtypes. For the pure S stars, a new system is introduced utilizing ZrO band strengths and the sodium D lines. Comparisons between the revised types and photometric colors have been made. In addition a new abundance index is proposed based on the relative strength of YO compared to ZrO and TiO. The revised system provides a connecting link between the M and C systems so that all red giants may be classified consistently from the strengths of TiO, ZrO and Na D.

The classification of the S stars is greatly complicated by the variety of S-type hybrids that have been found to exist. Although the S stars are defined by the presence at low dispersion of ZrO bands, some members of the class are similar to the M giants by exhibiting strong TiO, while others are weak-banded and display atomic-line spectra resembling carbon stars. Presumably, as a class the S stars are chemically intermediate between the M giants and the carbon stars, and thus, are difficult to classify because their molecular bands are severely affected by heavy-element and C/O abundance anomalies. The classification system devised by Keenan (1954) handles this problem by assigning temperature classes according to a weighted sum of ZrO and TiO band strengths such that the stronger bands are weighted more heavily. He also calculates an abundance parameter based on the ZrO/TiO ratio, although Keenan and McNeil (1976) have eliminated this index in the S notation and have replaced it by listing the TiO and ZrO strengths individually

A. G. Davis Philip and D. S. Hayes (eds.), The HR Diagram, 409–412.

with the temperature class assignment. The Keenan system, however, is inappropriate for the pure S and SC stars because TiO is absent and there is little similarity to the M giants. Because the C system for the carbon stars relies upon atomic-line criteria for the temperature types, it seems reasonable to assume that the same criteria should be available for the pure S and SC types. At low dispersion in the red, the Na D lines are quite useful, and a revised S-type system is introduced to improve the spectral types of the stars with Keenan abundance classes 8 and 9.

Image-tube spectrograms taken with the Warner and Swasey 91-cm reflector have been obtained for most of Keenan's original S stars and for some additional objects found in Stephenson's (1976) S star catalogue. The region 5450-7000 Å is observed at an inverse dispersion of 105 Å/mm with an estimated resolution of 5Å. For the pure S stars, the line spectra are visible but indistinct on these plates, and only the D lines were found to be useful as an atomic-line classification criterion. Absolute band strengths of TiO and ZrO were visually assigned on the Keenan systems, and Na D intensities were measured on a new system. The D-line system was later compared with that of the C stars so that the SC stars could be translated to comparable C types. It was found that an S star with D = 1 corresponds to type C5.

For the stars with strong TiO, Keenan's system is retained except for a minor change in the zero-point of the scale. Whereas the Keenan TiO strengths are the equivalent M type + 1, the TiO intensity here is taken to be just the M type in order to accommodate the higher dissociation energy of ZrO compared to TiO. When proceeding from the K giants to the S stars, ZrO should appear before TiO, and adding 1 to the M types would exaggerate the coolness of the stars. For the pure S stars, TiO is replaced by Na D in the computation. Since C types can be assigned by the D lines alone (Yamashita 1972), this substitution carries the S system to the C system for the weak-banded pure S and SC stars, where the D-line intensity is greater than the ZrO strength.

The classification then is

S-type = the stronger of ZrO and TiO + 1/2 the weaker of these

or

S-type = the stronger of ZrO and Na D + 1/2 the weaker of these.

For most stars there is little ambiguity about which criteria to use for the classification since the D lines are masked by TiO 5849 Å when TiO is strong. The warmer members of the class, however, may display TiO and Na D at the same time so that the spectral type is assigned at the discretion of the observer. Two cool

stars, AA Cyg and BI And, are unusual in that both TiO and Na D are quite strong, but TiO $\sim$ Na D so that the spectral types are stable.

Although different criteria are used for the TiO-strong and pure S types, comparisons of the new types with Piccirillo's (private communication) eight-color photometry of non-Mira S stars indicates that the types are assigned on a common temperature scale. The major improvement of the revised system is that the pure S stars are brought in line with the other S types. The most discrepant star is SU Mon, whose super-saturated D lines are comparable to those of the D-line carbon stars such as WZ Cas. Although some S stars display D lines approaching the intensity found in SU Mon, unlike this star their entire spectra indicate extremely low temperatures. For the long-period variables, the types appear to be well-behaved with phase.

Furthermore observations of the red YO bands were found to be inadequately explained by temperature differences between the various S types. The peculiar behavior of YO in M giants and S stars is well-known (e.g. Keenan 1966). These bands begin to appear in some M giants and are quite strong even at classification dispersion for the S stars. Their appearance in the M giants has been attributed to both incipient s-processing of Y as well as increasing C/O from the M giant to the S stage. In S stars of the same temperature class, YO appears weakly when TiO >> ZrO or ZrO >> TiO, and is strongest when ZrO $\sim$ TiO. For the SC stars, ZrO is present with strength less than 0.5, and YO and TiO are absent. It appears that the S stars can be divided into six abundance classes based on the comparison of YO to TiO and ZrO (Table I). The YO strengths are multiplied by 2 since they are not as strong as TiO and ZrO on these plates.

This pattern of behavior is predicted by increasing C/O from the TiO-strong stage to the pure S and SC types (e.g. Scalo and Ross 1976). Because the dissociation potential of YO lies between

TABLE I

DEFINITION OF THE ABUNDANCE INDEX

Index	Spectral Characteristics
1	TiO >> ZrO and YO
2	TiO $\geq$ ZrO $\geq$ 2 x YO
3	2 x YO $\geq$ ZrO $\geq$ TiO
4	ZrO $\geq$ 2 x YO > TiO
5	ZrO $\geq$ 2 x YO; TiO = 0
6	ZrO weak; YO, TiO = 0

that of TiO and ZrO, and YO is depleted by dioxide formation unlike ZrO, YO should have its highest concentration when the free-oxygen supply is small enough to reduce the TiO and YO_2 concentrations, but not so small that the ZrO content is reduced too. Although the behavior of YO is more complicated than this simple picture (Piccirillo, private communication), it is suggested that these crude divisions can be used as an abundance index. This parameter appears to be related to the C/O ratio, but because s-process enhancements influence the appearance of ZrO and YO, it should not be labelled as a C/O index. In particular, stars in abundance classes 2 and 3, where all molecular bands are strong, may have increased heavy-element abundances without C/O increases. It is a well-known fact that both s-process enhancements and C/O variations must be invoked simultaneously to explain both the molecular-band and atomic-line spectra of the S stars.

ACKNOWLEDGMENTS

The writer is pleased to acknowledge suggestions by John Piccirillo, William P. Bidelman and Philip C. Keenan, and is especially indebted to John Piccirillo for the use of his photometric observations prior to publication.

REFERENCES

Keenan, P.C. (1954). Astrophys. J., **120**, 484.

______. (1966). in IAU Symposium No. 24, Spectral Classification and Multicolor Photometry, K. Lodén, L. O. Lodén, and U. Sinnerstad, eds., Academic Press, New York, p. 26.

______, and McNeil, R. C. (1976). An Atlas of Spectra of the Cooler Stars, The Ohio State University Press, Columbus.

Scalo, J. M., and Ross, J. E. (1976). Astron. Astrophys., **48**, 219.

Stephenson, C. B. (1976). Publ. Warner and Swasey Observatory, Vol. **1**, No. 4.

Yamashita, Y. (1972). Ann. Tokyo Astron. Obs., 2nd Series, XIII, No. 3.

SOUTHERN HIGH-VELOCITY STARS

H.J. Augensen and W. Buscombe

Northwestern University

Using the model of the Galaxy presented by Eggen, Lynden-Bell and Sandage (1962), plane galactic orbits have been calculated for 800 southern high-velocity stars which possess parallax, proper motion, and radial velocity data. The stars with trigonometric parallaxes were selected from Buscombe and Morris (1958), supplemented by more recent spectroscopic data. Photometric parallaxes from infrared color indices were used for bright red giants studied by Eggen (1970), and for red dwarfs for which Rodgers and Eggen (1974) determined radial velocities. A color-color diagram based on published values of (U-B) and (B-V) for most of these stars is shown in Fig. 1. Plots of each of the orbital parameters apo- and peri-galacticum, eccentricity, change in radial angle per cycle, radial and azimuthal periods, and instantaneous orbital inclination to the galactic plane versus color indices (U-B), (B-V), and the ultraviolet excess δ(U-B) were also produced. In general, stars (especially F dwarfs) with greater space velocities and also greater components of speed perpendicular to the galactic plane have larger ultraviolet excesses, eccentricities and inclinations of the orbits around the galactic center (Fig. 2). Also, a comparison is made between the HR diagrams for the southern high-velocity star group and that of M3, a globular cluster, and again for M67, an old open cluster. The high-velocity star group is found to resemble an old open cluster more than a globular cluster (Fig. 3).

A. G. Davis Philip and D. S. Hayes (eds.), The HR Diagram, 413–416.

REFERENCES

Buscombe, W. and Morris, P.M. (1958). Mem. Mt. Stromlo Obs. 14.
Eggen, O.J. and Stokes, N.R. (1970). Astrophys. J. 161, 199.
Eggen, O.J., Lynden-Bell, D. and Sandage, A.R. (1962). Astrophys. J. 136, 748.
Rodgers, A.W. and Eggen, O.J. (1974). Publ. Astron. Soc. Pacific 86, 428.

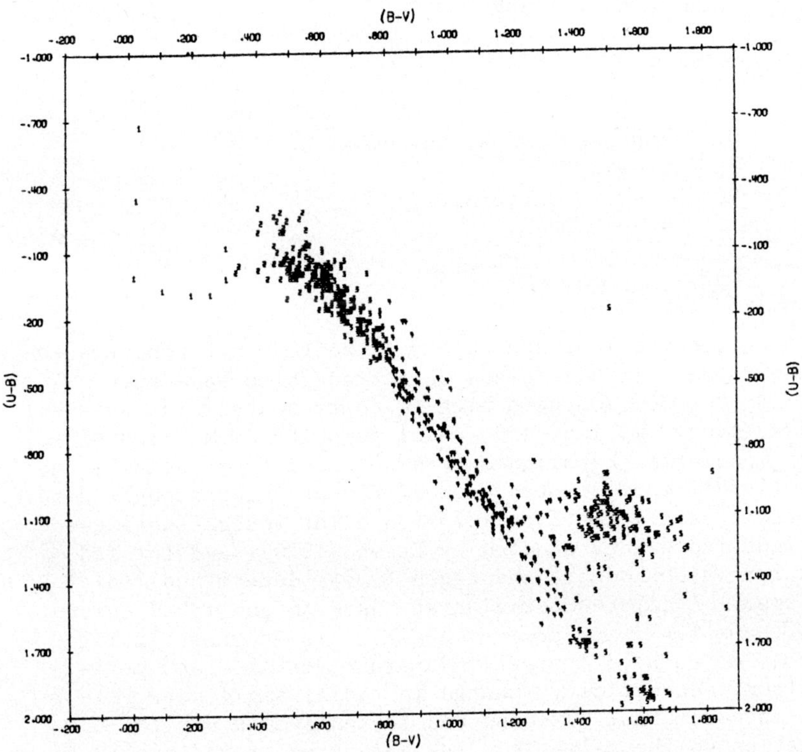

Fig. 1. A plot of (U-B) *versus* (B-V) for the southern high-velocity star group. Each symbol represents an approximate temperature class: 1 = A, 2 = F, 3 = G, 4 = K, 5 = M.

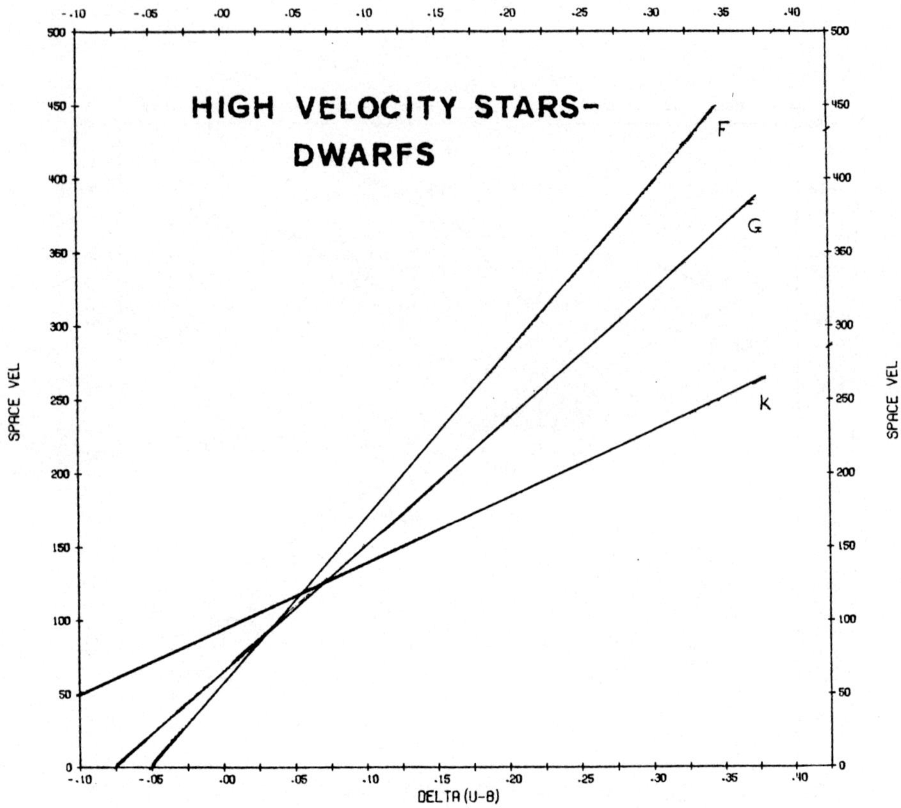

Fig. 2. A graph of the total space velocity versus the ultraviolet excess for F, G and K high-velocity dwarf stars. At least squares fit to the data is displayed for each group.

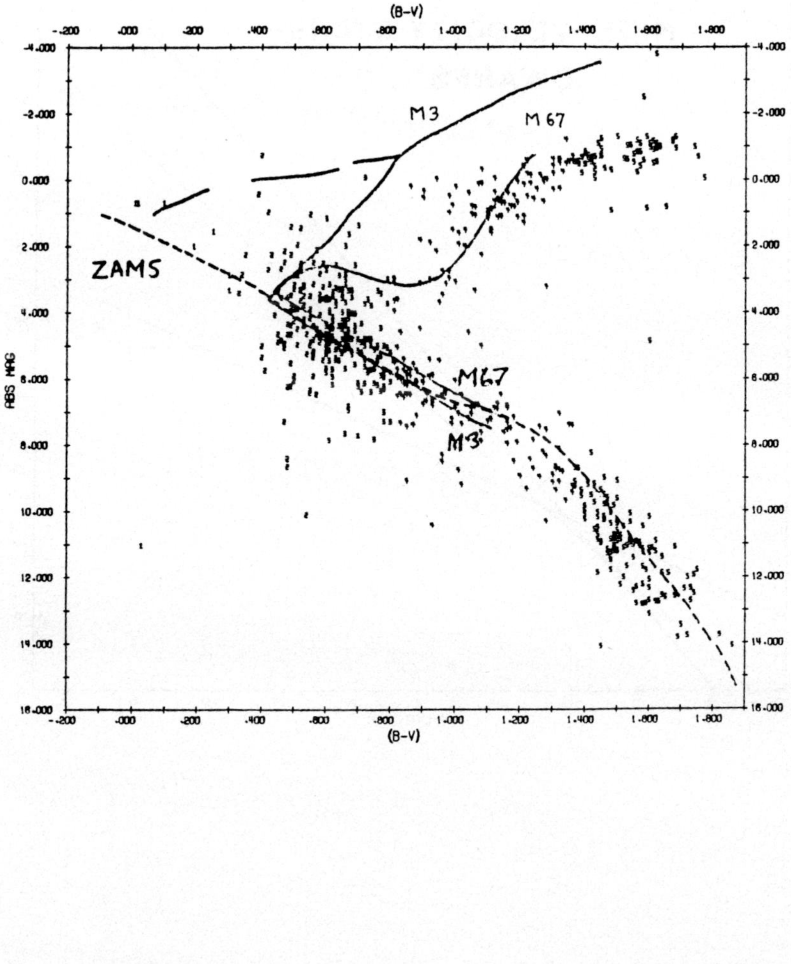

Fig. 3. The HR plot for the southern high-velocity star group. Superposed upon this diagram are sketches of the HR diagrams for M3, M67 and the zero-age main sequence.

ON G.P. KUIPER'S WORK ON THE STARS OF LARGE PROPER MOTION

William P. Bidelman

Warner and Swasey Observatory, Case Western Reserve University

ABSTRACT

The writer is preparing for publication a complete listing of the late G.P. Kuiper's spectral classifications of stars of large proper motion north of $\delta=-50^{\circ}$. This work, resulting from his observations made at the Yerkes and McDonald Observatories some 35 years ago, has been only partially published; it is estimated that useful unpublished classifications exist for the order of 1000 stars. This paper discusses the background of his program and the procedures to be followed in making these data generally available.

Kuiper's spectral classification program on the stars of large proper motion, which eventually comprised some 3500 objects observed during 1938-1944 at Yerkes and McDonald, represented a natural outgrowth of his earlier determination with the Lick Crossley reflector of the colors of all known stars of large parallax north of $\delta=-20^{\circ}$ for which no spectral class nor color index was known. On reporting its completion, Kuiper (1935) states: "Next in importance to the stars of (known) large parallax, in a statistical study of the near-by stars, are the stars with large proper motions (but without known parallax). Annual proper motions of 0''.50 and larger invariably indicate interesting stars; they are either in the immediate neighborhood of the Sun, or possess very high space velocities. The lower limit 0''.50 has been adopted in the present color program, but in future work the limit may well be taken as low as 0''.4 or 0''.3.

When Kuiper became a staff member at the Yerkes Observatory

A. G. Davis Philip and D. S. Hayes (eds.), The HR Diagram, 417–420.

in 1936 the feasibility of a systematic spectroscopic examination of a very large number of faint proper-motion stars was greatly enhanced by three factors: (1) he there had access to the extensive plate material used by Ross in his then largely completed proper-motion survey, which facilitated the preparation of finding charts for these stars; (2) he soon oversaw the modification of an old prismatic Brashear spectrograph into an instrument which gave the very low dispersion of 340 Å/mm at Hγ, enabling very faint stars to be observed; and (3) he decided, following the lead of W.W. Morgan who had just shown that M dwarfs could be very accurately classified in the visual spectral region (at considerably higher dispersion), to make extensive use of this region for the classification of the redder proper-motion stars. The new so-called "K" spectrograph utilized panchromatic film, bent to the proper focus over the entire spectral region recorded. Customarily two exposures of different densities were taken of each star, and the redder stars were classified in the visual region, the bluer ones in the photographic. With this efficient instrument he began, first with the 40-inch and then in many observing sessions with the McDonald 82-inch, including the first use of that telescope, his very important spectroscopic work on the stars with proper motions larger than approximately three-tenths of a second of arc.

The progress of his program is well recorded in Struve's several Yerkes Annual Reports of that period, as well as in Kuiper's own series of six papers dealing with his work (Kuiper 1938, 1939, 1940a, 1940b, 1942, 1943). Progress was rapid: the annual report for 1939-40 (Struve 1940) states that "Kuiper's program of stars of large proper motion is nearing completion. The spectra of more than 1500 stars have been determined..... The complete publication of the results is planned for the fall of 1940." But in the report for the next year (Struve 1941), we read that "Kuiper, in collaboration with Dr. W.J. Luyten of Minnesota, undertook an extensive program of observations of all new proper-motion stars within reach of the 82-inch, which had been discovered by the latter in his Bruce survey.... The total number of proper-motion stars for which Kuiper has determined spectral types since 1938 is about 2000." Finally, in the report for 1942-43 (Struve 1943) we learn that "Kuiper completed at McDonald the program, started at Yerkes in 1938, of determining accurate spectral types of stars with large proper motion, north of -50° in declination. A total of about 9000 spectra of 3200 stars was obtained during the six years. The spectra of some 300 stars for which Luyten provided the positions were also obtained by Kuiper. The results of this collaborative program will be published in the near future by Luyten and Kuiper." But alas, in the following years' report (Struve 1944) little is said of Kuiper's stellar work - instead the report notes his wartime discovery of an atmosphere on Saturn's largest moon, Titan. When we recall Kuiper's

many other enormously important contributions to stellar astronomy in addition to the work discussed here, we can only regretfully conclude that an era of remarkable personal achievement in that field had come to an end.

Unfortunately the promised comprehensive and collaborative catalogues containing the results of Kuiper's spectroscopic observations never eventuated. Though Kuiper provided magnitudes and spectral types for a great many of the entries in The General Catalogue of Trigonometric Stellar Parallaxes (Jenkins 1952), a substantial part of this work has never seen the light of day.

Some time after Kuiper's death and as a result of renewed interest in the characteristics of the stars of large proper motion aroused by the systematic Lowell Observatory survey (see Bidelman and Lee, 1975), the writer inquired of Dr. E.A. Whitaker of the Lunar and Planetary Laboratory of the University of Arizona as to whether any of Kuiper's unpublished proper-motion star material could be unearthed among his effects. When I learned that several notebooks that appeared to contain useful data had been found, I asked to have copies of this material sent to me, which, through the good offices of Drs. E. Roemer and H.A. Abt, eventually happened. What I finally obtained were copies of five large notebooks which were evidently themselves copies of Kuiper's original observing logs on which he had written his determination of the spectral type of every star observed. In many cases, several different films had been taken of the same star, with differing resultant classifications. Thus it was necessary to make up a card file containing each star (or pair, as many binaries were included) and to enter the individual classifications. Since Kuiper, being economical of clerical time, had only indicated the stars by name and right ascension (declinations not being given), finding the exact stars observed was at times a bit difficult. It is important to emphasize that I have not attempted to reclassify (or even inspect) Kuiper's original spectrograms; I think it very doubtful that any useful purpose would be served by doing this. However, I hope to cross-check my card file with one that he presumably also must have made up, if such can be found. In some cases it may be desirable to check his observing cards to determine which star was actually observed; these have also not yet been located.

The question of what to publish is a rather difficult one. In cases where Kuiper has determined more than one type an average will have to be taken (unless he has already done so). Further it is well known that Kuiper's classification scheme was not identical to that in general use today; thus I plan to provide a short table that will permit one to transform his spectral types to others scaled to the MK system for the K stars or to the new system recently established for the M dwarfs by Dr. P. Boeshaar.

Many of his stars have already, of course, been assigned objective-prism spectral types of varying quality, but Kuiper's types are undoubtedly better than most of these. Also, not surprisingly, slit-spectrograph types are available for a great many of Kuiper's stars, and there may seem little point in publishing his types if better ones are available. However, at times it is rather difficult to tell which types are better. My present inclination is to publish all of Kuiper's types, noting cases where they have been clearly or presumably superseded. A similar difficulty arises in the matter of magnitudes. It is not entirely clear where the magnitudes that Kuiper gives in his notebooks come from. Many were no doubt determined by the late Carl Seyfert, some may have been photometric determinations by Kuiper himself, but the majority are probably eye-estimates at the telescope. My thought is to give the V magnitude if it is available, and otherwise to use Kuiper's values.

In conclusion it should be pointed out that, though very extensive, the stars included in Kuiper's program do not represent a complete sample of proper-motion stars to any given limit in motion or magnitude. Later discovery lists such as those of Luyten or Giclas are far better in that respect.

Nonetheless, it is estimated that there must be contained in Kuiper's unpublished material of the order of 1000 stars for which his spectral types are better than those available anywhere else, and it thus seems certain that this somewhat belated contribution to our knowledge of the distribution of the stars in the HR diagram will prove a most valuable one. The writer is happy to have had a part in it.

REFERENCES

Bidelman, W.P., and Lee, S.-G. (1975). Astron.J. **80**, 239.
Jenkins, L.F. (1952). General Catalogue of Trigonometric Stellar Parallaxes, Yale University Observatory, New Haven.
Kuiper, G.P. (1935). Publ. Astron. Soc. Pacific **47**, 279.
Kuiper, G.P. (1938). Astrophys. J. **87**, 592
Kuiper, G.P. (1939). Astrophys. J. **89**, 548.
Kuiper, G.P. (1940a). Astrophys. J. **91**, 269.
Kuiper, G.P. (1940b). Astrophys. J. **92**, 126.
Kuiper, G.P. (1942). Astrophys. J. **96**, 315.
Kuiper, G.P. (1943). Astrophys. J. **97**, 275.
Struve, O. (1940). Publ. Amer. Astr. Soc. **10**, 104.
Struve, O. (1941). Publ. Amer. Astr. Soc. **10**, 205.
Struve, O. (1943). Publ. Amer. Astr. Soc. **10**, 358.
Struve, O. (1944). Astron. J. **51**, 82.

A B-STAR HIGH LUMINOSITY INDICATOR

I. Furenlid

Kitt Peak National Observatory

W. L. Sanders

New Mexico State University

In an effort to sort out BIa supergiants among B-stars for use in delineating spiral structure, a photometric discriminate has been sought in the following way. Wide slit 16Å/mm (flux) spectra were obtained of a sample of B stars over a suitable spectral and luminosity range. These spectra were digitized on the PDS Microdensitometer. Numerical filters of various widths were then slid along the spectra and numerous color-indices were formed and tested. An index $\ell_0 = u + v - 1.9\ (\ell)$ plotted against c_0, where ℓ is a 200Å halfwidth passband centered near 3840 Å and the other magnitudes are on the Strömgren 4-color system, succeeds in separating the Ia supergiants up to B1. For stars of spectral type B1 and earlier, confusion sets in and the separation is uncertain (although a fair percentage of success is also had for types B1 and B0). This luminosity discriminate is now in use for a large spiral structure program in which all known Be stars (and perhaps eventually all B stars) are being surveyed for Ia supergiants. Spectra for 2-dimensional classification are obtained for the Ia candidates.

A. G. Davis Philip and D. S. Hayes (eds.), The HR Diagram, 421. All Rights Reserved.

THE MAIN SEQUENCE AT G2 V

Johannes Hardorp

Department of Earth and Space Sciences

State University of New York at Stony Brook

1. ABSTRACT

About 3/4 of the sky were searched spectroscopically for stars whose ultraviolet spectra (3640 - 4100 Å) match that of the Sun, at 20 Å resolution. Down to the limit of the BSC, only two were found: HR 7504 and HR 2290. No G2 V star matches the Sun. Stars that do match have (B-V) = $0\overset{m}{.}66$. The search has a bearing on effective temperatures of G dwarfs and on metal-abundances relative to the Sun.

2. INTRODUCTION

When searching for a solar spectral twin to derive planetary albedos in 1974/75, I was surprised to find a G5 V MK standard (16 Cyg B = HR 7504 = HD 186427) match the solar energy distribution better than G2 V stars (Hardorp 1976). At the same time I discovered that a broad absorption feature near 3850 Å varied much more from star to star than did the energy distribution. This CN-band was subsequently used as the main criterion in a spectroscopic search for solar spectral twins in both hemispheres (Hardorp 1978).

3. OBSERVATIONS OF SOLAR SPECTRAL ANALOGS

77 G dwarfs were photoelectrically scanned with 20 Å resolution, using mainly 60 cm telescopes at ESO, Chile, and Mt. Hopkins, Arizona. Scans of the satellite Jupiter III and of the daylight sky represented the solar spectrum. The ratio spectra Sun/star of 9 of the 13 stars found most similar to the Sun are plotted in Fig. 1. HD 44594 = HR 2290 was subsequently calibrated from 3300 - 8500 Å by direct comparison with a black body (Hardorp and Tüg 1978), while HD 186427 will be compared to primary standards.

A. G. Davis Philip and D. S. Hayes (eds.), The HR Diagram, 423–426. All Rights Reserved.

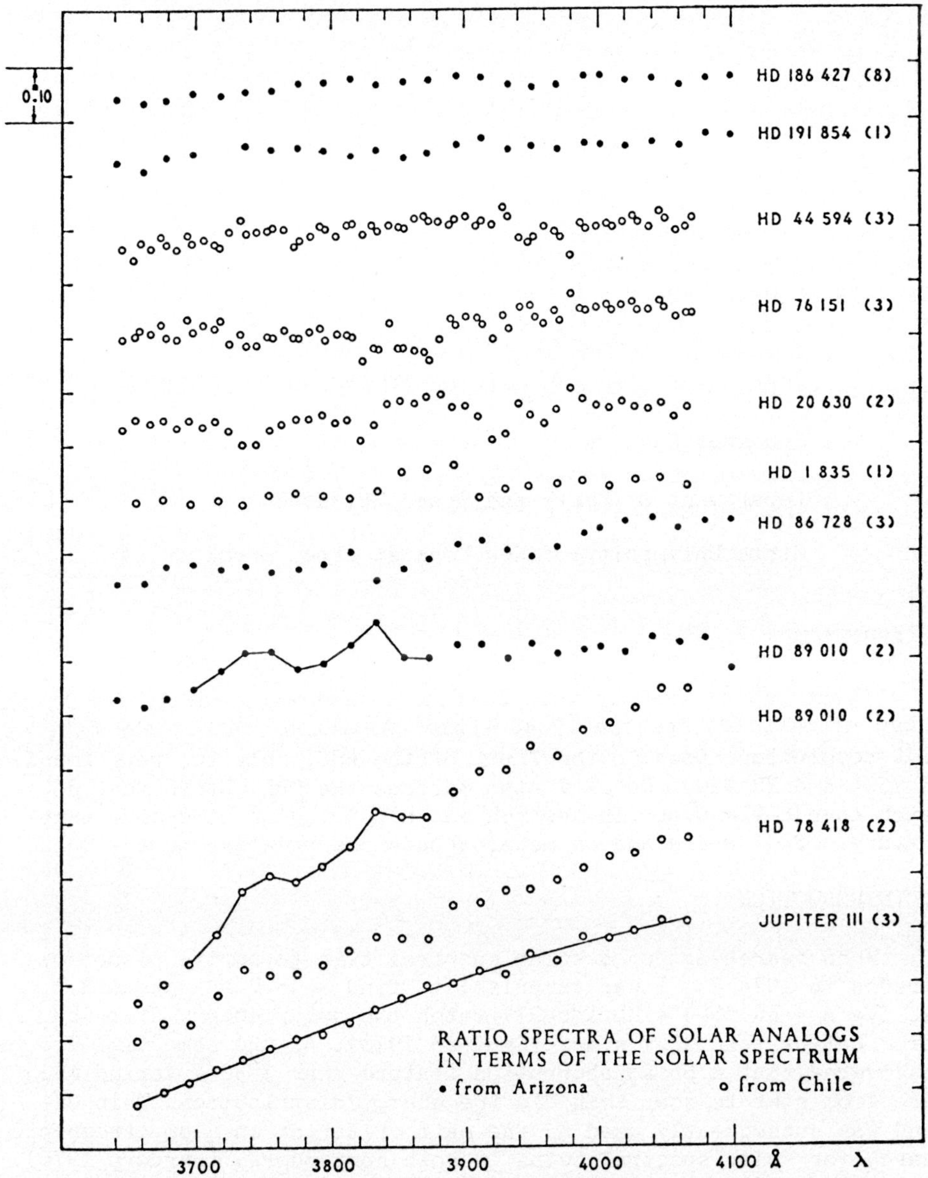

Figure 1. Ultraviolet spectra of stars found to be most similar to the Sun, in terms of the (tilted) spectrum of the daylight sky. Not corrected for extinction. The accuracy can be judged from the ratio Jupiter III/sky, which should be smooth in reality (in brackets, number of scans). HD 144585, 159222, 181655 were omitted from this graph for want of space. The three topmost stars-all of which have (B-V) = $0\overset{m}{.}66$ - cannot be distinguished from the Sun with the accuracy achieved here.

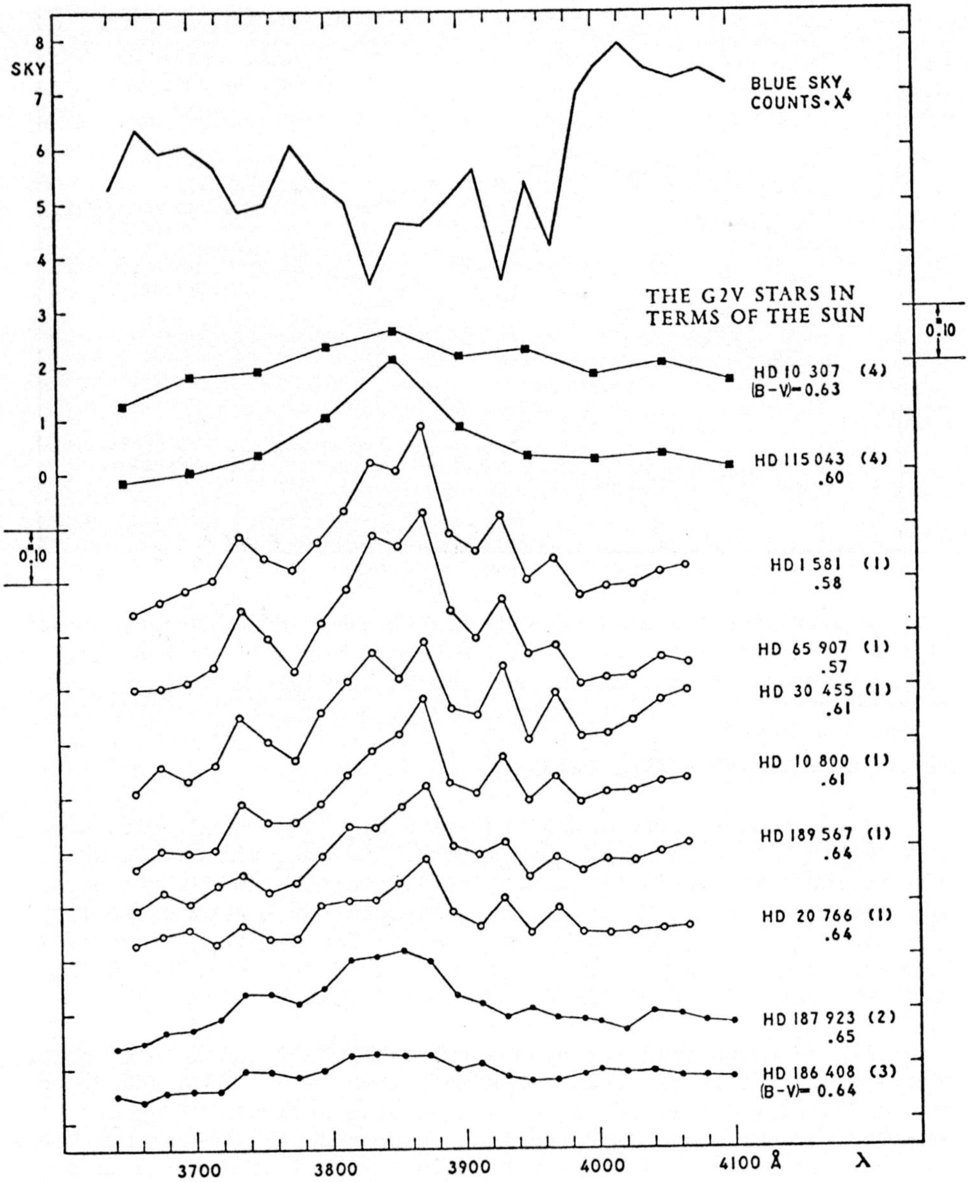

Figure 2: Ratio spectra star/sky of G2 V stars, whose types were taken from the Bright Star Catalogue. Not corrected for extinction. Squares: observed from Arizona with 50 Å resolution. Other symbols as in Fig. 1 (in brackets, number of scans). The daylight sky is displayed on a linear scale on top of the figure. The three G2 V stars omitted from this plot look similar to HD 65907. No G2 V star has ultraviolet absorption features as strong as the Sun.

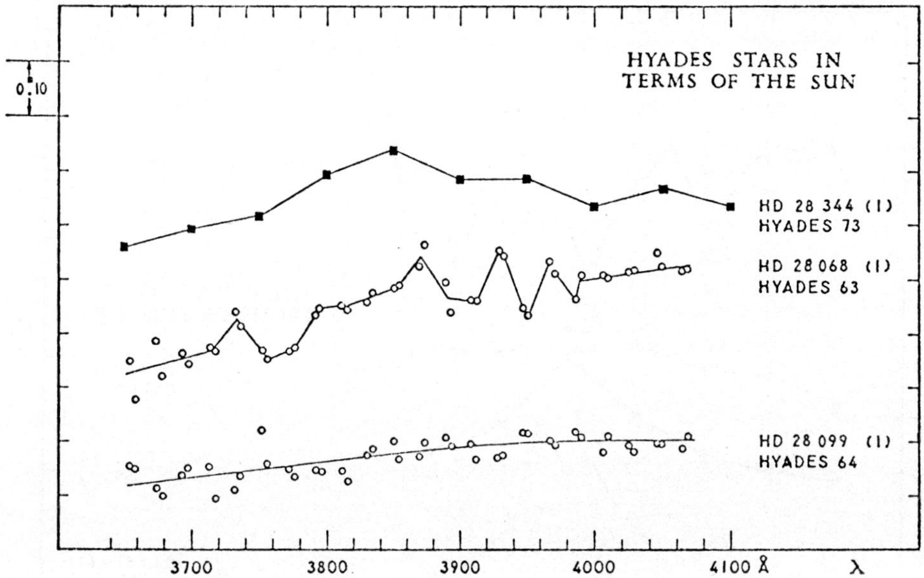

Figure 3. Within the limited accuracy obtained here, Hyades 64 is a solar spectral twin, while 63 and 73 are hotter than the Sun. Not corrected for extinction, symbols as in Fig. 2.

4. G2 V STARS AND HYADES DWARFS

13 G2 V stars were among the stars scanned, 10 of which are plotted in Fig. 2. None of them matches the Sun, all have weaker absorption features in the ultraviolet, even the two MK standards HD 10307 and 186408. MK types are obviously not a good criterion for preselection of solar spectral twins. The Geneva photometry was found much more useful for this purpose. THE SUN IS EITHER COOLER, MORE METALRICH OR OF DIFFERENT LUMINOSITY THAN ANY G2 STAR.

Fig. 3 shows that the Hyades dwarf HD 28099 could be of solar temperature, if it had solar abundances, while HD 28344 and 28068 are definitely hotter. Since most abundance analyses of Hyades dwarfs have been done assuming lower temperatures, these abundances may have to be revised. Energy distributions of Hyades G dwarfs should be measured to settle the question of their effective temperatures.

REFERENCES

Hardorp, J. 1976, Mitt. Astron. Ges. 38, 143.
Hardorp, J. 1978, Astron. Astrophys. in press
Hardorp, J., and Tüg, H. 1978 in preparation

OPTICAL AND INFRARED PROPERTIES OF THE NEWLY FORMED STARS IN CMa R1

William Herbst

Carnegie Institution of Washington

Rene Racine

University of Montreal

John W. Warner

University of Minnesota

UBVRIJHKL photometry and MK spectral types have been obtained for stars illuminating nebulae in the CMa R1 association. Five stars, including the two classical Herbig-emission-stars Z CMa and HD 53367, have (K-L) excesses and/or emission line spectra indicating the presence of circumstellar matter. A number of stars also have (V-K) indices larger than expected on the basis of their spectral types and (B-V) indices, suggesting that a steeper than normal extinction law applies to these stars. The color-magnitude diagram for the association is unusual; many stars over the entire spectral range of the association (B0 to A1) lie more than 1 magnitude above the ZAMS in V. The middle and early type B stars with this property are all variable stars and most have rotationally broadened spectra and/or shell indicators. No stars later than B5 are on the main sequence, making CMa R1 probably the youngest stellar group yet identified. It suggests an age of $\sim 3 \times 10^5$ years, compatible with the suggestion that a supernova explosion triggered star formation in this region. A comparison is made between the observations and theoretical models of pre-main sequence stars. A bolometric luminosity of $M_{bol} < -4\overset{m}{.}3$ is derived for Z CMa implying a core mass of $M > 10\ M_{\odot}$, and a spectral type of $\sim$B1, if the star is on the ZAMS.

A. G. Davis Philip and D. S. Hayes (eds.), The HR Diagram, 427.

LUMINOSITY CALIBRATION OF LATE-TYPE STARS BY STATISTICAL PARALLAX USING AGK3 PROPER MOTIONS

Keiichi Ishida

Tokyo Astronomical Observatory

Takao Mikami

University of Tokyo

1. INTRODUCTION

The luminosity of individual late-type stars is estimated mainly by spectral features in the MK spectral classification system or by means of the Wilson-Bappu effect. The MK spectral type is given for spectra with a dispersion of about 100 Å mm^{-1} which is within the reach of some Schmidt telescopes with objective prisms. On the other hand, the Wilson-Bappu effect is measured from spectra with a dispersion of about 10 Å mm^{-1}, obtained by coudé spectrographs. In either case the luminosities must be calibrated by knowledge of apparent magnitudes and distances for some stars.

Trigonometric parallaxes are one of the basic means of calibrating intrinsic luminosity. This most fundamental method can be applied successfully to main sequence stars of classes F5 through M, and with great care to giants of G5 to late K type. Zero-age main sequence fitting using cluster members is a usual means for calibrating high luminosity stars. The luminosity calibration of F, G, K and the later type giant stars depend primarily on the statistical parallax method.

Statistical parallaxes have been calculated for a large number of stars with AGK3 proper motions by making use of radial velocities in Wilson's (1963) and other catalogues. They have been published for carbon stars (Mikami 1975), M type stars (Mikami 1976), and F, G, and K type stars (Mikami 1977). The complete lists of those stars and their statistical material are available upon request from T. Mikami. Some results will be summarized briefly here.

A. G. Davis Philip and D. S. Hayes (eds.), The HR Diagram, 429–432.

2. CARBON STARS

There are 321 carbon stars classified in the C system in the present sample. The major source of spectral types is Yamashita (1972, 1975). It is found that their mean distance is over 1 kpc. Their mean distances can be derived from radial velocities and the first order differential galactic rotation equation. Mean visual absolute magnitudes thus obtained are $M_V=-1\overset{m}{.}3$ for C0-3, $-2\overset{m}{.}7$ for C4-5 and $-2\overset{m}{.}4$ for C6-7. Ninety stars have AGK 3 proper motions. Mean visual absolute magnitudes calculated with secular parallaxes are about $0\overset{m}{.}8$ fainter at each spectral subtype, although the difference is within the range of the probable error of the absolute magnitude derived by the secular parallaxes. The proper motion stars are subsamples of the radial velocity stars. It is likely that the stars with AGK3 proper motions are nearby stars and that, therefore, they are picked up more completely to fainter absolute magnitudes. It is suggested as a result that there exists a range of at least one magnitude of intrinsic dispersion in the visual absolute magnitude in each spectral subtype of carbon stars.

3. M-TYPE STARS

The sample includes 1490 M-type stars with luminosity classes in the MK or Mount Wilson system. A good correlation between the spectral types in the two systems was confirmed and they were combined in the analysis. There are 1116 giants with radial velocity data and 627 giants among 710 AGK3 proper motion stars. The mean distance of the giants is less than 1 kpc. Both p_U and p_T were calculated by the secular parallax method with radial velocity data and AGK3 proper motions. The mean visual absolute magnitudes obtained are listed in Table I. Mikami's values for M1 through M3 are $0\overset{m}{.}6$ brigher than those of Blanco (1965) which are mainly based on a distance modulus of $3\overset{m}{.}03$ of four giants in the Hyades based on the Wilson-Bappu effect. These values support Jung's (1970) values obtained by the statistical parallax method for stars in the Yale Catalogue of Bright Stars (Hoffleit 1964). Mikami's (1976) calibration of absolute magnitudes of the latest M-type giants are the first reliable ones available.

4. F, G AND K-TYPE STARS

In the sample of F, G and K-type stars there are 5384 stars with radial velocity data and 3574 stars with AGK3 proper motions. Most types are nearly equally divided into giants and dwarfs and the sample is also divided nearly equally into F, G and K star groups, though the majority of F stars are dwarfs and the majority of K stars are giants. Their mean distance is

TABLE I

MEAN VISUAL ABSOLUTE MAGNITUDES FOR M GIANT STARS

SP	Blanco (1965)	Jung (1970)	Mikami (1976)	Blanco -Mikami	n
M0	$-0^{m}.3$	$-1^{m}.2$	$-0^{m}.4$	$+0^{m}.1$	96(109)
M1	-0.5	....	-1.1	+0.6	60(67)
M2	-0.8	-1.5	-1.5	+0.7	126(114)
M3	-1.1	....	-1.6	+0.5	92(106)
M4	-1.0	-1.7:	-0.7	-0.3	119(123)
M5	-0.9:	....	-0.9	0.0	65(103)
M6	-0.9	....	+0.1	-1.0	44(86)
M7	-0.9:	....	+0.5	-1.4	25(48)
M8	....	....		....	(12)

from 230 pc for K giants to 30 pc for K dwarfs. Both p_U and p_T were calculated by the secular parallax method. The calculations were made for each Mount Wilson luminosity class and the results were converted to those in the MK luminosity class with FitzGerald's (1969) correlation tables. The resultant values of the absolute magnitude of dwarfs and subgiants confirm Blaauw's (1963) values. However, Mikami's values for K giants are $0^{m}.7$ brighter than Blaauw's (1963), and agree with those carefully derived from trigonometric parallaxes by Ljunggren and Oja (1966), as shown in Table II.

5. CONCLUDING REMARKS

AGK3 proper motions have been successfully used to derive

TABLE II

MEAN VISUAL ABSOLUTE MAGNITUDES FOR K GIANT STARS

Sp	Blaauw (1963)	Ljunggren & Oja (1966)	Mikami (1977)	Blaauw -Mikami	n
K0	$+0^{m}.8$	$+0^{m}.1$	$+0^{m}.1$	$+0^{m}.7$	148(212)
K1	+0.8	+0.8	+0.4	+0.4	114(171)
K2	+0.8	-0.1	+0.3	+0.5	113(172)
K3	+0.1	-0.2	-0.4	+0.5	116(180)
K4	+0.1	-0.9	-0.6	+0.7	130(187)
K5	-0.3	-1.4	-1.2	+0.9	160(251)

statistical parallaxes of late type giant stars. It seems that a revision is needed to the calibration formula of the Wilson-Bappu effect. Statistical parallaxes might depend on local peculiarities of the kinematical characteristics of field stars in the solar neighborhood. The consistency with results derived by other independent methods offer support for the results discussed here.

REFERENCES

Blaauw, A. (1963). Stars and Stellar Systems, Vol. 3, Basic Astronomical Data, K. Aa. Strand, ed., Univ. of Chicago Press, Chicago, p. 383.

Blanco, V.M. (1965). Stars and Stellar Systems, Vol. 5, Galactic Structure, A. Blaauw and M. Schmidt, eds., Univ. of Chicago Press, Chicago, p. 241.

FitzGerald, M.P. (1969). J. R. Astron. Soc. of Canada **63**, 251.

Hoffleit, D. (1964). Catalogue of Bright Stars, Yale University Observatory, New Haven.

Jung, J. (1970). Astron. and Astrophys. **4**, 53.

Ljunggren, B. and Oja, T. (1966). IAU Symp. 24, Spectral Classification and Multicolor Photometry, K. Lodén, L.O. Lodén and U. Sinnerstad, eds., D. Reidel, Dordrecht, p. 317.

Mikami, T. (1975). Publ. Astron. Soc. Japan **27**, 445.

Mikami, T. (1976). ibid. 29, No. 4.

Mikami, T. (1977). ibid. 29, No. 4.

Wilson, R.E. (1963). General Catalogue of Stellar Radial Velocities, Carnegie Institution of Washington.

Yamashita, Y. (1972). Ann. Tokyo Astron. Obs. **13**, 169.

Yamashita, Y. (1975). ibid. 15, 47.

FUNDAMENTAL PARAMETERS AND THE HELIUM CONTENT OF ECLIPSING SPECTROSCOPIC BINARIES

H. E. Jørgensen

Copenhagen Observatory

Today we know accurate masses, radii, surface gravities and luminosities or effective temperatures of a very small number of stars. There are several reasons for deriving accurate parameters of stars. I shall mention only four problems:

(i) checking stellar evolution calculations, particularly calculations of isochrones, using binaries,
(ii) derivation of helium content in stars of different age,
(iii) checking log g derived from photometry and stellar atmosphere calculations,
(iv) determination of accurate positions in the HR diagram to get an independent check on calibrations of photometric systems.

At the Copenhagen University Observatory we have carried out a program deriving fundamental parameters of eclipsing spectroscopic binaries. The spectroscopy is done by J. Andersen with the ESO 1.5m spectrographic telescope at La Silla (in some cases in collaboration with D.M. Popper). The photometry is done by a number of people in Copenhagen using our 50cm reflector at La Silla. Our telescope at La Silla is equipped with a simultaneous four channel spectrographic photometer. The four channels correspond to the Strömgren uvby photometric system. At present we have obtained approximately 40,000 accurate photometric observations of selected eclipsing binaries.

Our philosophy in this program is to study binaries with well separated components to avoid complications such as gas in the

A. G. Davis Philip and D. S. Hayes (eds.), The HR Diagram, 433-436.

system, strong deformation of the components, strong reflection effects, etc. Models of binaries with such complications are not very well developed and derived parameters of stars in such systems are therefore not very accurate. A binary with well separated components can be described using a relatively simple model. We use the model by Wood (1971).

Let us now ask the questions: What accuracy must we have to solve the problems (i) - (iv) already mentioned? We shall particularly consider the helium content problem. From measurements of emission lines from H II regions in the optical and radio region a present helium content in the interstellar medium is derived. From accurate photoelectric measurements of the helium line at 4026 Å Nissen (1976) has derived helium abundances in atmospheres of B stars. Both the emission line measurements and the 4026 Å measurements in the young stars seem to indicate a variation in the helium abundance from place to place. The abundance is normally found to be in the range $Y = 0.20 - 0.30$ and the variation from place to place amounts approximately to $\Delta Y = 0.10$. Also for very old objects such as globular clusters and halo planetary nebulae we may give an estimate of helium abundance. However the abundance of helium is, in this case, derived from stars in late evolutionary stages and we are ususally left with both the uncertainty of enrichment in the envelopes and the uncertainty in internal structure of horizontal-branch stars. Values in the range $Y = 0.2 - 0.4$ are often obtained.

At present we have little evidence for any general increase in helium abundance with time, although it has been proposed that enrichment of helium in the interstellar medium is correlated with the enrichment by heavier elements.

From this discussion it is obvious that the expected range in the abundance of helium from star to star is no more than $\Delta Y \sim 0.10$. To obtain further knowledge of an eventual enrichment of the interstellar medium we must therefore determine the helium abundance with an accuracy of approximately $\Sigma(Y)=0.03$. From stellar evolution calculations we can easily estimate that this corresponds to a typical set of uncertainties in the fundamental parameters: 2 per cent in mass, 1 per cent in radius, 2 per cent in effective temperature and 25 per cent in Z, the content of heavier elements.

Our results are given in Table I. The masses are in most cases determined to an accuracy of 2 per cent and the radii are known with sufficient precision. To derive effective temperatures we use [u-b] for B stars and the β index for the late A and F stars. In a few cases we still have no β index available and use (b-y), since the star is nearby. The m_1 index is used to determine [Fe/H] and from that the value of Z. An accuracy of 25 per cent in Z corresponds to an accuracy of $0^m.007$ in m_1, which normally is no problem.

TABLE I

FUNDAMENTAL PARAMETERS OF ECLIPSING SPECTROSCOPIC BINARIES

Binary	$M/M_\odot$	$R/R_\odot$	log g	Te	log Te	Z	Y	Age in yrs
CV Vel A	6.10±.04	4.10±.04	4.00±.01	18200	4.206±.01	.02	.28	33×10^6
B	5.99±.04	3.95±.04	4.02±.01	18060	4.257±.01			
ζ Phe A	3.85±0.2	2.84±.05	4.12±.02	14400	4.160±.02	.02	.28	63×10^6
B	2.5 ±0.1	1.85±.04	4.30±.02	12000	4.08 ±.02			
χ^2Hya A	3.61±.08	4.39±.04	3.71±.01	11700	4.07 ±.01	.02	.28	180×10^6
B	2.64±.05	2.16±.03	4.19±.02	11300	4.05 ±.01			
SZ Cen A	2.28±.02	3.62±.02	3.678±.006	8280	3.92 ±.02	.02	.30	450×10^6
B	2.32±.02	4.55±.02	3.478±.007	8000	3.90 ±.02			
HS Hya A B	1.31±.03	1.28±.01	4.34±.01	6770	3.83 ±.01	.014	.19	$\lesssim 10^9$:
RS Cha A	1.86±.02	2.17±.09	4.03±.04	7850	3.90 ±.01	.020	.24	1×10^9
	1.82±.02	2.34±.09	3.96±.03	7450	3.87 ±.01			
TV Cet A	1.39±.05	1.50±.01	4.23±.02	7000	3.845±.01	.014	.22	1.6×10^9
	1.27±.04	1.28±.01	4.33±.02	6582	3.818±.01			
RZ Cha $\overline{AB}$	1.51±.02	2.26±.02	3.91±.01	6580	3.82 ±.01	.020	.21	2.5×10^9
UX Men A	1.17±.03	1.32±.02	4.27±.02	6200	3.79 ±.01	.014	.16	6×10^9
B	1.11±.02	1.25±.02	4.29±.02	6130	3.79 ±.01			

The derived helium abundances in Table I are very close to 0.28 for the B stars and close to 0.22 for the F stars. Half of this difference (0.03) could be explained by the fact that Z for the F stars equals 0.016 and not 0.020 as adopted for the B stars. The other half (0.03) is most easily understood as a small error in the temperature scale or by errors in the calculated opacities. We notice, in particularly, the small variation in the helium content of F stars, where the spread is not more than $\Delta Y=0.03$ around the mean value of $Y=0.22$.

REFERENCES

Nissen, P.E. (1976). Astron. Astrophys. 50, 343.
Wood, D.B. (1971). Astron. J. 76, 701.

EXTENDING THE WHITE DWARF SEQUENCE: PLANS AND PROSPECTS

James Liebert

Steward Observatory, University of Arizona

1. INTRODUCTION AND SUMMARY

The recent discovery that the parallax star LP701-29 is a white dwarf has firmly extended the degenerate sequence below $M_V = +16^m$ (Dahn et al. 1978). As the search for white dwarfs extends to cooler and fainter stars, however, it becomes particularly important to develop a plan for selecting candidates among the many thousands of red proper motion stars. We begin by assessing the completeness of the known sample within 10 parsecs in the northern two thirds of the sky. Some color-dependent selection effects must be evaluated, however, since these may preferentially inhibit the discovery of cooler stars. A correction factor for the missing low velocity white dwarfs is estimated. Then, Green's(1977) recent determination of the number density of blue degenerates is used to normalize various theoretical luminosity functions, the benchmarks against which the current sample out to 10 pc can be compared. It is concluded that the sample may be approaching completeness in the northern sky for white dwarfs with tangential velocities (v_T) $\gtrsim$ 40 km/sec ($\mu \geq 1''.0$/yr.) and $M_{bol} < +15^m$. The implied luminosity function is thus consistent with that found by Sion and Liebert (1977). Below $M_{bol} = +15^m$ the different theoretical functions predict substantially different numbers.

It is proposed that the remaining search concentrate on finding more nearby stars with large proper motions, especially in the interval $15^m \leq M_{bol} \leq 17^m$, and then correcting for the missing stars of smaller motion. This approach offers the advantage of decoupling the degenerate star luminosity function problem from those posed by the disk M dwarfs and the background Population II dwarfs. Not that many faint stars of large proper motion remain

A. G. Davis Philip and D. S. Hayes (eds.), The HR Diagram, 437–440. All Rights Reserved.

unobserved in Luyten's catalogues, although some of them may have been missed. Some useful answers concerning the white dwarf density below the current M_{bol} limit should be forthcoming.

2. BARRIERS TO THE IDENTIFICATION OF COOL DEGENERATE STARS

2.1 Velocity Selection

All northern stars with $M_V > 13^m$ and $\pi \geq 0\overset{''}{.}1$ have proper motions in excess of 1"/yr and $v_T \geq 40$ km/sec. Yet many hot degenerates found on the basis of blue color have small space velocities. Thus the mean velocity characteristics of the known spectroscopic samples of blue and red stars differ greatly (Sion and Liebert 1977). There can be little doubt that the greater part of this velocity difference is due to velocity selection in the sample of cooler degenerates.

The cool white dwarf numbers can be corrected for the missing low velocity stars by assuming that the true velocity distribution is like that of old disk stars. The sample of non-emission late K and M dwarfs found independently of motion by the McCormick objective prism survey, and also appearing in Gliese's catalogue of nearby stars, was utilized for this purpose. See also Wielen(1974). Applying a correction is simple if it can be assumed that the northern 10 pc sample is nearly complete for $v_T \geq 40$ km/sec. Indeed, a spectroscopic and astrometric survey of catalogued stars with $\mu \geq 1''$ and $m_{pg} \leq 16^m$ is now nearly complete (see Greenstein's report to this Symposium, page 101). Degenerates with $M_{bol} \leq 15^m$ and $d \leq 10$ pc are no fainter than the red Palomar survey plate magnitude $(m_R) = 16^m$. Since 62% of the McCormick/Gliese stars have a lower velocity than 40 km/sec, the correction factor adopted is 2.6. For $\delta \geq -20^o$ there ought to be some $2.6 \times 9 = 23 \pm 8$ white dwarfs with $13 \leq M_{bol} \leq 15$ by this estimate.

2.2 Loss of Cool Degenerates in Composite Binary Systems

A cool white dwarf is particularly difficult to identify in an unresolved binary spectrum. Not only are such objects intrinsically faint, but they generally lack strong absorption features and may not show significant blue or red excesses relative to the other component. Since the most frequent companion is a K-M dwarf, blue degenerate stars in composite systems are often recognized. The sample of the nearest stars shows us, however, that there must be many degenerates which, regardless of color, will be hidden by bright companions like Sirius and Procyon. The simplest procedure for dealing with the bias effect on the luminosity function by close binaries is to remove them

altogether and calculate a luminosity function only for single and common proper motions stars. Then the total number density must be expressed by summing this single star function and then incrementing it by an uncertain binary contribution.

2.3 Other Selection Effects

Three additional problems discussed by Liebert (1978) in conjunction with this work are (3) line blanketing among cool degenerates, (4) the problem of correct spectroscopic identification of cool degenerates, and (5) the incompleteness of the proper motion surveys for faint stars of large proper motion.

3. PREDICTED AND OBSERVED NUMBERS OF NEARBY DEGENERATES

Green(1977) derives a total number density of blue stars of $1.43 \pm 0.28 \times 10^{-3}$ for $M_V \leq 12^m.75$ and a less certain $2.42 \pm 0.57 \times 10^{-3}$ for $M_V \leq 13^m.25$. These values lie within the range of previous estimates. For convenience here the geometric mean of these values was adopted for $M_V \simeq M_{bol} \leq 13^m$. For the 84% of the northern two thirds of the sky with measured Luyten Palomar fields, this density corresponds to 4.2 stars with $M_V \leq 13^m.0$. Using this number as the normalization for the restricted 10 pc sample, various theoretical calculations then provide predictions for the numbers of cooler stars.

TABLE I
EXPECTED AND OBSERVED NUMBERS
WITHIN RESTRICTED 10pc VOLUME

Intervals		Theoretical					2.6x
M_{bol}	M_V	SK	LVH	Sw	Mes	Observed	Observed
$<13^m$	$<13^m$	4.2	4.2	4.2	4.2	4	--
13-14	13.0-14.0	3.9	3.0	4.2	4.2	3	8
14-15	14.0-15.3	7.4	3.2	4.9	8.4	4	10
15-16	15.3-17.2	14.1	2.9	5.6	16.8	2	5
16-17	17.2-20.	26.3	2.2	4.9	33.6	0	0

SK = Shaviv and Kovetz (1976) for $0.6M_\odot$
LVH = Lamb and Van Horn (1975) for $1.0M_\odot$
Sw = Sweeney (1976) for $0.51M_\odot$
Mes = Mestel cooling relation

The reasons for the different predictions for $M_{bol} > 15^m$ and other possible effects are reviewed in Liebert (1978).

It is proposed that a survey of large proper motion stars fainter than $m_R = 16^m$ would find nearly all low luminosity degenerates within 10 pc and with $v_T \geq 40$ km/sec. There are only 29 of these in Luyten's catalogues. An evaluation of the completeness of the surveys is necessary, but there are reasons for believing that this is not a critical problem. The different theoretical predictions of Table I may be tested within a relatively short time. In fact, the failure to find any lower luminosity stars in work so far on this Luyten sample (Liebert and Strittmatter 1978) indicates that there are actually fewer stars in the last M_{bol} interval than predicted by the continuous function of Shaviv and Kovetz (1976).

REFERENCES

Dahn, C., Hintzen, P., Liebert, J., Stockman, H. and Spinrad, H. (1978). Astrophys. J., in press.
Green, R. (1977). Ph.D. Thesis, California Institute of Technology.
Lamb, D.Q., and Van Horn, H.M. (1975). Astrophys. J., **200**, 306.
Liebert, J. (1978). in preparation.
Liebert, J., and Strittmatter, P. (1978). in preparation.
Shaviv, G., and Kovetz, A. (1976). Astron. Astrophys., **51**, 383.
Sion, E., and Liebert, J. (1977). Astrophys. J., **213**, 468.
Sweeney, M. (1976). Astron. Astrophys., **49**, 375.
Wielen, R. (1974). in Highlights of Astronomy, **3**, ed. G. Contopoulos, D. Reidel Publishing Co., Dordrecht, Holland, p. 395.

PECULIARITIES IN SOME CLUSTER COLOR-MAGNITUDE DIAGRAMS

Bernard J. McNamara and W. L. Sanders

New Mexico State University

There has existed for some years a generally accepted conceptual scheme of stellar evolution as reflected in HR diagrams. Such evolutionary sequences are normally only apparent in stellar clustering where, in the first order at least, coeval formation is assumed. Such observational evidence of evolutionary sequences is directly dependent upon the identification of cluster membership. Only in very recent years have the necessary high precision and full membership based independently upon proper motion become available for a significant sample of clusters. The result of these studies for some sample cluster C-M diagrams are discussed and some possible difficulties with our evolutionary scheme are evidenced.

A. G. Davis Philip and D. S. Hayes (eds.), The HR Diagram, 441.

THE STRÖMGREN FOUR-COLOR SYSTEM AND THE LOG T_{eff}, LOG G DIAGRAM

Lawrence J. Relyea

Dudley Observatory

A. G. Davis Philip

Dudley Observatory and Union College

1. INTRODUCTION

Strömgren (1966) set up the four-color system of photometry as a means of studying early-type stars in the spectral range O through F. In this spectral range one of the important measures that can be made of a stellar spectrum is the magnitude of the Balmer jump. For stars of a given (b-y) color the discontinuity is very large for stars of high luminosity and small for low luminosity stars. For very hot stars (O and B stars) this luminosity effect becomes too small to be used as a luminosity indicator, but because of this insensitivity to luminosity changes the magnitude of the Balmer discontinuity becomes a temperature indicator (see Philip and Newell 1974 and Davis and Shobbrook 1977). Photoelectric measures of the strength of the $H\beta$ line can be used in a similar, but reverse, manner. For the B and O stars the index is a measure of luminosity; for late A and F stars the index is a good temperature index. In the spectral range A0-A3 both the Balmer jump and the Beta index are functions of temperature and luminosity. For a bibliography of the many uses of Strömgren four-color and $H\beta$ photometry see the paper prepared by Philip and Perry (1977).

Kurucz (1977) has computed a grid of model atmospheres for O-G stars with the input parameters of T_{eff} (5500K to 10000K), log g (4.5 to 2.0) and [M/H] (solar abundance, 1/10 and 1/100 solar abundance). By means of this grid, it is now possible to transform a set of intrinsic four-color indices into the fundamental parameters log T_{eff} and log g for stars in the spectral range

A. G. Davis Philip and D. S. Hayes (eds.), The HR Diagram, 443-446.

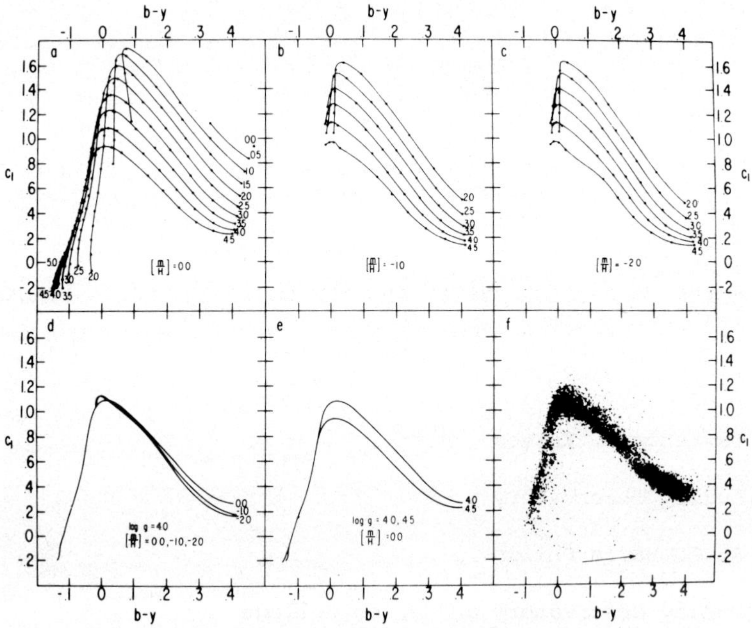

Fig. 1. The (c_1) vs. (b-y) diagrams for (a) models with [m/H] = 0.0, (b) models with [m/H] = -1.0, (c) models with [m/H] = -2.0, (d) models with log g = 4.0, three separate abundances, (e) models with log g = 4.0 and 4.5, [m/H] = 0.0, and (f) stars (not dereddened) in the HMC with the direction of reddening indicated by an arrow.

A-F and luminosity classes V-III of both Pop I and II, provided one learns the approximate metal abundance of the star by means of some other investigation (see Relyea and Kurucz 1977).

2. THEORETICAL COLORS

A theoretical photometric system was defined in Relyea and Kurucz (1977), who used the photometric calibration of Vega (Hayes and Latham 1975) to calibrate the system. Transformation coefficients between the theoretical colors and the observed colors for Vega (Crawford and Barnes 1970) were determined by convolving the filter transmission curves of the uvby filters from Matsushima (1969) with the fluxes predicted by Kurucz's atmospheric models.

The transformation between the theoretical colors, $(c_1)_o$ and $(b-y)_o$, and the observed colors was checked by comparing the calculated log T_{eff} and log g values for 28 stars (considered as standards because spectroscopic measures had been made from which log T_{eff} and log g were calculated) with the published values (see Philip, Miller and Relyea 1976). The rms errors of the calculated values relative to the published values were $\pm$ 250 K in

T_{eff} and ± 0.2 in log g. A small discrepancy in log g's calculated for F-type stars (noted in Relyea and Kurucz 1976) was corrected by moving the grid so that the ZAMS of Crawford (1975) fell on the log g = 4.3 grid line in the c_1, (b-y) diagram. The grid of log g values was corrected at log g = 2.0, 2.5, 3.0, 3.5, 4.0 and 4.5 by moving the grid up, along the temperature lines by the amount which corrects the ZAMS. The theoretical m_1 indices, calculated for log g's of 4.0 and 4.5, do not agree with the distribution of points in an observational m_1 versus (b-y) diagram for stars with (b-y) > $0^m.05$, so this grid was not used for the calculation of T_{eff} and log g.

The $(c_1)_o$, $(b-y)_o$ grid for [M/H] = 0, -1.0 and -2.0 is shown in Fig. 1 abc (from Relyea and Kurucz 1978). The effects of changes in [M/H] and log g are shown in Fig. 1 de. The distribution of 5183 stars in the Hauck Mermilliod catalogue (uncorrected for interstellar reddening) is shown in Fig. 1 f. Philip and Relyea (1977) have prepared large scale grids for use in transforming dereddened $(c_1)_o$ and $(b-y)_o$ indices in Θ_{eff} and log g. To use the grids, one must first deredden the star, and then knowing the population type and the metallicity, plot the point in the $(c_1)_o$, $(b-y)_o$ grid and read off the Θ_{eff} and log g. If the star is peculiar, luminosity class I or II, or outside the limits $0^m.0 < (b-y)_o < 0^m.4$ then Θ_{eff} and log g can not be determined.

Code *et al.* (1976) give T_{eff} for some early-type main sequence stars in their Table 6 and a relation between T_{eff} and $(B-V)_o$ in Table 7. The T_{eff}, $(b-y)_o$ relation for log g = 4.3 is very close to the Code relation and does as well or better than similar relations determined by Palmer (1977), Olson (1974), or Breger (1975).

3. APPLICATIONS

Four-color observations of any normal Population I or II stars in the $(b-y)_o$ color interval of $0^m.0$ to $0^m.4$ can be transferred into the log T_{eff}, log g plane. Astronomers who calculate evolutionary tracks present their calculations in such diagrams, so it is possible to make direct comparisons between theory and observation. Some comparisons of this sort will be described in more detail in another paper at this Symposium (Philip 1978).

REFERENCES

Breger, M. (1975). In *Multicolor Photometry and the Theoretical HR Diagram*, A.G.D. Philip and D.S. Hayes, eds., Dudley Observatory Report No. 9, 31.

Code, A.D., Davis, J., Bless, R.D. and Hanbury Brown, R. (1976). Astrophys. J. **203**, 417.
Crawford, D.L. (1975). Astron. J. **80**, 955.
Crawford, D.L. and Barnes, J.V. (1970). Astron. J. **75**, 978.
Davis, J. and Shobbrook, R.R. (1977). Mon. Not. Roy. Astron. Soc. **178**, 651.
Hauck, B. and Mermilliod, M. (1975). Astron. and Astrophys. **22**, 235.
Hayes, D.S. and Latham, D.W. (1975). Astrophys. J. **197**, 593.
Kurucz, R.L. (1978). in preparation.
Matsushima, S. (1969). Astrophys. J. **158**, 1137.
Olson, E.C. (1974). Publ. Astron. Soc. Pacific **86**, 80.
Palmer, L. (1977). Astron. J. **82**, 158.
Philip, A.G.D. (1978). In IAU Symposium No. 80, The HR Diagram, A.G.D. Philip and D. S. Hayes, eds., Reidel, Dordrecht, p. 209.
Philip, A.G.D., Miller, T.M. and Relyea, L.J. (1976). Dudley Observatory Report No. 12.
Philip, A.G.D. and Newell, B. (1974). In Multicolor Photometry and the Theoretical HR Diagram, A.G.D. Philip and D.S. Hayes, eds., Dudley Observatory Report No. 9, p. 161.
Philip, A.G.D. and Perry, C.L. (1977). In Vistas in Astronomy, A. Beer, ed. (in press).
Philip, A.G.D. and Relyea, L.J. (1978). in preparation
Relyea, L.J. and Kurucz, R.L. (1978). Astrophys. J. Suppl. (May issue).
Strömgren, B. (1966). In Annual Reviews in Astronomy and Astrophysics, Vol. **4**, L. Goldberg, ed., Annual Reviews Inc., Palo Alto, p. 433.

ANGULAR DIAMETERS OF COOL STARS MEASURED BY THE METHOD OF LUNAR OCCULTATIONS

Nathaniel M. White

Lowell Observatory

Given the angular diameter and total energy flux of a star, its effective temperature can be determined. Until recently, there were few well-determined angular diameters of cool stars, and effective temperature calibrations rested on angular diameter measurements of several not necessarily "normal" stars. The number of measured angular diameters of stars cooler than about spectral type G is steadily being increased by using the method of photoelectric observations of lunar occultations.

Angular diameter measurements of 42 stars are compiled in the Table. These are the results of many authors; most have been published since 1970. Excluding several geometric occultations (i.e., no diffraction pattern), the mean angular diameter measured has been about 5 milliseconds of arc, with an average error of about 0.8 milliseconds of arc. (Methods of error estimates vary among authors.) This is an indication of the unique power of the lunar occultation technique.

Thirteen of the 42 stars have more than one independent observation, with μ Gem having eight observations. Since the observational data are recorded in a fraction of a second, are not repeatable, and therefore are unique for a given site and time, duplication of observations at various sites is extremely important and is obviously possible.

As the lack of known cool star angular diameters is being alleviated systematically through the lunar occultation technique, the lack of homogeneous photometry is becoming more important. Photometry reducible to absolute fluxes and covering a wide wavelength range for the stars listed in the Table will provide the

A. G. Davis Philip and D. S. Hayes (eds.), The HR Diagram, 447–450.

STELLAR ANGULAR DIAMETERS FROM LUNAR OCCULTATIONS

BS	Name	Spectral Type[1]	m_V[1]	Milliseconds of Arc[2] ϕUD	ϕLD	Central Wavelength[3]	Ref. Code[4]
224	63 δ Psc	K5 III	4.ᵐ44	4.2 ± 1.		≃.5	7
601		gM2	6.11		2.6 ± 0.5	≃.5	2
867	45 Ari	M6⁻ III	5.94		9.5 ± 2.0	≃.75	1
				9.88 ± 0.2	10.27 ± 0.21	1.64	24
				11.16 ± 0.99	11.38 ± 1.01	3.45	24
	+24°571	K5	6.77	1.8			5
1977	Y Tau	C6,4	Var	8.39 ± 0.24	8.80 ± 0.25	1.64	24
				7.98 ± 0.23	8.38 ± 0.24	1.64	24
2013	BD+27°888	gG7	5.52	2.4 ± 1.2		≃.5	7
2047	χ^1 Ori	G0 V	4.41		1.6 ± 0.5	≃.70	12
2063	U Ori	M7.0-M9.0	Var	14.30 ± 0.54		1.64	24
				15.45 ± 0.33	15.45 ± 0.33	2.10	24
2286	μ Gem	M3 III	2.97	11.8 ± 2.4	13.2 ± 2.4	.456	10
				12.3 ± 1.4	13.5 ± 1.4	.820	10
				12.1 ± 1.8		V	27
				12.0 ± 0.7		V	27
				11.8 ± 0.9		y	27
				12.3 ± 1.8		V	27
				12.3 ± 0.34	13.70 ± 0.41	.45,.52,.58	23
				14.2 ± 1.3	15.81 ± 1.45	V	20
2473	ε Gem	G8 Ib	3.08	5.6 ± 0.6		≃.5	7
				1.8			5
2533	87 B Gem	gK5	5.62	2.8			5
2671	R Gem	S3,9e-S6,9e	6.5	3.7			5
2938	74 Gem	M0 III	5.07	2.78 ± 0.27		.75	28
3094	1 Cnc	gK3	5.81	2.1 ± 0.6		.6	7
	V Cnc	S2.9e	7.5-	2.6 ± 0.7	2.8 ± 0.8	.845	16
	+13°1994	M4 III	6.61	3.84 ± 0.55		1.64	24
				4.0 ± 0.5		.75	28
3541	X Cnc	C5,4	6.22	7.9 ± 0.8	9.3 ± 0.8	.69	4
3550	60 Cnc	gK5	5.56	3.3 ± 0.4		.75	28
3882	R Leo	gM8e	5.0-	67 ± 5		≃.70	19
3950	π Leo	M2 III	4.71	5.2 ± 0.5	5.9 ± 0.5	R	25
3980	31 Leo	K4 III	4.43	2.8 ± 0.6		≃.55	7
				3.1 ± 0.2		≃.5	13
4127	46 Leo	gM2	5.54	5.6 ± 1.1		.51	21
4432	87 Leo	K4 III	5.07	3.7 ± 0.4	4.1 ± 0.4	.547	9
				2.9 ± 0.5		y	28
4517	ν Vir	M1 III	4.02		5.65 ± 0.05		22
4902	ψ Vir	M3 III	4.91		6.1 ± 0.3	.63	26
					6.5 ± 0.3	.69	12
				4.92 ± 0.38	5.11 ± 0.39	1.64	24
5301		gM3	4.91		3.6 ± 0.5	≃.80	2
				4.65 ± 0.33	4.79 ± 0.34	1.64	24
				5.0 ± 1.		b	28
5824	42 Lib	K4 III	4.95	2.2 ± 0.3		V	15
6134	α Sco	M1 Ib	1.08	41. ± 1.			11
	IRC-20 321	M6		4.37 ± 0.25		2.2	24
	V774 Sgr	M5	12.-	5.65 ± 0.73		1.64	24

STELLAR ANGULAR DIAMETERS FROM LUNAR OCCULTATIONS (continued)

BS	Name	Spectral Type[1]	m_v[1]	Milliseconds of Arc[2] ϕ_{UD}	ϕ_{LD}	Central Wavelength[3]	Ref. Code[4]
	IRC-20 445	M1		4.0 ± 1.0		2.40	24
6861	70 B Sgr	gM5	6.25	3.2 ± 0.8		V	15
6913	22 λ Sgr	K2 III	2.85	4.4 ± 0.3			18
7023		M4	6.49		8.5 ± 1.5	≈.80	3
7150	ζ^2 Sgr	K1 III	3.52		3.0 ± 1.0	≈.71	1
				3.7 ± 0.5	4.2 ± 0.5	R	25
	SAO 162049	Ma	6.7		2.8 ± 0.9	≈.8	2
7776	β Cap	gKo	3.07	3.05 ± 0.12	3.17 ± 0.17	1.64	24
7900	ν Cap	M2 III	5.10	4.1 ± 0.5	4.7 ± 0.5	.694	8
				8.6 ± 2.		.762	7
8318	47 Cap	gM3	6.20	2.8 ± 0.6		≈.5	15
8698	λ Aqr	M2 III	3.74	7.4 ± 0.4	8.2 ± 0.4	V	17
8834	φ Aqr	M2 III	4.22	4.9 ± 0.8		V	21
9004	19 Psc	C6,2	5.30	9. ± 1.		R	14
				8.0 ± 1.	8.9 ± 1.0	.71μ	6
					10.2 ± 2.5	.69μ	8

NOTES TO TABLE

[1]Spectral types and magnitudes are from the reference listed or the Bright Star Catalog.

[2]ϕ_{UD} and ϕ_{LD} indicate the fitted model used, either a uniform disk or limb-darkened disk approximation, respectively. Nearly all limb-darkened fits used the fully darkened assumption, i.e., β = 1.0.

[3]The approximate central wavelengths are in micron units, with the letters being the usual notation for standard photometric systems.

[4]The five angular diameters with reference code 28, White, N. M. (1978), have not been published previously.

---

best estimate of their total flux output. Since the effective temperature is a function of the fourth root of the total flux, the error in effective temperature determinations will be reduced to mostly that related to the angular diameter measurement, that is, about 5% in the effective temperature.

The purpose of this compilation is to draw attention to these stars (some of which do not even have good MK spectral types), in order to encourage their detailed observational and theoretical investigation.

REFERENCES TO TABLE

1. Africano, J. L., Cobb, C. L., Dunham, D. W., Evans, D. S., Fekel, F. C., and Vogt, S. S. (1975). Astron. J. **80**, 689.
2. Africano, J. L., Evans, D. S., Fekel, F. C., and Ferland, G. J. (1976). Astron. J. **81**, 650.
3. Africano, J. L., Evans, D. S., Fekel, F. C., and Montemayor, T. (1977). Astron. J. **82**, 631.
4. Bartholdi, P., Evans, D. S., Mitchell, R. I., Silverberg, E.C., Wells, D. C., and Wiant, J. R. (1972). Astron. J. **77**, 256.
5. Beavers, W. I., and Eitter, J. J. (1974). Bull. Amer. Astron. Soc. **6**, 216.
6. de Vegt, Chr. (1974). Astron. Astrophys. **34**, 457.
7. de Vegt, Chr. (1976). Astron. Astrophys. **47**, 457.
8. Dunham, D. W., Evans, D. S., Silverberg, E. C., and Wiant, J. R. (1975). Astron. J. **78**, 199.
9. Dunham, D. W., Evans, D. S., and Sandmann, W. H. (1974). Astron. J. **79**, 483.
10. Dunham, D.W., Evans, D.S., and Vogt, S.S. (1975). Astron. J. **80**, 45.
11. Evans, D. S. (1957). Astron. J. **62**, 83.
12. Evans, D. S., Fekel, F. C., and Silverberg, E. C. (1977). Astron. J. **82**, 828.
13. Glass, I. S., and Morrison, L. V. (1976). Monthly Notices Roy. Astron. Soc. **175**, 57.
14. Lasker, B. M., Bracker, S. B., and Kunkel, W. E. (1973). Publ. Astron. Soc. Pacific **85**, 109.
15. Harwood, J.M., Nather, R.E., Walker, A.R., Warner, B., and Wild, P.A.T. (1975). Monthly Notices Roy. Astron. Soc. **170**, 229.
16. McGraw, J. T., and Angel, J.R.P. (1974). Astron. J. **79**, 485.
17. Nather, R. E., McCants, M. M., and Evans, D. S. (1970). Astrophys. J. **160**, L181.
18. Nather, R. E. (1972). Ph.D. dissertation, Univ. of Cape Town.
19. Nather, R. E., and Wild, P.A.T. (1973). Astron. J. **78**, 628.
20. Nelson, M. R. (1975). Astrophys. J. **198**, 127.
21. Poss, H. L. (1971). in Highlights of Astronomy, Vol. **2**, C. de Jager, ed., D. Reidel, Dordrecht, p. 692.
22. Rakos, K. D. (1971). in Highlights of Astronomy, Vol. **2**, C. de Jager, ed., D. Reidel, Dordrecht, p. 675.
23. Ridgway, S. T., Wells, D. C., and Carbon, D. F. (1974). Astron. J. **79**, 1079.
24. Ridgway, S. T., Wells, D. C., and Joyce, R. R. (1977). Astron. J. **82**, 414.
25. Vilas, F., and Lasker, B. M. (1977). Publ. Astron. Soc. Pacific **89**, 95.
26. Walker, A.R. (1975). Monthly Notices Roy. Astron. Soc. **173**, 29P.
27. White, N. M. (1974). Astron. J. **79**, 1076.
28. White, N. M. (1978). In preparation.

TWO-DIMENSIONAL SPECTRAL CLASSIFICATION OF M STARS BY NARROW-BAND PHOTOMETRY

Robert F. Wing

Ohio State University

Nathaniel M. White

Lowell Observatory

1. INTRODUCTION

To place a star on the HR diagram by means of spectroscopic data alone--i.e., without knowledge of the parallax--one must separate the spectroscopic effects of temperature and luminosity as cleanly as possible. Here we discuss the use of the red and near-infrared bands of TiO, VO, CN, and CaH as classification criteria for M stars. Our eight-color system of filter photometry (Wing 1971; White 1971a) has provided data on the first three molecules.

2. TiO: THE PRIMARY TEMPERATURE INDICATOR

The great temperature sensitivity of the TiO bands is well known. Numerical comparisons of the photometric TiO indices that have been given on various systems indicate that the eight-color system measures TiO in the most favorable way, namely, with a filter centered at 7120 Å (full width at half power: 60 Å) within the (0,0) band of the γ system, together with a 50 Å continuum filter at 7540 Å. This combination provides good sensitivity throughout the range K4 to M8. The limit at M8 is imposed by the growth of VO bands, such that there are no good continuum regions left anywhere shortward of 1 μ.

Within the useful range of the eight-color TiO index, both the TiO filter and the continuum filter are contaminated by lines of the CN molecule; these filters were, however, chosen to give the

A. G. Davis Philip and D. S. Hayes (eds.), The HR Diagram, 451–454.

best separation of the overlapping band systems of TiO and CN. In a recently completed study of 128 M supergiants (White and Wing 1978), we explain the procedure used for correcting our TiO index for CN contamination and show that spectral types based on the corrected TiO index are on the same scale as those of Morgan and Keenan (1973) for supergiants as well as giants.

Even when the luminosity sensitivity of CN is removed from the TiO index, there remains the question of whether the TiO bands themselves are luminosity-sensitive. Wing (1973a) found an appreciable difference between the TiO-color relations for giants and dwarfs; dwarfs have stronger TiO bands than giants of the same temperature in the range K4 to M2, while the reverse is true in the later types. These differences never exceed about 30% in band strength, however, and seem very small in terms of the enormous difference in absolute magnitude. The relations for giants and supergiants appear to be nearly the same, but here it is hard to make a precise statement because of the difficulty of correcting the color temperatures of the supergiants for reddening.

Abundance differences must also affect the TiO index; however, these effects appear to be surprisingly small. Several hundred M giants, including known metal-deficient stars, have been observed on the eight-color system, and the tightness of the relation they define between band strength and color temperature indicates that the TiO band strength depends only weakly on the metallicity.

The Mira variables present a special problem; different TiO bands often indicate different spectral types. The discrepancy is usually in the sense that the (0,0) band is anomalously strong relative to high-excitation bands and to the color temperature (Wing 1974).

3. VO: A TEMPERATURE INDICATOR IN VERY COOL STARS

The VO bands in the visible region (Keenan 1963) and in the near-infrared (Cameron and Nassau 1955) have served as temperature criteria for late M stars, and the bands in the one-micron region serve the same purpose on the eight-color system (Wing and Lockwood 1973) for stars cooler than M7. Their temperature sensitivity is demonstrated by their behavior in Mira variables, in which they not uncommonly vary from invisible to very strong, their greatest intensity occurring at the time of minimum light and lowest color temperature.

Unfortunately, since most of our information about VO comes from the Miras, it is difficult to evaluate the sensitivity of the VO bands to luminosity or abundance effects. Boeshaar (1976) was able to detect VO bandheads in several of the coolest M dwarfs,

which have TiO types of M5 or M6, and it appears that VO/TiO ratio is nearly the same in dwarfs as in giants. The supergiants tell a different story: Although most M supergiants are too warm to show VO bands, there do exist stars that have all the earmarks of high luminosity together with strong VO bands. Examples of this phenomenon are VX Sgr, NML Cyg, and S Per at minimum. Wing (1973b, 1974) has suggested that high VO/TiO ratio of the stars is a luminosity effect.

4. CN: A LUMINOSITY INDICATOR IN HIGH-LUMINOSITY STARS

The strong positive luminosity effect shown by the CN bands has been used often for the recognition of supergiants in low-resolution spectral surveys. However, a considerable spread in CN strength occurs at a given temperature and luminosity among G and K giants (Griffen and Redman 1960) and similarly among M giants (Wing 1967). This spread is presumably attributable to abundance differences, and it limits the usefulness of CN as a luminosity indicator, although giants can almost always be distinguished from dwarfs and supergiants.

Among the M supergiants the situation is more favorable, and the CN measurements on the eight-color system are capable of reproducing the MK luminosity subdivisions Ib, Iab, and Ia, which are based upon atomic line ratios (White 1971a, 1971b; White and Wing 1978). Abundance effects seem to be smaller among supergiants than among giants, perhaps because the M supergiants have more homogeneous population characteristics.

The infrared CN bands are quite insensitive to temperature in the late K and M stars. Whereas Griffin and Redman (1960) found that the bands of the violet system pass through a maximum in the early K's (for any given luminosity class) and fade to invisibility by about M0, the bands of the red system weaken much more gradually and are still measurable in middle-M giants. A single calibration of the eight-color CN index in terms of luminosity class is at least valid throughout the interval K4 to M4 (White 1971b; White and Wing 1978).

5. CaH: A LUMINOSITY INDICATOR IN LOW-LUMINOSITY STARS

Although M dwarfs can be recognized by their lack of CN bands, it is obvious that CN cannot be used to define finer luminosity subdivisions among low-luminosity stars. For this purpose we turn to CaH, which has bands in the red region and because of its low dissociation energy is very sensitive to gas pressure.

Measurements of CaH not only distinguish dwarfs from giants

(Jones 1973), but also allow the separation of subdwarfs from normal main sequence stars (Mould 1976; Boeshaar 1976; Wing, Dean, and MacConnell 1976). The recent narrow-band photometry of TiO and CaH by Mould and McElroy (1978) has shown that subdwarfs belonging to the old disk population occupy an intermediate position in the (CaH, TiO) plane between the old disk main sequence stars and the halo subdwarfs.

The temperature dependence of the CaH bands, as indicated by the model-atmosphere calculations of Mould (1976), is rather strong. However, for classification purposes, the CaH bands must be viewed as a luminosity indicator; if they are used for temperature classification, subdwarfs are assigned spectral types that are too late to be consistent with their color temperatures (Wing, Dean, and MacConnell 1976).

REFERENCES

Boeshaar, P.C. (1976). Ph.D. thesis, Ohio State University.

Cameron, D.M. and Nassau, J.J. (1955). Astrophys. J. **122**, 177.

Griffin, R.F. and Redman, R.O. (1960). Mon. Not. R. Astron. Soc. **120**, 287.

Jones, D.H.P. (1973). Mon. Not. R. Astron. Soc. **161**, 19.

Keenan, P.C. (1963). In Basic Astronomical Data, K.Aa. Strand, ed., Univ. of Chicago Press, Chicago, p. 78.

Morgan, W.W. and Keenan, P.C. (1973). Ann. Rev. Astron. Astrophys. **11**, 29.

Mould, J.R. (1976). Astrophys. J. **207**, 535.

Mould, J.R. and McElroy, D.B. (1978). Astrophys. J., in press.

White, N.M. (1971a). In Proceedings of the Conference on Late-Type Stars, G.W. Lockwood and H.M. Dyck, eds. (KPNO Contr. No. 554), p. 163.

White, N.M. (1971b). Ph.D. thesis, Ohio State University.

White, N.M. and Wing, R.F. (1978). Astrophys. J., in press.

Wing, R.F. (1967). Ph.D. thesis, Univ. of California, Berkeley.

Wing, R.F. (1971). In Proceedings of the Conference on Late-Type Stars, G.W. Lockwood and H.M. Dyck, eds. (KPNO Contr. No. 554), p. 145.

Wing, R.F. (1973a). In Spectral Classification and Multicolor Photometry, Ch. Fehrenbach and B.E. Westerlund, eds., D. Reidel, Dordrecht, p. 209.

Wing, R.F. (1973b). Bull. Amer. Astron. Soc. **5**, 443.

Wing, R.F. (1974). Highlights of Astronomy **3**, 285.

Wing, R.F., Dean, C.A. and MacConnell, D.J. (1976). Astrophys. J. **205**, 186.

Wing, R.F. and Lockwood, G.W. (1973). Astrophys. J. **184**, 873.

INDEX

1. Name Index - An underlined page number following an individual's name indicates the first page of an article by that individual. The names of people mentioned in references were not indexed.

2. Object Index - Stars with HD numbers are listed in order by HD number. Non-HD stars are listed alphabetically by constellation or by BD number. Other objects are listed by NGC number or alphabetically under their category.

3. Subject Index - Underlined numbers refer to subjects mentioned in the titles of articles.

Frank Edmondson addressing Banquet.

NAME INDEX

OBJECT INDEX

STARS (cont'd)

HD Stars (cont'd)

STARS (cont'd)

HD Stars (cont'd)

STARS (cont'd)

SUBJECT INDEX

A